Fluid Flow in Porous Media

Fundamentals and Applications

Highly Recommended Titles

The Imperial College Lectures in Petroleum Engineering
Volume 1: An Introduction to Petroleum Geoscience
by Michael Ala
ISBN: 978-1-78634-206-5

The Imperial College Lectures in Petroleum Engineering
Volume 2: Reservoir Engineering
by Martin J Blunt
ISBN: 978-1-78634-209-6

The Imperial College Lectures in Petroleum Engineering
Volume 3: Topics in Reservoir Management
by Deryck Bond, Samuel Krevor, Ann Muggeridge, David Waldren and Robert W Zimmerman
ISBN: 978-1-78634-284-3

The Imperial College Lectures in Petroleum Engineering
Volume 4: Drilling and Reservoir Appraisal
by Olivier Allain, Michael Dyson, Xudong Jing, Christopher Pentland, Marcel Polikar and Sander Suicmez
ISBN: 978-1-78634-395-6

The Imperial College Lectures in Petroleum Engineering
Volume 5: Fluid Flow in Porous Media
by Robert W Zimmerman
ISBN: 978-1-78634-499-1

Fluid Flow in Porous Media

Fundamentals and Applications

Liang Xue
Xiaozhe Guo
Hao Chen
China University of Petroleum, Beijing, China

World Scientific

NEW JERSEY • LONDON • SINGAPORE • BEIJING • SHANGHAI • HONG KONG • TAIPEI • CHENNAI • TOKYO

Published by

World Scientific Publishing Co. Pte. Ltd.
5 Toh Tuck Link, Singapore 596224
USA office: 27 Warren Street, Suite 401-402, Hackensack, NJ 07601
UK office: 57 Shelton Street, Covent Garden, London WC2H 9HE

Library of Congress Cataloging-in-Publication Data
Names: Xue, Liang (Petroleum engineering teacher), author. | Guo, Xiaozhe, author. | Chen, Hao (Petroleum engineering teacher), author.
Title: Fluid flow in porous media : fundamentals and applications / Liang Xue, Xiaozhe Guo, Hao Chen, China University of Petroleum, Beijing, China.
Description: First edition. | Singapore ; a Hackensack, NJ : World Scientific Publishing Co. Pte. Ltd., [2020] | Includes bibliographical references and index.
Identifiers: LCCN 2020023713 | ISBN 9789811219528 (hardcover) | ISBN 9789811219535 (ebook) | ISBN 9789811219542 (ebook other)
Subjects: LCSH: Oil reservoir engineering. | Petroleum--Migration. | Fluids--Migration.
Classification: LCC TN870.57 .X84 2020 | DDC 622/.3382--dc23
LC record available at https://lccn.loc.gov/2020023713

British Library Cataloguing-in-Publication Data
A catalogue record for this book is available from the British Library.

地下储层渗流力学
Originally published in Chinese by Petroleum Industry Press

For any available supplementary material, please visit
https://www.worldscientific.com/worldscibooks/10.1142/11805#t=suppl

Desk Editor: Shaun Tan Yi Jie

Typeset by Stallion Press
Email: enquiries@stallionpress.com

This book is dedicated to our teachers and colleagues whom we have learned from, and to the students whom we have worked with.

Preface

This book focuses on the fundamental theory of fluid flow in porous media and its applications to petroleum engineering. It can be used as a textbook for senior undergraduate students and graduate students in petroleum engineering. When teaching the course of flow mechanics in porous media, we find that most textbooks only cover the classical flow theory, such as steady state flow, transient flow and multiphase flow. But with the development of unconventional energies, such as coal bed methane, shale gas and tight oil, the classical flow theory cannot explain the flow characteristics in such reservoirs. Therefore, we decide to summarize these special flow theories and write this book to complement the current classical flow theory in porous media. In addition, the flow mechanics in porous media provides the quantitative basis for other problems in petroleum engineering, and this book shows how the flow theory can be applied to these problems.

This book covers the fundamentals of fluid flow theory in porous media. It is organized as follows: Chapter 1 introduces the basic concepts of reservoir rock and fluid; Chapter 2 introduces the basic law of flow in porous media and how we can build a mathematical model to describe the flow system; Chapter 3 introduces the classical flow theory, which includes single-phase flow under the steady state and transient conditions, and multiphase flow; Chapter 4 introduces the flow in complex porous media, such as fractured porous medium, anisotropic medium, the deformable medium, and the fractal medium; Chapter 5 introduces the physicochemical fluid flow theory; Chapter 6 introduces the flow theory of unconventional energy sources, such as shale gas, tight oil, coal bed methane, shale oil, natural gas hydrate, geothermal, water-soluble gas; and Chapter 7 introduces how these flow theories can be applied to reservoir numerical simulation, modern well testing, production engineering, and reservoir engineering. In each chapter, the key concepts have been demonstrated concisely and the derivations of equations have been presented logically. The solutions of some practical problems are given so that the readers can get an idea on how to actually use these equations.

We would like to thank all our colleagues in the College of Petroleum Engineering at the China University of Petroleum — Beijing for their enormous help during the preparation of the book manuscript.

Liang Xue

About the Authors

Liang Xue is an associate professor in the College of Petroleum Engineering at China University of Petroleum (Beijing). He teaches an undergraduate course named *Flow in Porous Media*. His research interests include numerical simulation of flow in porous and fractured media, uncertainty quantification, and machine learning. Xue holds a PhD degree in hydrology from University of Arizona, USA, in addition to a Master's degree in hydrology and a Bachelor's degree in environment engineering, both from China University of Geosciences (Beijing).

Xiaozhe Guo is an associate professor in the College of Petroleum Engineering at China University of Petroleum (Beijing). He received his Bachelor's degree in petroleum engineering from Daqing Petroleum Institute, and Master's and PhD degrees in oil & gas field development engineering from China University of Petroleum (Beijing).

Hao Chen is an associate professor in the College of Petroleum Engineering at China University of Petroleum (Beijing). He received both his Bachelor's and PhD degrees in petroleum engineering from China University of Petroleum (Beijing).

Contents

List of Tables

List of Figures

Chapter 1

Basic concepts

The characteristics of liquid and rocks and the physical properties of relevant parameters in subsurface reservoirs are shown in this chapter. It provides simple conceptual understanding and specific numerical references to understand the flow mechanisms in porous media. In addition, we analyze the characteristics of the mechanical phenomena, flow regimes and various pressures, which form the foundation to extract the physical law and establish the mathematical model of flow mechanics.

1.1 Reservoirs

1.1.1 *Introduction*

The oil, gas, water and other fluids that we recover are all from geological formations, which are composed of rocks formed in different geological ages. Some of these formations have few pores to store the fluids and usually form flow barriers, such as mudstone or shale. Some other formations have developed sufficiently large pores that can form the space to store the fluids, and are referred to as reservoirs. Sandstone and carbonate rock are common reservoirs. A typical oil or gas production system is shown in Figure 1.1.

The scale of the reservoirs should be large enough so that it can be cost effective for industrial development. The depth of the reservoir ranges from several hundred meters to several kilometers. A common classification in terms of depth is that: if the depth is less

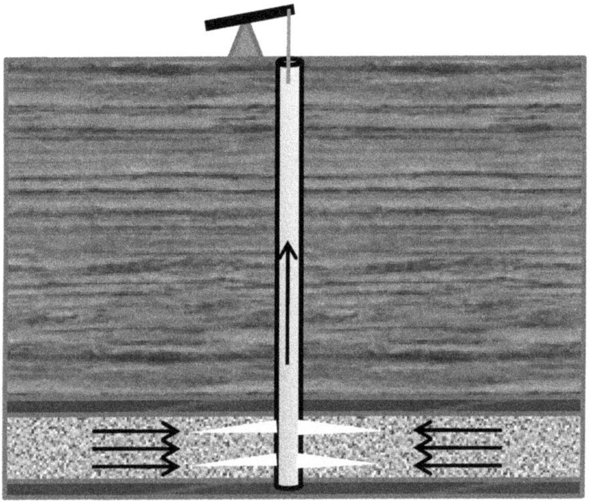

Figure 1.1. Schematic diagram of oil or gas production system

than 1500 m, it can be called a shallow layer; if the depth is between 1500–2800 m, it can be called a middle-shallow layer; if the depth is between 2800–4000 m, it can be called a deep layer; and if the depth is larger than 4000 m, it can be called an ultra-deep layer.

The reservoir space to store the fluids is rock pore. The pore diameter can be as small as only a few nanometers and as large as several millimeters. On average, the pore diameter is on the magnitude of micrometers. Porosity is used to describe how well the pore develops, and is less than 5% in shale reservoirs and less than 10% in tight reservoirs. For conventional reservoirs, the porosity is mainly around 20%. For high porosity reservoirs, it can reach to more than 30%.

When producing oil and gas, it is necessary to connect the reservoir through the wellbore, so that the fluid in the reservoir pore flows into the wellbore along the pore channel, and then flows to the surface. Therefore, the role of pores has two aspects: reservoir space and flow channel. Permeability, measured in D (Darcy), is the parameter to describe flow capacity, and it has a wide range of values. The permeability in shale reservoirs can be 0.001 μD–0.01 mD, that in tight reservoirs is lower than 0.01–0.1 mD, in low permeable reservoirs it is 0.1–10 mD, and in high permeable reservoirs it is more than 500 mD, even up to dozens of D.

1.1.2 *Concepts*

Several basic concepts relating to reservoirs are introduced in this section.

Formation: *formation is a general term for the stratified rocks that have certain burial depth and geological time.*

Reservoir: *reservoir is a stratum composed of reservoir rocks. It provides the storage site and migration pathway for oil and gas.*

Porous media: *porous media is the continuous solid phase that contains the interconnected pores, fractures or caves of different sizes and shapes.*

Geological structure: *geological structure is the characteristics of a spatial extension of strata. Usually, anticline, fault block and monocline are favorable structures for oil and gas.*

Trap: *trap is a geologic structure that can prevent the migration of oil and gas and force them to accumulate. A trap is composed of reservoir rock, cap rock and source rock, as shown in Figure 1.2.*

Petroleum reservoir: *petroleum reservoir is the part of the reservoir occupied by oil and gas with the same pressure system in a single trap, as shown in Figure 1.2.*

Oil and gas field: *the summation of all the oil and gas reservoirs within a certain continuous oil and gas production area on the ground surface.*

Porosity: *porosity, denoted as* ϕ, *is the ratio of pore volume,* V_p, *to total rock volume in the rock,* V_f.

Permeability: *permeability is the ability of a rock to allow fluid to pass, and generally refers to the absolute permeability of rocks. The symbol is* K. *The common unit is mD, and its international unit is* m^2.

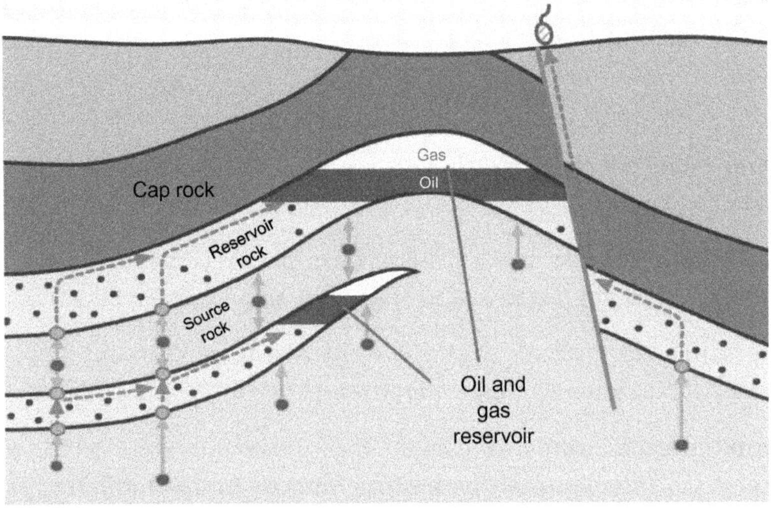

Figure 1.2. Schematic diagram of oil and gas reservoir

Effective thickness: *effective thickness is the part of thickness in a reservoir with industrial oil production capability. The unit is m. The effective thickness depends on the physical properties of the reservoir. Usually, the effective thickness is less than 1 m for thin reservoirs, and more than 5 m for thick reservoirs. For some reservoirs, the effective thickness can reach 30 m or even 60 m.*

Single-porosity medium: *single-porosity medium is a medium with only one pore structure.*

Dual-porosity medium: *dual-porosity medium is a medium with two pore structures, such as the fractured porous medium and the vuggy porous medium.*

Triple-porosity medium: *triple-porosity medium is a medium with three pore structures, such as vuggy fractured porous medium.*

Closed boundary: *closed boundary is a boundary without energy supply or fluids passing through it, such as fault, pinch-out and unconformity.*

Supply boundary: *supply boundary is a boundary with energy supply and fluids passing through, such as bottom water and edge water.*

1.1.3 *Analysis*

1.1.3.1 *Pore structure of reservoirs*

There are three different basic pore types in reservoirs: the inter-granular pore, fracture and vug. The combination of these pore types yields 7 pore structures.

(1) The inter-granular pore structure

The inter-granular pore structure mainly exists in sandstone petroleum reservoirs, as shown in Figure 1.3(a). There are cements (the black background with white points) and inter-granular pores (the white part) between the clastic grains (shadow area with parallel lines). The inter-granular pores are the fluid storage space and flow channels, which are also the main object of flow mechanics in porous media. If there is no special explanation on the pore structures, it generally refers to this type of structure. The conventional reservoirs of continental deposit in China are mostly of the inter-granular pore structure.

(2) The fracture structure

The fracture structure generally exists in the carbonate reservoir, as shown in Figure 1.3(b). The fractures (white part) are distributed in the reservoir along a certain direction, or form an interwoven fracture

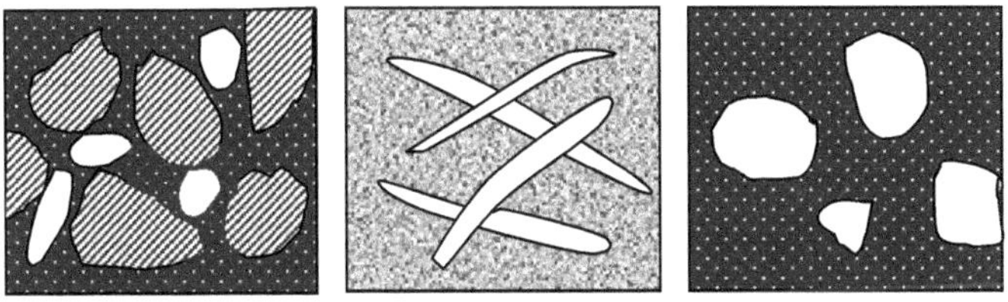

(a) Intergranular pore (b) crack (c) karst cave

Figure 1.3. Schematic diagram of single pore structures

network with several groups of different directions. In this case, the fractures are the main storage space and flow channels for oil and gas. Most of the marine sedimentary reservoirs in the Middle East and America are of this type.

(3) The vug structure

The vug structure mostly exists in carbonate reservoir, as shown in Figure 1.3(c), because of the dissolution of the carbonate rock. The size of a vug pore is much larger than an inter-granular pore, and the flow characteristics in vug pores is substantially different from that in the inter-granular pores. The flow in vug pores cannot be described by Darcy's law, therefore it is not taken as the research object in this book.

(4) The fractured pore structure

The fractured pore structure is the most common and important type of dual medium. A dual medium generally refers to a reservoir with dual structure of fractures and inter-granular pores. It takes inter-granular pores as the main storage space and fractures as the main flow channel. This type of pore structure usually occurs in tight sandstone reservoirs and in some shale reservoirs. Fractures are originated from two aspects: natural fractures and hydraulic fractures. The tight oil and shale gas in unconventional reservoirs need the massive hydraulic fracturing to create a fracture network, and the fractured pore structure can be used to build the flow model.

(5) The vuggy fracture structure

This type of pore structure can be found in carbonate reservoirs, where both dissolution vugs and fractures exist with different sizes, irregular shapes and disorderly distribution. The vugs are the main reservoir spaces for storing oil and gas. Fractures connect these vugs and form the main flow channels. As mentioned before, the flow characteristics in vugs is out of the scope of this book.

(6) The vuggy pore structure

Inter-granular pores can be enlarged due to the effect of mineral dissolution, and they form the so-called secondary pores. If the dissolution process continues, the secondary pores can grow to become vugs. The co-existence of vugs and pores yields a vuggy pore structure.

(7) The vuggy fractured pore structure

All the vugs, fractures and inter-granular pores co-exist in the same reservoir. This type of pore structure can be found in a fully developed carbonate reservoir.

1.1.3.2 *Geometry of reservoirs*

A reservoir of oil and gas is a trap formed by the strata with a certain thickness and a specific distributed structure. If vertical flow in the reservoir is not considered (due to the fact that the vertical thickness of the reservoir is often small relative to the horizontal plane size), the reservoir can be projected onto the plane to study the flowing characteristics of fluids. Take the anticline structure as an example; the planar projection of a reservoir is shown in Figure 1.4.

If the outer boundary is closed, that is, there is no energy supply outside the boundary when the reservoir pressure decreases, the boundary is a closed boundary as shown in Figure 1.4(a). If the outer boundary is connected to a large volume of aquifer, there is always energy supply outside the boundary when the reservoir

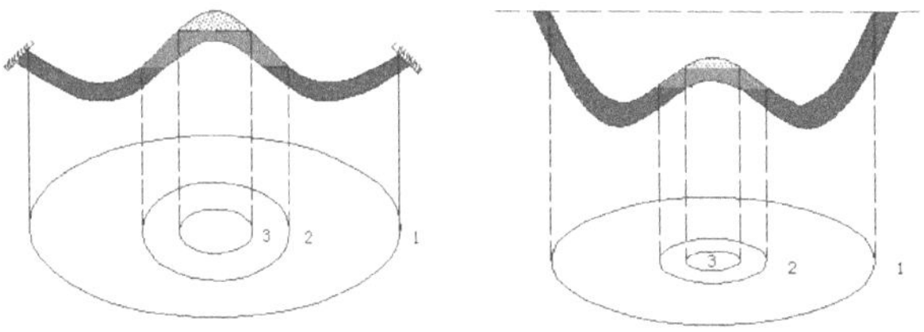

(a) Closed boundary (b) Supply boundary

Figure 1.4. Planar projection of a reservoir (Line 1 is outer boundary of reservoir; Line 2 is outer boundary of oil and water; Line 3 is outer boundary of oil and gas)

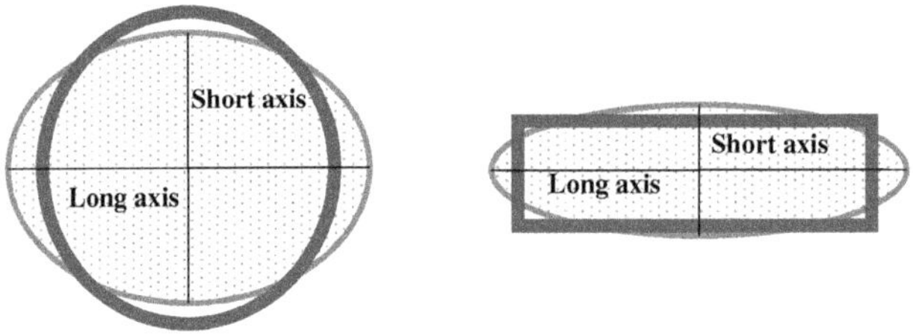

Figure 1.5. Geometry of reservoirs

pressure decreases, and the boundary is a supply boundary as shown in Figure 1.4(b). These two kinds of outer boundaries are often used in flow models.

Due to the irregular geometry obtained from the projection of the actual reservoir on the plane, for convenience, it is often simplified into two kinds of regular geometry: rectangular shape and circular shape. If the ratio of long axis to short axis after projection is less than 3, it is simplified as a circular reservoir, which is a geological model of radial flow, as shown in Figure 1.5(a). If the ratio of long axis to short axis after projection is greater than 3, it is simplified as a rectangular reservoir, which is a geological model of linear flow, as shown in Figure 1.5(b). These two simplified flow regimes are widely used in the research of flow mechanics in porous media.

1.2 Fluid properties

1.2.1 *Introduction*

The fluid in reservoirs can be divided into two types: liquid and gas. Liquid mainly refers to oil and water. In addition to the main components of natural gas, some inorganic gases such as nitrogen, carbon dioxide and hydrogen sulfide are contained in the gas. In percolation mechanics, the gas component is not specially described, but generally refers to the natural gas of hydrocarbon.

An underground reservoir is in a state of high temperature and pressure, and the physical properties of the fluid stored in it are very different from those on the ground: liquid water dissolves little to none of the gas, and its physical properties are relatively similar to those on the ground; liquid oil can dissolve a large amount of natural gas, and its viscosity, density, formation volume factor and so on change greatly under the influence of pressure change; the gas has strong compressibility and pressure sensitivity. In addition, the continuous exchange of dissolved and free natural gas in oil makes the change in physical properties of natural gas more complicated.

The main physical parameters describing fluid properties are: viscosity, density, formation volume factor, compressibility, solution gas-oil ratio, etc.

1.2.2 *Concepts*

Several basic concepts relating to reservoir fluids are introduced in this section.

Viscosity: *viscosity is the resistance caused by internal friction in the fluid flow process. The symbol is μ. The common unit is $mPa \cdot s$ or cp. The international unit is $Pa \cdot s$.*

Newtonian liquid: *a Newtonian liquid is a liquid conforming to Newton's internal friction law. It is expressed as the ratio of shear stress per unit area to velocity gradient in any point of fluid. That is to say that the shear stress is linearly proportional to the shear rate. The equation is:*

$$\tau = \mu\dot{\gamma} \tag{1.1}$$

where τ is shear stress, Pa; μ is viscosity, $Pa \cdot s$; and $\dot{\gamma}$ is shear rate, 1/s.

Density: *density is the mass per unit volume of fluid at a certain temperature and pressure. The symbol is ρ. The common unit is g/cm^3 or kg/m^3. The international unit is kg/m^3.*

Relative oil density: *relative oil density is the ratio of oil density to the pure water density ($\rho_w = 1\ g/cm^3$) at 1 atm and $4°C$ under the specified temperature and pressure. The symbol is γ_o. It is unitless, and is also known as specific gravity.*

API degree: *API degree is another expression of relative oil density in Europe and America. The equation is:*

$$API = \frac{141.5}{\gamma_o} - 131.5 \tag{1.2}$$

where API represents the API degree of oil; and γ_o *is relative oil viscosity.*

The API of water is 10, the API of oil is inversely proportional to γ_o *and is proportional to the solubility of gas in oil.*

Relative gas density: *relative gas density is the ratio of the density of natural gas to the density of dry air at the normal temperature and pressure (20° C, 1 atm), standard condition (0° C, 1 atm) or the same temperature and pressure. The symbol is* γ_g*. It is unitless.*

Formation volume factor of oil: *it is the ratio of the volume of oil underground to that after degassing on the ground. The symbol is* B_o*. It is unitless.*

Formation volume factor of water: *it is the ratio of the volume of equal amounts of formation water underground to that on the ground. The symbol is* B_w*. It is unitless.*

Formation volume factor of natural gas: *it is the ratio of the volume of a certain amount of natural gas underground to that on the ground in standard condition. The symbol is* B_g*. It is unitless.*

Elastic compressibility of liquid: *elastic compressibility of liquid is the rate of underground liquid volume with the change of the pressure. Namely, it is the volume change of liquid per unit volume under unit pressure difference. If the pressure decreases, the volume of liquid will expand, the density will decrease, and the elastic energy will release. The equation is:*

$$C_L = -\frac{1}{V_L}\frac{dV_L}{dp} \tag{1.3}$$

where C_L is the elastic compressibility of liquid with the common unit as MPa^{-1} and the international unit as Pa^{-1}; and V_L is the volume of liquid, m^3.

Compressibility factor of natural gas: *At the same temperature and pressure, compressibility factor of natural gas is the ratio of the volume of an equivalent amount of the real gas to the ideal gas. The symbol is Z. It is unitless.*

Gas Isothermal Compressibility Factor: *namely natural gas compressibility. It is the volume change of natural gas with pressure. It has the same definition as the elastic compressibility of liquid. Combined with state equation, the formula is:*

$$C_g = -\frac{1}{V_g}\left(\frac{dV_g}{dp}\right)_T = \frac{1}{p} - \frac{1}{Z}\frac{\partial Z}{\partial p} \tag{1.4}$$

where C_g is gas isothermal compressibility factor with the common unit as MPa^{-1} and the international unit as Pa^{-1}; V_g is the volume of gas, m^3; T is temperature, K; and Z is gas compressibility factor.

Solution gas-oil ratio: *solution gas-oil ratio is the volume of natural gas under the standard condition dissolved in unit volume or unit mass of oil under the reservoir condition. The symbol is R_s. The unit is m^3/m^3.*

Oil pour point: *oil freezing point refers to the critical temperature from flowing state to non-flowing state in the oil cooling process. It is related to the wax content, bitumen resin content and light oil content in crude oil. The unit is $^\circ C$, between $-56\sim50^\circ C$. High pour point oil is more than $40^\circ C$.*

Saturation pressure: *also named bubble point pressure. It is the pressure at which the first bubbles separate from the liquid phase during the pressure reducing process when the temperature is constant. For unsaturated oil, when the pressure is higher than saturation pressure the flow is single-phase liquid flow. When the pressure decreases to saturation pressure, the gas dissolved in oil begins to separate and forms the free gas. The phase state changes* *to a two-phase gas-liquid state. The symbol is P_b. The common unit is MPa. The international unit is Pa.*

Unsaturated reservoir: *unsaturated reservoir is the oil reservoir with unsaturated natural gas under the condition of pressure and temperature of the original reservoir. Namely, the original reservoir pressure is higher than saturated pressure. The oil reservoir is always liquid phase, before the pressure decreases to saturated pressure.*

Total salinity of formation water: *total salinity of formation water is equal to the sum of positive and negative ions of mineral substance in unit volume of water. The common unit is mg/L or ppm (one millionth of g/m^3).*

Hardness of formation water: *hardness of formation water refers to the concentration of divalent cations, such as calcium and magnesium, in formation water. The common unit is mg/L or ppm.*

1.2.3 *Analysis*

1.2.3.1 *Chemical composition of fluids*

At normal temperature and pressure, the state of the fluid is dependent on the number of the carbon atoms, C_n. $C_1 \sim C_4$ is in gaseous state, which is the major component of natural gas; $C_5 \sim C_{16}$ is in liquid state, which is the major component of crude oil; more than C_{17} is in solid state, which is the major component of wax.

Under the formation condition, methane is the main component in natural gas; its molar concentration can reach $70 \sim 98\%$, while ethane content is about 10%, and other contents are less. The compositions of non-hydrocarbon gases vary greatly, in which hydrogen sulfide content is generally less than 5%, while others such as CO_2 content can be as high as 70% and nitrogen content can be as high as 60%.

Under the formation condition, the content of light oil C_1 is more than 60%, and content of more than C_5 is less than 20%; the content of C_1 in black oil (black oil refers to non-volatile oil components, such as conventional reservoirs, heavy oil reservoirs, etc.) is less than 50%, and more than C_5 takes up to 40%.

Under the formation condition, water often contains a lot of metal salts, especially potassium salt and sodium salt, thus it is also called brine. The common cations are Na^+, K^+, Ca^{2+}, Mg^{2+}, and the common anions are Cl^-, SO_4^{2-}, HCO_3^-, NO_3^-, etc. The salinity ranges from hundreds to hundreds of thousands of mg/L. When the chemical reaction occurs, it may produce chemical precipitates and affect the oil and gas development performance.

1.2.3.2 *Physical properties of formation oil*

(1) Density of formation oil

A large amount of natural gas is dissolved in underground crude oil so that its density is less than that of degassed crude oil on the ground surface. When the formation pressure is less than saturation pressure, the solution gas-oil ratio increases and the oil density decreases with increase of formation pressure. When the formation pressure is larger than saturation pressure, the solution gas-oil ratio has no change, but the volume of oil is compressed and the density increases. The higher the formation temperature, the higher the dissolved gas-oil ratio and the smaller the density of oil. Figure 1.6 shows how the formation oil density changes with the pressure.

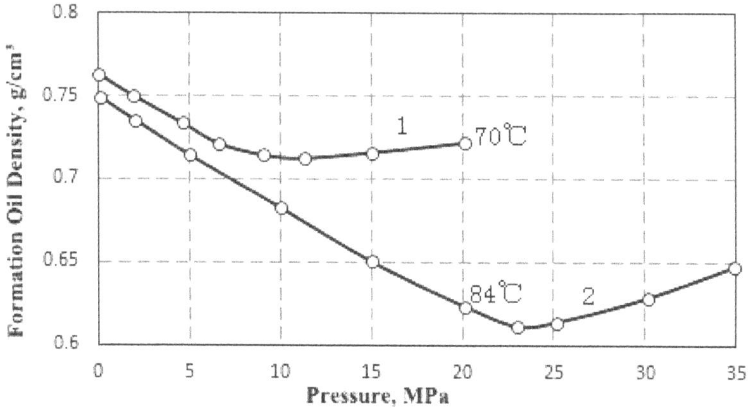

Figure 1.6. Formation oil density versus pressure

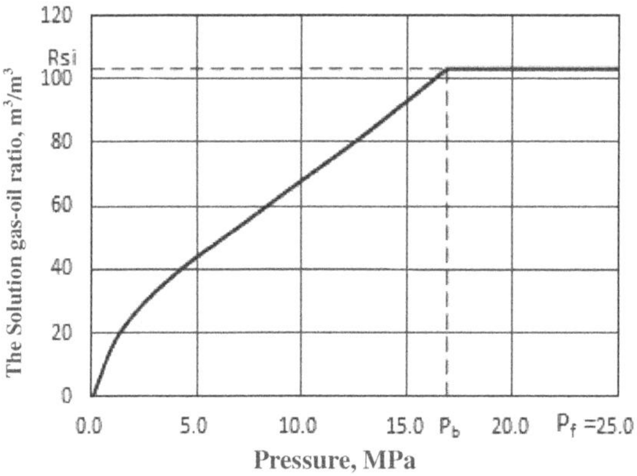

Figure 1.7. Solution gas-oil ratio versus pressure

(2) Dissolved gas-oil ratio

When the formation pressure is higher than saturation pressure (P_b), the solution gas-oil ratio doesn't change with pressure. When the formation pressure is less than saturation pressure, the solution gas-oil ratio increases with the increase of pressure. The solution gas-oil ratio varies greatly in different reservoirs, and generally heavy oil reservoir $<$ conventional oil reservoir $<$ low permeability oil reservoir. Its values range from dozens to hundreds. The change of solution gas-oil ratio with pressure is shown in Figure 1.7.

(3) Formation volume factor

Formation volume factor is the ratio of the volume of fluid at the reservoir condition to the volume of fluid at standard conditions. When the formation pressure is higher than saturation pressure (P_b), with the increase in pressure, the formation oil is compressed and the volume factor decreases. When the formation pressure is less than saturation pressure, with the increase in pressure, more gas is dissolved into oil, the volume of oil increases and then the oil formation volume factor (B_o) increases, as shown in Figure 1.8.

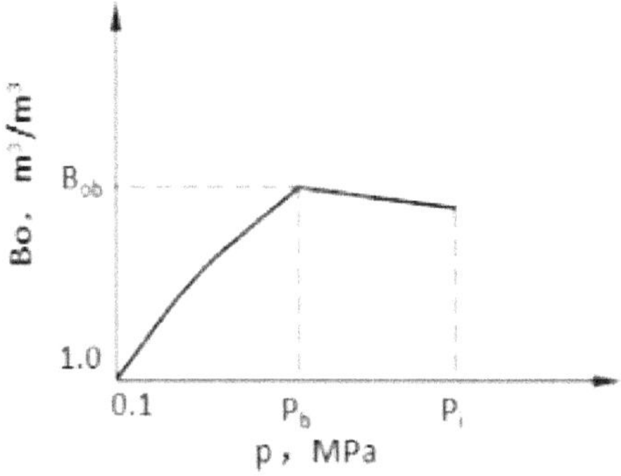

Figure 1.8. Oil formation volume factor versus pressure

Generally, the formation volume factor increases with the increase in solution gas-oil ratio, because the influence of underground solution gas and thermal expansion is far more obvious than the volume change caused by compression. The oil formation volume factor ranges from 1 to 15 in most of the cases.

(4) Elastic compressibility

The elastic compressibility of formation oil depends on the solution gas-oil ratio, reservoir temperature and pressure. The larger the solution gas-oil ratio, the larger the elastic compressibility; the higher the temperature, the larger the elastic compressibility; and the higher the pressure, the lower the elastic compressibility.

The compressibility of degassed oil is $(4 \sim 7) \times 10^{-4}\,\text{MPa}^{-1}$, and the compressibility of formation oil is $(10 \sim 140) \times 10^{-4}\,\text{MPa}^{-1}$.

(5) Viscosity

The viscosity of oil is sensitive to the variation of temperature, especially heavy oil. The viscosity-temperature curve is the key data to calculate the effects of temperature on viscosity in thermal recovery, as shown in Figure 1.9.

Because the solution gas-oil ratio varies under different pressures, the viscosity of oil varies greatly under different pressures, as shown

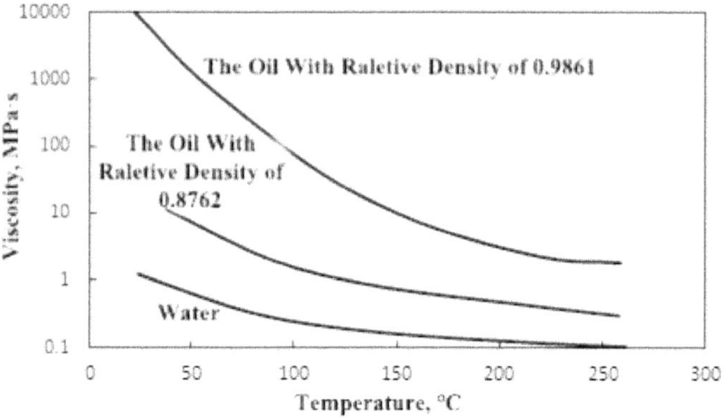

Figure 1.9. Viscosity of formation oil versus temperature

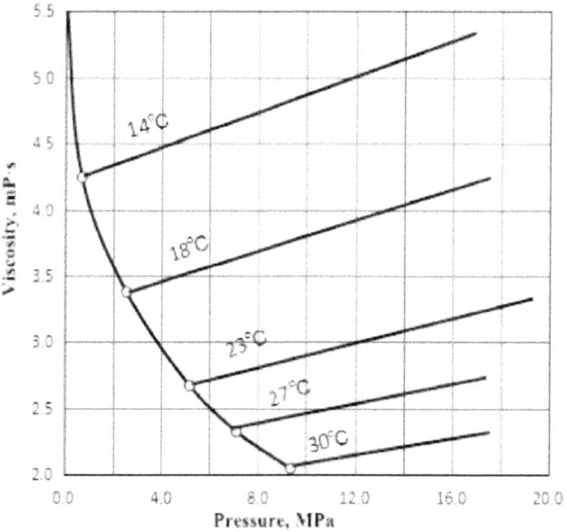

Figure 1.10. Viscosity of formation oil versus pressure

in Figure 1.10. When the pressure is higher than saturation pressure, the increase in pressure causes the elastic compression of oil. The higher the density, the higher the internal frictional resistance and the higher the oil viscosity. When the pressure is less than saturation pressure, with the decrease in pressure, the solution gas escapes continuously from the oil and the oil viscosity increases rapidly.

The viscosity of heavy oil in the reservoir condition is larger than 50 MPa · s, and the relative density is larger than 0.92.

1.2.3.3 *Physical properties of formation water*

The solubility of natural gas is relatively low in formation water. It generally does not exceed $2\,m^3/m^3$ under a pressure of 10 MPa. Even under higher pressures, it is also less than $5\,m^3/m^3$. The temperature has little effect on the solubility.

Normally, the elastic compressibility of formation water is less than that of formation oil, which is roughly $(3.4 \sim 5.0) \times 10^{-4}\,MPa^{-1}$. The water formation volume factor is small, which ranges from roughly $1.01 \sim 1.06$. Temperature has great influence on the viscosity of formation water, while pressure has little effect. Therefore, a reservoir with a depth of 2000 m has a temperature of 50°C and a viscosity of about 0.3 mPa · s of formation water as shown in Figure 1.11.

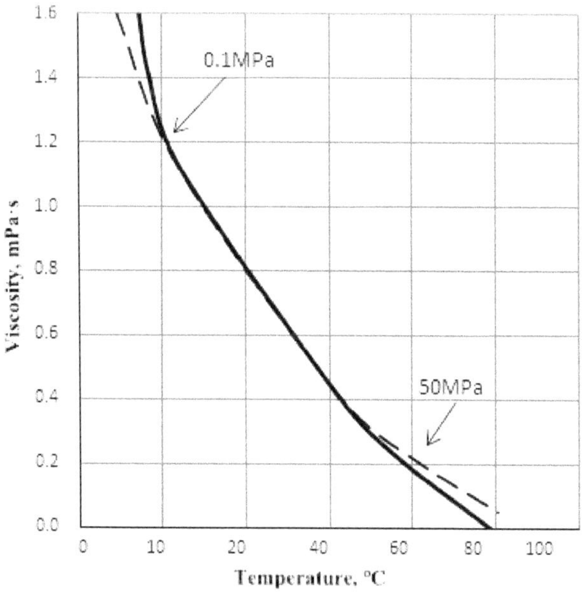

Figure 1.11. Viscosity of formation water versus temperature and pressure

1.2.3.4 *Physical properties of natural gas*

(1) Density

The density of natural gas can be obtained from the gas state equation:

$$\rho_g = \frac{pM}{ZRT} \tag{1.5}$$

where ρ_g is the density of natural gas, kg/m^3; M is molecular weight of natural gas, kg/kmol, computation of natural gas components; R is universal gas constant, its value is 0.008314 MPa $\cdot$ $m^3/(kmol \cdot K)$; T is temperature, K; and p is pressure, MPa.

Under the standard condition, the density of air is $\rho_{asc} = 1.293\,kg/m^3$. Under normal temperature and pressure, it is $\rho_a = 1.205\,kg/m^3$. Under the formation condition, the density of natural gas can reach $200 \sim 300\,kg/m^3$.

The molecular weight of dry air is about 29 kg/kmol, so the relative density of natural gas is:

$$\gamma_g = \frac{M}{29} \tag{1.6}$$

where γ_g is the relative density of natural gas. Its value is between $0.55 \sim 0.8$.

(2) Compressibility factor

Under the formation condition, compressibility factor of natural gas is the lowest when the pressure is less than 10 MPa. With the increase in pressure, the compressibility factor increases gradually. And when the pressure exceeds 35 MPa, compressibility factor may be greater than 1. The compressibility factor of natural gas increases with the increase of temperature, and the range of variation becomes narrower. For example, the compressibility factor of natural gas is 0.783 when the relative density is 0.8, the formation temperature is 65°C and the pressure is 12 MPa; the compressibility factor of natural gas is 0.81 when the relative density is 0.65, the formation temperature is 50°C and the pressure is 20 MPa; the compressibility factor of natural gas is 0.92 when the relative density is 0.65, the formation temperature is 60°C and the pressure is 30 MPa. Under normal temperature and pressure, the compressibility factor of natural gas is 1.

(3) Formation volume factor

The formation volume factor of natural gas can be expressed by the formula:

$$B_g = \frac{273 + t}{293} \frac{p_{sc} Z}{p} \tag{1.7}$$

where t is centigrade degree, °C and p_{sc} is the pressure under the standard condition, generally 0.1 MPa.

Supposing that the compressibility factor of natural gas $Z = 0.8$ and the temperature $t = 65$°C are both invariant, the volume factor of natural gas decreases with the increase in pressure, as shown in Figure 1.12.

When the pressure is less than 10 MPa, the decrease of formation volume factor is dramatic; when the pressure is more than 10 MPa, the formation volume factor changes slightly; the formation volume factor of liquid is about 100 times bigger than that of natural gas.

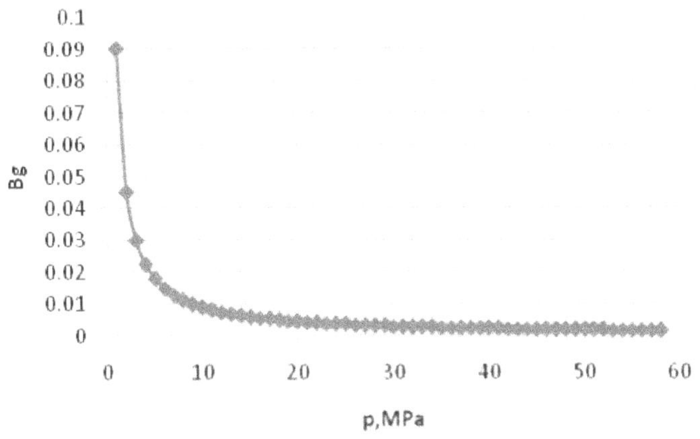

Figure 1.12. Relationship between pressure and volume factor of natural gas (the temperature is 65°C)

(4) Isothermal compressibility

Compared with liquids, the compressibility of gas is much larger. For example the isothermal compressibility of natural gas is $1123 \times 10^{-4}\,\text{MPa}^{-1}$, which is about 300 times greater than that of formation water.

(5) Viscosity

The viscosity of natural gas is much less than that of oil or water. Under normal temperature and pressure, the viscosity of natural gas is $0.005 \sim 0.015\,\text{mPa}\cdot\text{s}$. Under the formation condition, the viscosity of natural gas can be calculated by a semi-empirical formula method:

$$
\begin{aligned}
\mu_g &= 10^{-4} K \exp(X \rho_g^Y) \\
K &= \frac{2.6832 \times 10^{-2}(470 + M)T^{1.5}}{116.1111 + 10.5556M + T} \\
X &= 0.01009 \left(350 + \frac{54777.7}{T} + M\right) \\
Y &= 2.447 - 0.2224X
\end{aligned}
\tag{1.8}
$$

where ρ_g is the density of natural gas, g/cm^3.

Let $p = 20\,\text{MPa}$, $\gamma_g = 0.65$, $T = 323\,\text{K}$, then we obtain the viscosity of natural gas as $0.022\,\text{mPa}\cdot\text{s}$. This calculated viscosity of natural gas is an order of magnitude of $10^{-2}\,\text{mPa}\cdot\text{s}$, which is about one percent of the viscosity of formation water.

1.2.3.5 *Phase behavior*

The phase diagram of a typical multicomponent hydrocarbon system is shown in Figure 1.13.

Envelope: the solid line in phase diagram, that is $\text{aEC}_\text{T}\text{BCGC}_\text{p}\text{Ib}$ line;

Dew point line: $\text{aEC}_\text{T}\text{BC}$ line;

Bubble point line: $\text{CGC}_\text{p}\text{Ib}$ line;

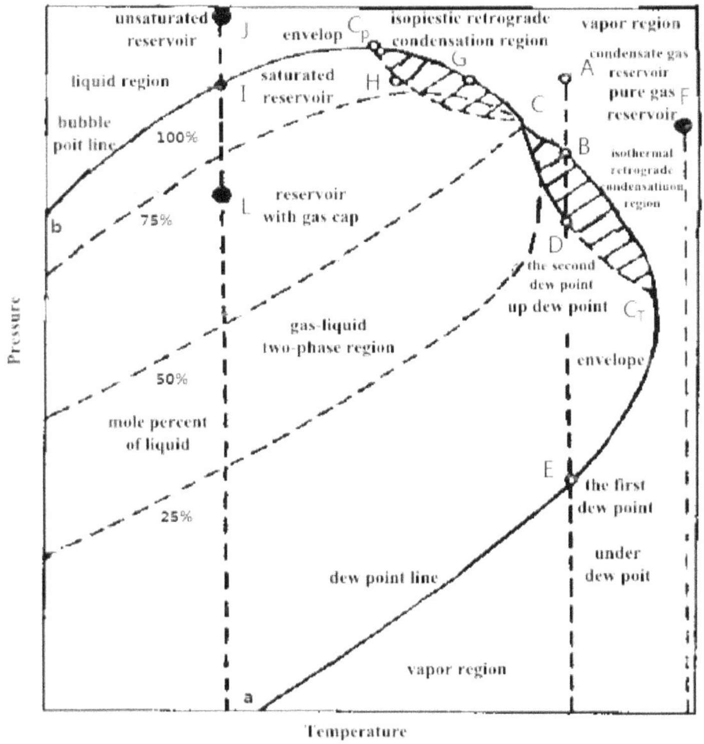

Figure 1.13. Phase diagram of multicomponent hydrocarbon system

The region within this envelope is the two-phase region; the region outside the bubble point line is the liquid region;

Point C is critical point;
C_T is critical gas-condensate temperature;
C_P is critical gas-condensate pressure;

The region of CGC_PHC is isopiestic retrograde condensation region; the region of CBC_TDC is isothermal retrograde condensation (the reservoir is generally isothermal).

Point J is unsaturated reservoir;
Point I is saturated reservoir;
Point L is the reservoir with gas cap;

Point F is pure gas reservoir;
Point A is condensate gas reservoir;
Point E is the first dew point;
Point D is the second dew point;

The original state of point A is pure gas, and then pressure decreases with temperature unchanged. The liquid increases from B to D and the liquid vaporizes from D to E. After E, it is pure gas.

1.3 Rock properties of reservoirs

1.3.1 *Introduction*

Reservoir rocks constitute the porous medium on which the oil and gas depend. They have the characteristics of storage capacity, permeability, large specific surface and complex pore structure. The distribution and flow of fluids in them are affected by the properties of rocks.

The underground hydrocarbon reservoirs are mainly sedimentary rock strata, including clastic rock and carbonate rock. Clastic rock reservoirs, including various types of sandstone, conglomerate, sandy conglomerate and mudstone, take inter-granular pores as the main storage space and flow channel. Carbonate rock is dominated by dolomite and limestone, with fractures and vugs in them.

This section mainly illustrates the characteristics of rocks by taking sandstone of clastic reservoir as an example. Sandstone can be classified according to particle size and specific surface area: For common sandstone, particle size is in the range of $0.25 \sim 1\,\text{mm}$ and the specific surface area is less than $950\,\text{m}^2/\text{m}^3$; for silty sandstone, particle size is in the range of $0.1 \sim 0.25\,\text{mm}$ and the specific surface area is in the range of $950 \sim 2300\,\text{m}^2/\text{m}^3$; for shaly sandstone, particle size is in the range of $0.01 \sim 0.1\,\text{mm}$ and the specific surface is higher than $2300\,\text{m}^2/\text{m}^3$.

There are many physical quantities to describe the properties of rocks, and the main factors closely related to flow mechanics in porous media are: porosity, permeability, saturation, capillary force, elastic compressibility and relative permeability curve.

1.3.2 *Concepts*

Several basic concepts relating to reservoir rocks are introduced in this section.

Particle size: *particle size is the size of the rock particle, which is expressed by its diameter. The symbol is d. The unit is mm or* μm.

Median grain diameter: *the median grain diameter is the grain diameter value at the 50 percent point (by weight) on the grain-size distribution curve. The symbol is* M_d. *The unit is mm or* μm.

Sorting coefficient: *sorting coefficient is the ratios of the particle size values at the 25 percent and 75 percent point (by weight) on the cumulative distribution curve. The closer the value is to 1, the better the reservoir. The symbol is* S_0.

Specific surface: *the specific surface is a ratio of total surface area of rock pores to the rock volume. In* $1 m^3$ *siltstone, the total surface area of pores can reach* $20{,}000 m^2$ *and the specific surface is* $20{,}000 m^{-1}$. *The specific surface area is a key property of the porous medium, and it also illustrates the resistance of fluid flowing through the medium is large.*

Median pore radius: *the median pore radius is the pore radius value at the 50 percent point (by weight) on the cumulative distribution curve of pore radius. The symbol is* R_{50}. *The unit is* μm. *The main range is* $0.1 \sim 100\ \mu m$.

Tortuosity: *the tortuosity is the ratio of actual length of flow pathway to the straight line length through the porous medium, that is, the actual length of the particle trajectory moving along the pathway when the fluid particle passes through the unit distance of the medium. The symbol is* τ. *The range is about* $1 \sim 1.4$.

Elastic compressibility of rocks: *it is the reduction of pore volume per unit volume of rock when the formation pressure decreases by unit* *pressure. The equation can be written as:*

$$C_f = \frac{1}{V_f}\frac{dV_p}{dp} \tag{1.9}$$

where C_f is elastic compressibility of rocks with the common unit as MPa^{-1} and the international unit as Pa^{-1}; V_f is the volume of rocks, m^3; and V_p is the volume of pores, m^3.

Fluid saturation: *fluid saturation is the ratio of the fluid volume to the total pore volume in the reservoir. The symbol is $S_l(S_o, S_w, S_g)$.*

Initial water saturation: *initial water saturation is the ratio of initial water volume to the total pore volume in the reservoir rock, before the reservoir is developed. The symbol is S_{wi}.*

Irreducible water saturation: *irreducible water saturation is the ratio of the unmovable water volume to the pore volume in the reservoir rocks. In other words, it is impossible to contain 100% oil in the reservoir rock. The symbol is S_{wc}. The lower the permeability of the reservoir, the higher the irreducible water saturation.*

Residual oil saturation: *residual oil saturation is the ratio of the volume of oil that cannot be recovered to pore volume in the reservoir rock after the displacement effect of displacing fluid. The symbol is S_{or}.*

Remaining oil saturation: *remaining oil saturation is the ratio of oil volume that cannot be reached by the displacing fluid and remains in the reservoir pores to the total pore volume after oil production.*

Phase: *phase is a homogeneous material with the same properties, such as liquid phase, gas phase, oil phase, water phase, solid phase, etc.*

Composition: *composition is the various components that constitute the phase. For example, the paraffins, naphthenes and aromatics in the oil phase.*

Wetting Phenomena: *when the immiscible two-phase liquids (such as oil and water) contact with a rock surface, one tends to spread* *along and the other tends to adhere to the rock surface. This is called wetting phenomena. If the oil phase tends to spread on the rock surface, the rock is oil wet. If the water phase spreads, the rock is considered to be water wet.*

Capillary pressure: *capillary pressure is the pressure difference across the interface between two immiscible fluids. The symbol is* p_c. *The unit is MPa.*

Absolute permeability: *absolute permeability is the ability of a single-phase liquid passing through the rock. The symbol is* K. *The common unit is mD, and the international unit is* m^2.

Effective permeability: *effective permeability is the ability of one phase to flow through the rock in the presence of other fluid phases.*

Relative permeability: *relative permeability is the ratio of effective permeability to absolute permeability of each phase fluid when multiphase liquids co-exist. The symbol is* K_r.

Oil displacement efficiency: *oil displacement efficiency is the ratio of the displaced oil volume to the initial oil volume in the reservoir. The equation can be written as:*

$$E_D = \frac{1 - S_{wi} - S_{or}}{1 - S_{wi}} \tag{1.10}$$

where E_D *is the oil displacement efficiency;* S_{wi} *is the initial water saturation; and* S_{or} *is the residual oil saturation.*

Recovery factor: *recovery factor is the ratio of the total produced oil to the total amount of oil in the reservoir. The symbol is* E_R. *Generally, it is used to evaluate the contribution of a specific recovery technology. It is referred to as the recovery percentage when the reservoir is abandoned.*

Recovery percentage: *the recovery percentage is the ratio of produced oil to the total amount of oil in the reservoir at a certain production time when the oil is producing.*

Original oil in place: *the original oil in place is the total volume of oil stored in the reservoir before oil production. The equation can be written as:*

$$N_{ooip} = \varphi A h S_{oi} / B_o \tag{1.11}$$

where N_{ooip} *is the original oil in place,* m^3; φ *is the rock porosity;* A *is the reservoir area,* m^2; h *is the reservoir thickness, m;* S_{oi} *is the initial oil saturation;* B_o *is the formation volume factor of oil.*

Reserve abundance: *the reserve abundance is the ratio of original oil in place to the reservoir area. The unit is* $10^4\ m^3/km^2$ *or* ton/km^2. *High abundance is generally more than* $300 \times 10^4\ ton/km^2$, *and low abundance is less than* $100 \times 10^4\ ton/km^2$.

1.3.3 *Analysis*

(1) Porosity

When the sorting coefficient of sandstone is close to 1, the porosity is maximized, which has almost nothing to do with the particle size. It is shown as Figure 1.14. It is clear that sandstone porosity is related to shale content, which is shown as Table 1.1.

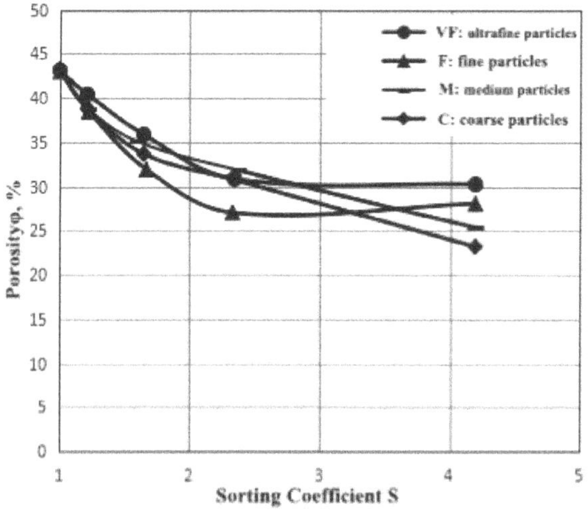

Figure 1.14. Relationship between sorting coefficient and porosity

Table 1.1. Relationship between sandstone porosity and shale content

Shale content, %	<2	$2 \sim 5$	$5 \sim 10$	$10 \sim 15$	$15 \sim 20$
Porosity, %	$28 \sim 34$	$29 \sim 31$	$25 \sim 30$	<25	<20

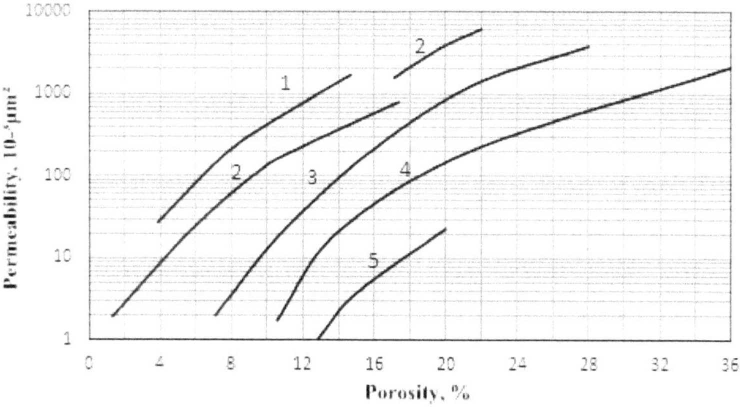

Figure 1.15. Relationship between permeability, porosity and particle size (1: coarse grains; 2: medium grains; 3: fine grains; 4: muddy grains; 5: clay grains)

(2) Permeability

Generally, permeability and porosity have a high positive correlation. Fine grain size, small pore radius and high specific surface correspond to low permeability, which is shown as Figure 1.15.

The larger the sorting coefficient, the lower the permeability.

The greater the compaction, the larger the effective cover, and the lower the permeability.

There are many quantitative calculation methods of permeability, among which the formula based on the capillary tube bundle model is:

$$K = \frac{\varphi \times r^2}{8\tau^2} \tag{1.12}$$

where K is permeability, μm^2; φ is porosity; r is rock pore radius, μm; and τ is the tortuosity.

Let $r = 1\,\mu\text{m}$, $\varphi = 0.25$, $\tau = 1.2$, then $K = 0.0217\,\mu\text{m}^2 = 21.7\,\text{mD}$.

(3) Elastic compressibility of rocks

When the pore pressure decreases, the volume of rock matrix expands, and the pore volume shrinks. The elastic compression of reservoir rocks forms the driving force for oil displacement. The value is generally in the range of $(1 \sim 2) \times 10^{-4}\,\text{MPa}^{-1}$.

(4) Saturation

In sandstone reservoirs, the lower the permeability, the higher the irreducible water saturation, the less the mobile oil, and the higher the residual oil saturation. The relationship between the irreducible water saturation and permeability is shown in Figure 1.16.

(5) Capillary force

The rock pore can be simplified to a capillary tube, and the capillary force can be produced by the wetting function of two-phase interface

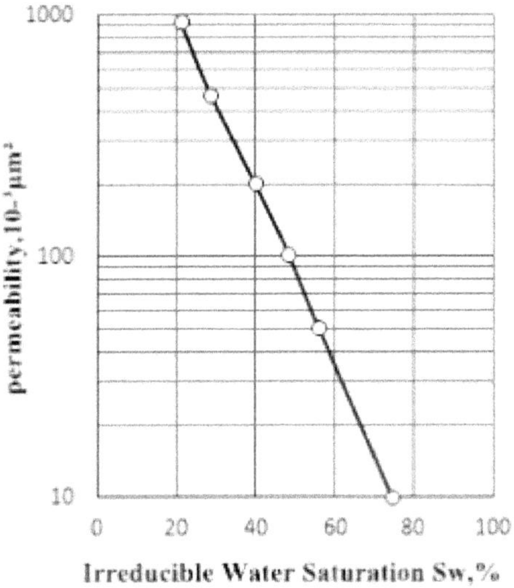

Figure 1.16. Relationship between irreducible water saturation and permeability of sandstone

in the capillary. The equation of capillary force in the oil-water system is:

$$p_c = \frac{2\sigma_{wo} \cos\theta_{wo}}{r} \tag{1.13}$$

where p_c is the capillary force, Pa; σ_{wo} is the oil-water interfacial tension, mN/m; θ_{wo} is the contact angle (oriented in the direction of water); and r is the capillary radius, mm.

Let $r = 0.001$ mm, $\sigma_{wo} = 24$ mN/m (under atmospheric pressure water-gas interfacial tension is 72 mN/m), $\theta_{wo} = 60°$, then $p_c = 24132.8$ Pa $= 0.024132$ MPa; let $r = 0.01\,\mu$m $= 0.00001$ mm, then $p_c = 2.4132$ MPa.

It can be seen that when the pore radius is large, the capillary force is small. And when the pore radius is small (such as low-permeability, tight reservoir and shale), the capillary force is large.

The oil-water two-phase capillary force is the function of water saturation. Mercury intrusion method and semi-permeable diaphragm method are the common experimental methods, and the experimental results need to be converted into underground capillary. The capillary pressure curves in experimental and reservoir conditions are shown in Figure 1.17.

The capillary pressure curve can be used to calculate the water saturation of the oil-water transition zone in addition to establishing the relationship between phase and phase pressure in multiphase flow.

(6) Relative permeability

Liquid is generally more than two-phase (oil-water, oil-gas, gas-water) in the real reservoir, and the physical properties of liquid make the flow characteristics of various fluids quite different. For example, the fluid consistency of oil is larger than that of water, so the flow resistance of water is smaller and the flow velocities of these two phases are different. Therefore, the relationship between phase permeability and relative permeability needs to be clear, which requires the relative permeability curves. It is shown in Figure 1.18.

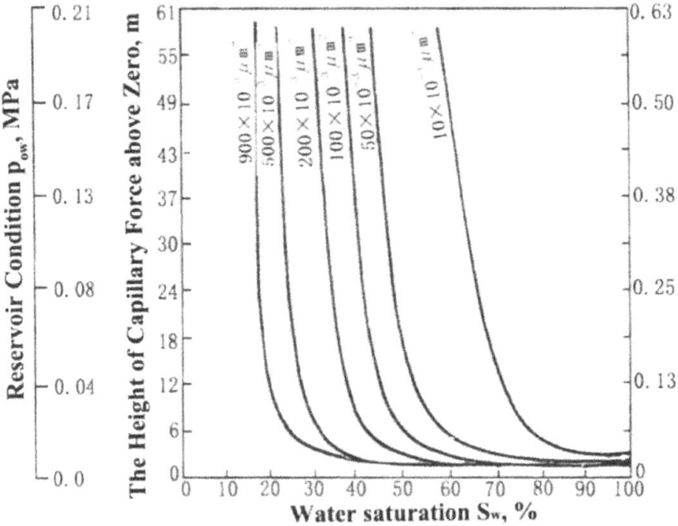

Figure 1.17. Conversion of capillary pressure curves in experimental and reservoir conditions

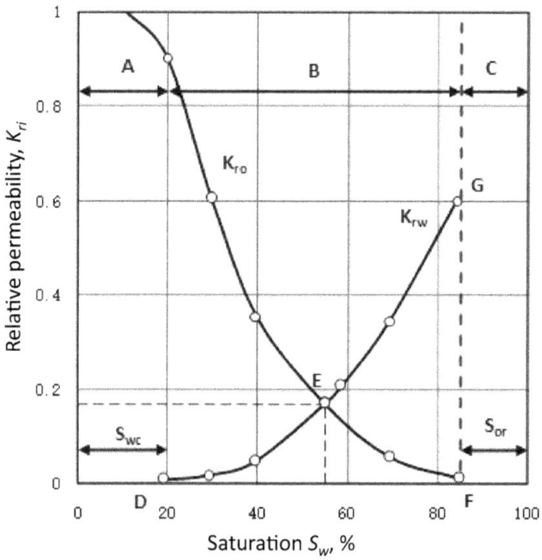

Figure 1.18. Typical oil-water relative permeability curves

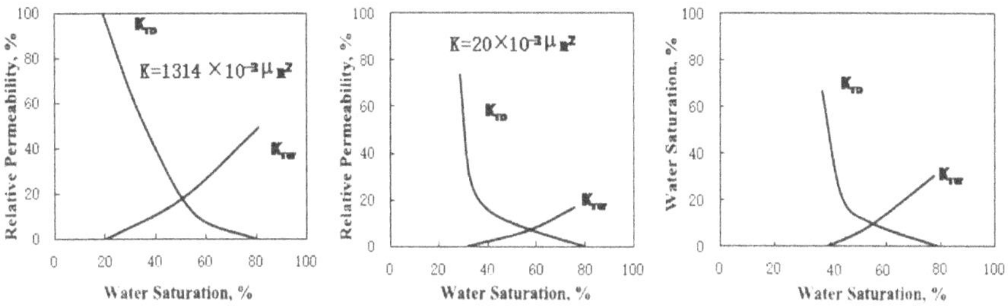

Figure 1.19. Oil-water phase permeability of different sandstones: (a) large pore, good connectivity, high permeability; (b) small pore, low permeability; (c) large pore, low connectivity

Two curves: curve K_{ro} and curve K_{rw}.

Three regions: single-phase oil flowing region A, oil-water concurrent flowing region B, single-phase water flowing region C.

Four characteristic points: point D is the irreducible water saturation S_w, point F is residual oil saturation S_{or}, point E is crossover point, point G is relative saturation of water in residual oil.

Relative permeability reflects the pore characteristics of rocks, and different rocks show different phase permeabilities. The phase-permeability curves of different rocks are shown in Figure 1.19.

The two-phase flow zone of sandstone with high permeability and large pores is wide, and the irreducible water saturation is low, as shown in Figure 1.19(a);

The endpoints of K_{ro} and K_{rw} of sandstone with small pores and low connectivity are both low, as shown in Figure 1.19(b);

The irreducible water saturation of sandstone with large pores and low connectivity is high, and the two-phase flow zone is narrow, as shown in Figure 1.19(c);

The sum of K_{ro} and K_{rw} is less than 1 in the two-phase flow, and the value of $K_{ro} + K_{rw}$ is minimum at the crossover point. Relative permeability is generally low at the crossover point in the case of low permeability.

1.4 Mechanical analysis of flow process

1.4.1 *Introduction*

There is energy in the reservoirs. If the energy reaches an equilibrium state, the fluid will not flow. If the energy does not reach an equilibrium state, the fluid will flow along the direction of weakening energy, which tends to balance the energy. Therefore, fluid flow is essentially a process of energy rebalance.

Fluids in the reservoirs have flow ability themselves; however, fluid flow needs the driving force to overcome flow resistance under the formation conditions. Mechanical analysis is necessary to analyze the resistance and the driving force.

The main forces in fluid flow are: hydrodynamic force, gravity, viscous force, elastic force, capillary force and inertia force.

The flow process is the result of comprehensive interaction of various forces. In the process of oil and gas reservoir exploitation, the fluid will be driven by some main driving force, which determines the driving mechanisms of the reservoirs. Regardless of the supply of artificial energy, the natural energy driven in reservoirs can be divided into five driving mechanisms: hydraulic drive, elastic drive, solution gas drive, gas cap drive and gravity drive.

1.4.2 *Concepts*

Force: *force is the action of an object on another. The force cannot exist separately from the object. In dynamics, it is equal to the product of mass and acceleration of an object. The international unit is Newton (N). It means the force used to make the acceleration of an object of 1 kg mass reach* $1\,m/s^2$.

Pressure: *pressure is the force per unit area, namely,* N/m^2. *The common unit is MPa. The international unit is Pa.*

Formation pressure: *formation pressure is the pressure acting on the fluid contained in the formation. It is also called fluid pressure, pore pressure or reservoir pressure. The common unit is MPa. The international unit is Pa.*

Hydrostatic pressure: *hydrostatic pressure is the pressure produced by gravity of the stationary water column that increases with increasing depth.*

Pressure coefficient: *pressure coefficient refers to the ratio of measured formation pressure to hydrostatic pressure calculated at the same depth.*

Formation pressure gradient: *formation pressure gradient is the increasing rate of formation pressure with depth. The common unit is MPa/100 m. When the pressure coefficient is 1, the formation pressure gradient can be recorded as 1 MPa/100 m.*

Temperature gradient: *the temperature gradient is the increasing rate of formation temperature with depth under the upper layer of ground surface (20 ∼ 130 m). The common unit is °C/100 m.*

Hydrodynamic force: *when the water supply and discharge areas both exist in the underground strata, the fluid flows under the result of head pressure, which is the main driving force. Generally, the value is characterized by pressure difference or pressure gradient.*

Inertia force: *inertia force is a force of inertia caused by the mass of an object, whose magnitude depends on mass and acceleration.*

Water drive: *fluid flow is driven by the pressure energy generated by hydrodynamics of edge water and bottom water connected with the outside.*

Elastic drive: *elastic drive mainly depends on the elastic force of rocks and liquids as the main driving force.*

Solution gas drive: *dissolved gas drive mainly depends on the elastic force of dissolved gas separated from oil as the main driving force.*

Gas cap drive: *pressure drive mainly depends on the elastic force of free gas in the gas cap as the main driving force.*

Gravity drive: *gravity drive mainly depends on the gravity of liquids as the main driving force.*

Rigidity: *Rigid reservoirs or rigid fluids, as opposed to elastic terms, are reservoirs or fluids considered to be incompressible.*

Elasticity: *elastic reservoirs or elastic fluids are considered to produce the elastic force when the pressure decreases.*

1.4.3 *Analysis*

1.4.3.1 *Pressure and temperature*

Pressure and temperature are the major factors affecting the frequent variation of each reservoir, and each factor is the potential energy that can be used. With their changes, the reservoir pores also change. The most important change is the change of the fluid volume in the pores. Therefore, many physical properties of rocks and fluids are all the function of these two parameters. The rebalance and redistribution of the energy in reservoir production can also be regarded as the redistribution of pressure and temperature. And the force on fluids is essentially the reflection of pressure and temperature.

(1) Pressure

The pressure distribution and time-varying regularity of reservoirs' points are the most important contents of flow mechanics, which are also the concerns of oil-gas development. The pressure distribution of each point of the reservoir and its change rule with time are the most important contents of the study of flow mechanics, and also the problem concerned by oil and gas development.

The physical quantity that describes the formation pressure is usually the formation pressure coefficient. Under general geological conditions, the formation pressure and hydrostatic pressure are equivalent, that is, the pressure coefficient is approximately equal to 1, but there are also special geological conditions. According to the formation pressure coefficient, the reservoir can be divided into three categories: abnormal high pressure is the case when the pressure coefficient is greater than 1.2; abnormal low pressure is the case when the pressure coefficient is less than 0.7; and normal pressure is between them.

(2) Temperature

It is generally considered that reservoirs are in isothermal conditions except during thermal recovery.

The temperature gradient is the common physical quantity used to describe the formation temperature. Geothermal gradients are different in different regions of crust, which is generally from $0.9 \sim 5.2°C/100\,m$. Geothermal gradient is closely related to the accumulation of oil-gas reservoirs, and the range of thermal gradient is $2.5 \sim 4.0°C/100\,m$.

1.4.3.2 *Force*

(1) Hydrodynamic force

As shown in Figure 1.20, the reservoir has an open water head connecting to the atmosphere. Under the action of head pressure, oil flows to well A and well B, which generates the hydrodynamic force. It is the main driving force of fluid flow. If the hydraulic head is constant in the open atmospheric area, the pressure of the well will be reduced due to the oil production and the pressure difference will be increased, that is to say, the hydrodynamic force will be increased, so the energy of the driving oil will be increased.

Especially if the reservoir is horizontal strata, the fluid needs more hydrodynamic force (i.e. pressure difference) to flow.

(2) Gravity

As shown in Figure 1.20, the G1 part flowing into well A and the G2 and G3 parts flowing into well B are influenced by the gravity

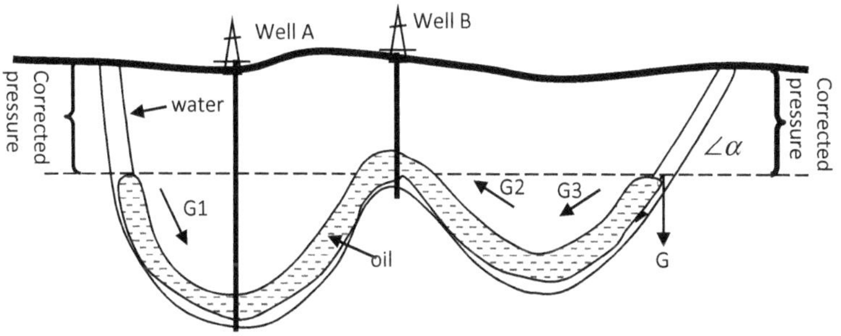

Figure 1.20. Schematic diagram of the effect of hydrodynamic force and gravity

effect. Gravity can be expressed as driving force or resistance in the fluid flow process. In the G1 and G3 parts, gravity is considered as the driving force, and in the G2 part, gravity is considered as resistance. The component of gravity along the direction of fluid flowing is $\rho g \sin \alpha$.

(3) Viscous force

When the fluid flows, there will always be internal friction, which always blocks the fluid flowing, so the viscous force is resistance. Its value is related to the viscosity of the fluid. The greater the viscosity, the greater the flow resistance, the more the energy consumed, and the more difficult the oil production. For example, compared with conventional reservoirs, the heavy oil reservoirs are more difficult to develop with higher cost.

(4) Inertia force

When the fluid flows, the inertia force will act if the velocity magnitude or direction changes. The pore size in reservoirs is irregular and the pores are connected by the smaller pore throat. The velocity of fluid flow is changing almost all the time in it, the inertia force also exists.

The magnitude of inertia force is positively proportional to v^2 (square of velocity). When the velocity of fluid flow is small, the inertia force can be neglected when compared with other forces, such as conventional reservoirs with middle-low permeability and heavy oil reservoirs; however, when the velocity of fluid flow is large, inertia force increases exponentially with the increase of v^2. Compared with other forces, it cannot be neglected any more, such as the flow of gas in reservoirs.

The inertia force in fluid flow is shown as resistance and energy loss.

(5) Elastic force

The oil-gas reservoirs are under high pressure, which is hundreds of times higher than the surface atmospheric pressure. The fluids and rocks in the reservoirs both contain elastic energy and manifest themselves in three aspects:

Liquid elastic force: oil and water are compressible and their magnitude can be quantified by elastic compressibility. When the formation pressure decreases, the expansion of liquid produces the elastic force and pushes the liquid to flow to the bottom of wells.

Gas elastic force: gas has greater compressibility than liquids. When the formation pressure decreases, the gas expands more dramatically. The gas elastic force comes from free gas with gas cap or the solution gas escaped from oil. If the solution gas is still dissolved in the oil, it will only appear as the elastic effect of oil. Only when the gas escapes from the oil and becomes free gas, can the stronger gas elastic force take effect.

Rock elastic force: the elastic expansion of rock due to pressure decrease results in the shrinkage of pore volume, which produces the elastic force to drive fluids.

The release of elastic force must rely on the decrease of pressure. Reservoir development in the early stage often relies on the nature energy and the reservoir pressure will be depleted, hence the elastic force plays an important role at this time. In the middle and later stages of reservoir development, the elastic energy is almost released and the change of reservoir pressure can be neglected. At this stage, it can be considered that there is no elastic energy and the reservoir is rigid.

The elastic force is generally the driving force for oil production.

(6) Capillary pressure

The reservoir can be regarded as being composed of numerous tiny capillary tubes. When the two-phase fluid flows, the interface of two-phase will bend, which produces the capillary force, as shown in Figure 1.21.

The capillary pressure can be expressed as:

$$p_c = p_o - p_w \tag{1.14}$$

where p_c is capillary force, Pa; p_o is the pressure of oil phase, Pa; and p_w is the pressure of water phase, Pa.

Figure 1.21. Schematic diagram of capillary force action

If the wettability is water-wet, p_c is positive, and capillary pressure is the driving force when oil is displaced by water. If the wettability is oil-wet, p_c is negative, and capillary pressure is the resistance when oil is displaced by water. Usually, the capillary force is small, but the capillary force plays a key role in low permeability reservoirs, such as the imbibition of tight oil or shale gas.

1.4.3.3 *Driving mechanisms*

When a fluid moves in a porous medium, various energies work at the same time. However, there will be a more prominent force that plays the major role to drive the fluid flow in different periods or reservoir conditions. Therefore, the driving modes are classified according to the action of this force.

(1) Water drive

Water drive relies on the hydrodynamic pressure difference to drive fluid flow. If the hydrodynamic pressure remains unchanged due to the existence of a large adjacent natural aquifer in the process of reservoir development and the elastic forces of rocks and fluids in the reservoirs are negligible, the water drive can be called rigid water drive at this time.

In many oil and gas reservoirs, there is no strong and continuously stable energy supply source. Some reservoirs are even closed. It is necessary to create an artificial energy supply source through injecting water. In this case, it is called water flooding instead of water drive.

The water drive could exist in the whole development period of oil and gas reservoirs, and its recovery factor can reach $40 \sim 60\%$.

(2) Elastic energy drive

When fluids flow from high pressure to low pressure, the elastic energy of the fluids is released. Elastic energy drive here refers to the driving mechanisms where only the elastic force of liquids and rocks is the key driving force. In this case, the reservoir is an unsaturated reservoir and the formation pressure in the process of development is higher than saturation pressure.

Generally, the recovery factor of elastic drive is low (2–5%).

(3) Solution gas drive

In the process of elastic drive, when the formation pressure is lower than saturation pressure, solution gas escapes from oil to form free gas. It has much more elastic energy than liquids and rocks. At this time, the elastic force of solution gas becomes the dominant driving force.

The solution gas escaped from oil under certain pressure difference is limited. Besides, the escaping of the solution gas will also greatly increase the viscosity of oil, which is not conducive to oil recovery. Therefore, many reservoirs do not utilize or effectively utilize solution gas drive, and usually terminate or avoid solution gas by supplying additional formation energy.

In reservoirs with higher solution gas-oil ratio, the recovery factor of solution gas flooding can reach $10 \sim 20\%$.

(4) Gas cap drive

Gas cap drive occurs in the oil or oil-gas reservoirs with a gas cap. When exploiting such reservoirs, we usually avoid the gas cap and take the lead in exploiting the oil below and use the elastic force of the free gas of the gas cap to maintain the formation energy. Sometimes, artificial gas caps are also used to supplement formation energy.

The energy of gas cap drive depends on the size of gas caps and the recovery factor of reservoir with strong gas cap energy can reach $20 \sim 30\%$.

(5) Gravity drive

Only when other energies have been exhausted and the reservoir is obviously dipped, the gravity drive can take effect under the conditions of low fluid viscosity and high permeability.

1.5 Basic concepts of flow

1.5.1 *Introduction*

Fluid flow in the pores of reservoirs under the various forces has high flow resistance and thus the flow velocity is slow. The flow in pipes is relatively rapid. That is the key difference between the fluid flow in porous media and in pipes.

In addition to understanding about the concepts of reservoirs, fluids, rocks and energy, it also requires understanding the flow regimes, inner and outer boundaries and various pressures.

1.5.2 *Concepts*

Initial formation pressure: *the initial formation pressure is the formation pressure measured before the reservoir is developed. The symbol is* p_i.

Current formation pressure: *the current formation pressure is the average formation pressure of different periods in the process of reservoir development. The symbol is* p_R.

Supply pressure: *the supply pressure is the pressure on the supply boundary of the reservoir. The symbol is* p_e.

Bottom hole pressure (BHP): *the bottom hole pressure is the pressure measured in the middle of the production wells under the normal production condition. The symbol is* p_w.

Corrected pressure: *the corrected pressure is the pressure on each point in the formation converted on the same horizontal plane, which essentially represents the total energy of the fluid at that point. The symbol is* p_r.

Linear flow: *linear flow usually occurs in the rock core during the oil displacement test in the lab or between the well arrays in the field.*

Radial flow: *radial flow occurs around the bottom of an injection or production well.*

Homogeneous reservoir: *in homogeneous reservoirs, the physical properties at different locations do not change.*

Tubing pressure: *the tubing pressure is the pressure measured in the tubing at the wellhead. The symbol is p_{wh}. The unit is MPa.*

Casing pressure: *the casing pressure is the pressure measured at the annular space between the tubing and casing. The symbol is p_t. The unit is MPa.*

1.5.3 *Analysis*

1.5.3.1 *Flow regimes*

(1) Linear flow

As shown in Figure 1.22, linear flow refers to the fluid flow along one direction (x direction). The supply boundary has supply pressure p_e, and the production side has the bottom hole pressure p_w. The reservoir thickness is h, the reservoir width is B, and the reservoir length is L. The flow rate is Q.

If the reservoir is homogenous and the vertical flow caused by gravity is neglected, the flow parameters on all points in the reservoir are the same. In this case, the research object can concentrate on the flow characteristics along the x direction in the section view diagram.

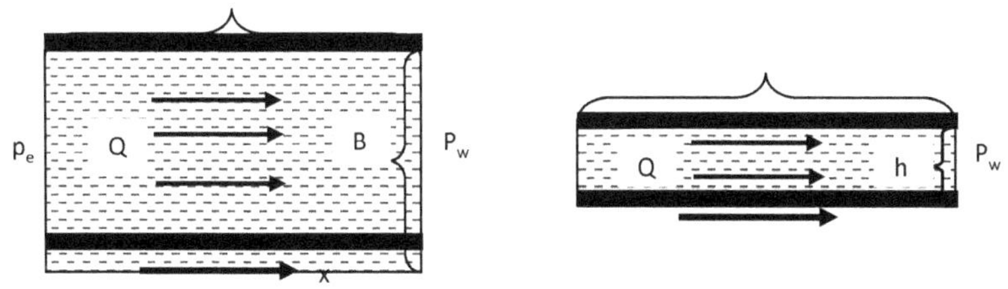

Figure 1.22. Schematic diagram of linear flow

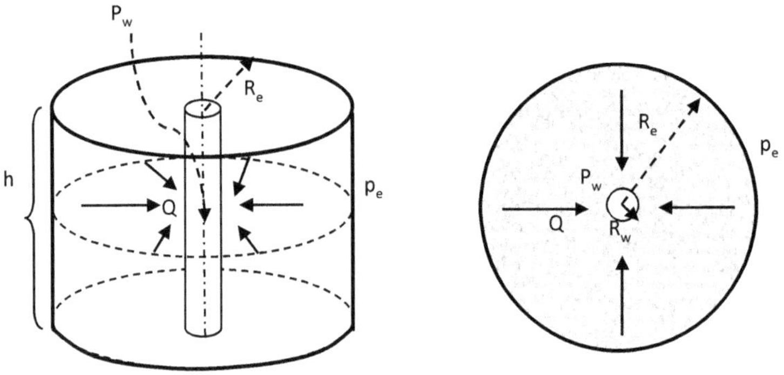

Figure 1.23. Schematic diagram of radical flow

Linear flow is a simple flow regime, and it is also a typical representation of one-dimensional flow.

(2) Radical flow

As shown in Figure 1.23, in the 3D case, the reservoir is a homogeneous cylinder with thickness h and radius R_e. There is a well in the center of the reservoir (completely penetrates the reservoir), the radius of well is R_w, and the bottom hole pressure is p_w. In the 2D case, the fluid flow in (or out) the well is Q. This flow regime is named as *radical flow*.

Radical flow is the most common flow regime in actual oil field development. Flow around wells can be regarded as radical flow. Radical flow can be regarded as two-dimensional flow in the Cartesian coordinate system defined by (x, y). For simplicity, the Cartesian coordinate system is converted into polar coordinates (r, θ).

(3) Spherical flow

As shown in Figure 1.24, there is a well in a thick reservoir. This well is different from planar radical flow that penetrates the reservoir completely. It only produces oil on the top of the reservoir. Under these conditions, the flow near the bottom of the well works as a point sink, and the supply boundary has a semi-spherical shape and the fluid flows from the semi-spherical surface to the central point.

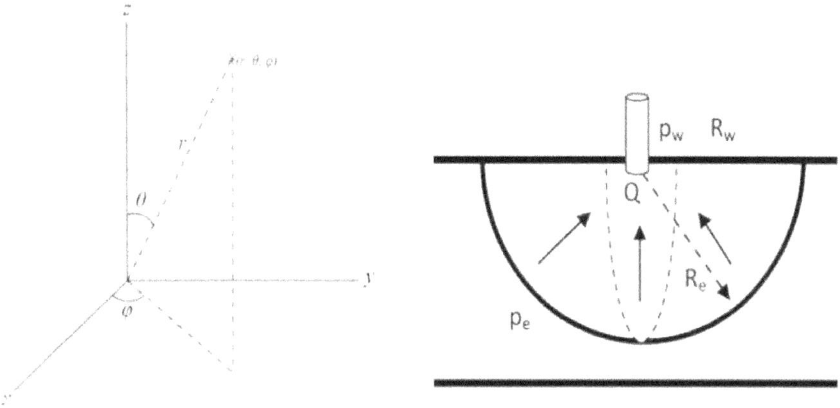

Figure 1.24. Schematic diagram of spherical flow

This flow regime is called *spherical flow* and it can be better characterized in the sperical coordinate system (r, θ, φ).

This flow form is not common in oil fields, and thus it is not emphasized in this book.

1.5.3.2 *Various pressures*

(1) Original formation pressure

Once the reservoir is put into development, the original state of the reservoir will be broken. Thus, it is difficult to measure the original formation pressure. The original formation pressure can only be measured in the initially drilled exploration wells, and in other cases it is generally obtained by indirect methods. The relationship curve between the original formation pressure and depth (also known as pressure gradient curve) can be obtained by the first group of exploration wells. By using this curve, the original formation pressure can be estimated from its well depth.

(2) Current formation pressure

The current formation pressure is an index used to measure the formation energy at different development stages. The current formation pressure in a certain area can be obtained by measuring the bottom hole pressure of the wells that have been shut in for a long time and by the well testing method. The well testing method will be introduced in a future chapter.

(3) Supply pressure

Whether the boundary is open or closed can be judged by the supply pressure, which is determined by the pressure head in the hydrodynamic system.

When the supply pressure is stable, the pressure is rigid aquifer pressure. It is also possible that the pressure changes over time. If the supply boundary of a single well is large, the current formation pressure will be regarded as the supply pressure.

(4) Bottom hole pressure

When the well has been shut in for a long time and the bottom hole pressure is stable, its value is current formation pressure.

When the well begins to produce oil, the pressure difference between the bottom hole pressure and formation pressure is formed. That is:

$$\Delta p = p_R - p_w \tag{1.15}$$

where Δp is the pressure difference, MPa; p_R is the formation pressure, MPa; and p_w is the bottom hole pressure, MPa.

The fluid flow is driven by various forces. The bottom hole pressure can be measured directly by pressure gauge and estimated by the relationship between the flow rate and pressure difference. It can also calculate the pressure loss of the multiphase flow in the wellbore by the tubing pressure and casing pressure.

(5) Corrected pressure

For the convenience of research, the energy in the flowing system needs to be converted on a unified plane so that the energy at different locations can be compared.

As shown in Figure 1.25, if the horizontal plane where point B is located is set as the datum level and point A is projected, then the

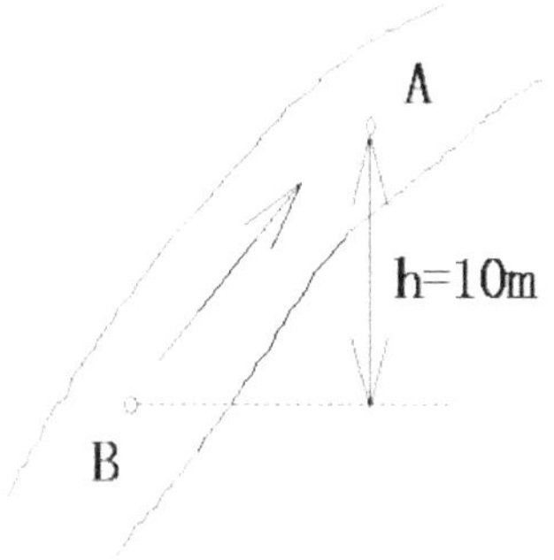

Figure 1.25. Schematic diagram of flow direction obtained from corrected pressure

corrected pressure of point A is:

$$p_{rA} = p_A + \rho g h \tag{1.16}$$

where p_{rA} is the corrected pressure of point A, Pa; p_A is the actual pressure of point A, Pa; ρ is the density, kg/m^3; g is acceleration of gravity, 9.8 N/kg; and h is the height, m.

1.5.3.3 *Inner and outer boundaries*

Fluid flow in a reservoir is always in a certain region, and the boundary is an important part of it. Boundaries can be divided into two categories according to their location, that is, the outer boundary and the inner boundary.

For the outer boundary, the external geometry of the reservoir is known, and the two types of outer boundaries are supply boundary and closed boundary.

For the inner boundary, it refers to the production side in the linear flow regime and the wall of the wellbore in the radical flow regime. It is also divided into two categories: constant pressure production and constant flow rate production. Constant pressure production is a production state in which the bottom hole pressure on the production side remains unchanged. Constant flow rate production is a production state in which the volume flow of a well remains unchanged. The wells are also divided into production well and injection well.

Table 1.2. Standard industrial oil flow

depth, m	Minimum oil flow rate, t/d	Minimum gas flow rate, 10^4 m^3/d
<500	0.3	0.05
500–1000	0.5	0.1
1000–2000	1.0	0.3
2000–3000	3.0	0.5
3000–4000	5.0	1.0
>4000	10.0	2.0

Flow rate of a production well can also be named the production rate, which is a most important index in oil field production. The industrial oil flow of exploratory wells is also expressed by the flow rate of the wells. As shown in Table 1.2, the standard industrial oil flow indicates the minimum oil flow rate at a certain depth. Only if the production rate is above this level, can the oil production be considered as being industrially worthy.

Example

As shown in Figure 1.25, there are two points in the reservoir, $h = 10\,\text{m}$, $p_A = 9.35\,\text{MPa}$, $p_B = 9.5\,\text{MPa}$, relative oil density $\gamma_o = 0.85$. What is the direction of the oil flow?

Solution:

Taking the horizontal plane at point B as the datum level, the corrected pressure of the two points is calculated as:

$$p_{rB} = p_B = 9.5\,\text{MPa}$$

$$p_{rA} = p_A + \gamma_o \rho_w g h = 9.35 + 0.85 \times 1 \times 10^3 \times 9.8 \times 10 \times 10^{-6}$$
$$= 9.433\,\text{MPa}$$

$p_{rB} > p_{rA}$, the oil should flow from point B to point A.

Chapter 2

Basic theory

This chapter mainly introduces the basic law of flow in porous media: Darcy's law, and other concepts such as flow velocity, flow resistance and non-linear flow. The composition and establishment of the mathematical model to describe the flow process will be introduced, which provides the basis to solve the flow problem quantitatively.

2.1 Basic law of flow

2.1.1 *Introduction*

Fluid flow in the porous medium is affected by various forces, and its essence is to consume energy and produce fluid through the wellbore. The relationship between energy and flow rate becomes the most important problem in flow mechanics through porous media. Darcy's law is the most basic law that describes this relationship. This law and the rules extended from it become the main subject of flow mechanics in porous media.

2.1.2 *Concepts*

Darcy's law: *Darcy's law describes the linear relationship between the flow rate Q and pressure difference Δp when the fluid flows in porous media, which can also be named as linear flow law.*

Linear relationship: *if the relationship between two variables can be expressed by a linear function, that is, $y = ax + b$, then it is also called linear relationship.*

Non-linear relationship: *the relationship between the two variable quantities cannot be expressed by a linear function.*

Darcy Flow: *Darcy Flow is the flow that satisfies the linear relationship described by Darcy's law.*

Non-Darcy Flow: *non-Darcy Flow is the flow that has a non-linear relationship between flow rate and pressure difference and cannot be described by Darcy's law.*

Flow area: *the flow area is the cross-sectional area of fluid flowing through porous media, including the rock grains area and the pore area on the cross section. The symbol is A. The unit is m^2. The flow area of linear flow regime is Bh and that of radical flow regime is $2\pi rh$.*

Specific weight: *the specific weight is the weight of unit volume fluid. The symbol is γ. The formula is $\gamma = \rho g$. The unit is N/m^3.*

Productivity index: *the productivity index is the oil recovery per unit time under the unit pressure difference. The symbol is J. The common unit is $m^3/d/MPa$.*

Darcy velocity: *the Darcy velocity is the flow rate passing through a unit cross sectional area of the porous medium, which is an imaginary apparent flow velocity.*

Real flow velocity: *the real flow velocity is the flow rate passing through a unit pore area.*

Reynolds number: *Reynolds number is the ratio of inertia force to viscous force when the fluid flows in the porous medium. The symbol is R_e. The Reynolds number for air flowing through an airplane is 5,000,000, for air flowing through a seagull's wing it is 100,000, for flow through a circular pipe it is 2320, for blood flowing in the brain it is 100 and for oil and water flowing in the reservoirs it is less than 0.2.*

Laminar: *the fluid is said to be in a laminar state when it flows with low velocity in tube, where any material point moves smoothly in a straight line that is parallel to the axis of the tube. The velocity* *of fluid is maximum at the center of the tube and minimum near the tube wall. The Reynolds number for laminar flow is generally less than 2000. And it is larger than 4000 for turbulent flow. The Reynolds number of fluid between 2000 and 4000 is transition flow.*

2.1.3 *Analysis*

2.1.3.1 *Darcy's law*

Darcy's law was proposed by French engineer Henry Darcy through experiments in 1856. A schematic diagram of the experimental device is shown in Figure 2.1.

If there is a stable flow rate at outlet *c*, then the total hydraulic head at the cross section 1-1 in Figure 2.1 is:

$$H_1 = Z_1 + \frac{p_1}{\gamma} \tag{2.1}$$

The total head at the cross section 2-2 is:

$$H_2 = Z_2 + \frac{p_2}{\gamma} \tag{2.2}$$

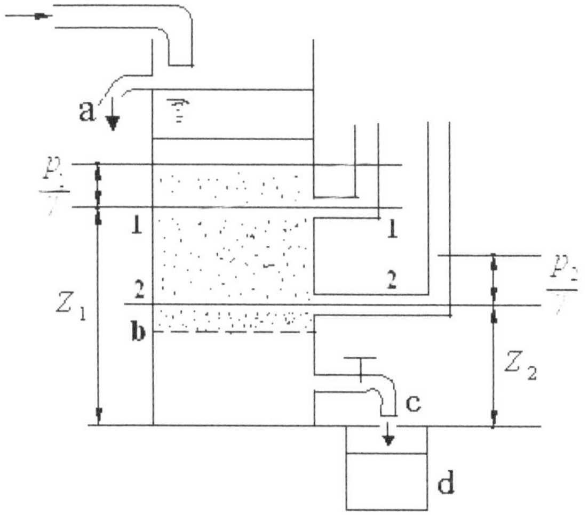

Figure 2.1. Schematic diagram of experimental device of Darcy's law

The head difference between the two cross sections is:

$$\Delta H = \left(Z_1 + \frac{p_1}{\gamma}\right) - \left(Z_2 + \frac{p_2}{\gamma}\right) \tag{2.3}$$

The pressure difference is:

$$\Delta p = \gamma \Delta H \tag{2.4}$$

The relationship between flow rate and pressure difference can be measured by adjusting the valve at the outlet c, and it is shown in Figure 2.2.

The first part of the curve from the origin shows that the relationship is linear. When the flow rate is large, the non-linear relationship starts to show up.

For the linear relationship segment, Darcy studied the influencing factors of flow. If the experimental sand box is placed horizontally, Darcy's law can be obtained as:

$$Q = \frac{K}{\mu} A \frac{\Delta p}{L} \tag{2.5}$$

where Q is the flow rate, cm^3/s; K is the permeability, D; A is the cross sectional area of flow, cm^3; Δp is the pressure difference between the two sections, 10^{-1} MPa; μ is fluid viscosity, mPa · s; and L is the distance between the two sections, cm.

In this equation, the permeability K was introduced as a fitting coefficient to describe the characteristics of porosity and permeability

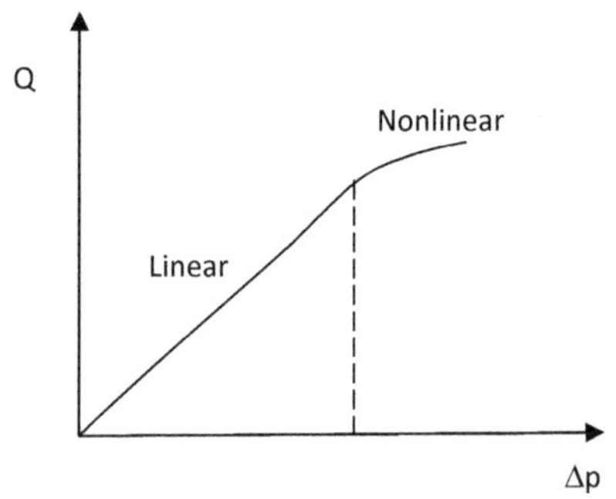

Figure 2.2. Relationship between flow rate and pressure difference

when Darcy analyzed the experiment. To honor his important contribution, the unit of permeability K is named as *Darcy*. The symbol is D, which is also μm^2.

For oil and gas reservoirs, the unit of permeability, D, is too large, so the common unit is mD or $10^{-3}\ \mu m^2$.

In Darcy's equation, the units of the parameters have various forms; the main forms are experimental unit, engineering unit and international unit. They are shown in Table 2.1.

The flow resistance is defined as:

$$R = \frac{\mu L}{KA} \tag{2.6}$$

Then Darcy's equation can be written as the ratio of the driving force of flow to flow resistance:

$$Q = \frac{\text{Driving force}}{\text{Flow resistance}} = \frac{\Delta p}{R} \tag{2.7}$$

The productivity index is defined as:

$$\mathrm{J} = \frac{Q}{\Delta p} = \frac{\mathrm{KA}}{\mu \mathrm{L}} \tag{2.8}$$

If Q is the production rate of oil, then J is the oil productivity index. If Q is the production rate of liquid, then J is the fluid productivity index. If Darcy's law is expressed in the form of productivity index, then it can be written as:

$$Q = J \cdot \Delta p \tag{2.9}$$

Table 2.1. Units of parameters in Darcy's law

Parameter	Experimental unit	Engineering unit	International unit
K	D	mD	m^2
μ	$mPa \cdot s$	$mPa \cdot s$	$Pa \cdot s$
A	cm^2	m^2	m^2
Δp	atm	MPa	Pa
L	cm	m	m
Q	cm^3/s	cm^3/s	m^3/s

2.1.3.2 *Darcy velocity*

Generally, in flow mechanics through porous media, the flow velocity refers to imaginary flow velocity, that is, velocity of fluid passing through per unit cross sectional area of rock.

According to the definition, Darcy velocity can be written as:

$$v = \frac{Q}{A} = \frac{K}{\mu}\frac{\Delta p}{L} \tag{2.10}$$

Define x direction as the one along the direction of fluid flow, and write $\Delta p/L$ as the changing rate along x direction, i.e., pressure gradient. It can be expressed by using the differential form:

$$grad(p) = -\frac{dp}{dx} \tag{2.11}$$

The Darcy velocity is:

$$v = -\frac{K}{\mu}\frac{dp}{dx} \tag{2.12}$$

This is the differential form of Darcy's law, which is also the equation of motion to establish the mathematical model of flow in future chapters.

In the pores of reservoir rocks, the flow velocity of material point, v_φ, is the real velocity, and the relationship between the real and Darcy velocity is:

$$v_\varphi = \frac{v}{\varphi} \tag{2.13}$$

2.1.3.3 *Non-linear flow*

(1) High-speed non-linear flow

In Figure 2.2, the non-linear segment occurs due to the higher flow velocity caused by the increased inertia force, which makes the flow not satisfy Darcy's Law and the non-linear relationship show up.

The flow mechanics through porous media is a branch of flow mechanics. In this book we mainly study the characteristics of fluid flow with low velocity in underground reservoirs.

In flow mechanics, Reynold number (R_e) is an important physical quantity judging the state of fluid flow. The Reynold number in porous media is:

$$R_e = \frac{v_\varphi \sqrt{K}\rho}{1750\mu\varphi^{\frac{3}{2}}} \tag{2.14}$$

where v_φ is the real flow velocity, cm/s; K is the permeability, μm^2; ρ is the density, g/cm^3; μ is the viscosity, $mPa \cdot s$; and φ is the porosity.

Darcy flow is in the laminar flow segment, where the Reynold number is generally less than 0.2 or 0.3.

If the flow velocity is large and Reynold number is larger than the critical value, it is named as high-speed non-linearity. In this case, the flow law can be expressed in exponential and binomial forms.

For the exponential form, it can be written as:

$$Q = C(\Delta p)^n \tag{2.15}$$

For the binomial form, it can be written as:

$$\Delta p = aQ + bQ^2 \tag{2.16}$$

(2) Low-speed non-linear flow

There is an immobile layer on the interface between the fluid and rocks. When the pores are small and the permeability is low, the flow velocity is small, the immobile layer becomes thicker and the additional resistance becomes larger. The essence is that the flow channel becomes smaller, the permeability also becomes smaller, and thus the flow shows a non-linear relationship. In low-speed non-linear flow, apparent permeability can be introduced to replace actual permeability or the threshold pressure gradient can be introduced. Since there is no consensus on how to deal with this issue and is still an ongoing research topic, we will not discuss it in this book.

2.1.3.4 *Example*

(1) In Darcy's experiment, if the rock core is placed horizontally, the diameter of core is 25 mm, the length is 40 cm, the permeability is

250 mD, the viscosity of liquid is 5 mPa · s, and the pressure difference at two ends is 3 atm. Please calculate the flow rate Q (cm^3/min) and Darcy velocity $v(\mu\text{m/s})$.

Solution:

$$Q = \frac{K}{\mu} A \frac{\Delta p}{L}$$
$$= \frac{250 \times 10^{-15}}{5 \times 10^{-3}} \times 3.14 \times (25 \div 2 \times 10^{-3})^2 \times \frac{3 \times 10^5}{40 \times 10^{-2}} \times 60 \times 10^6$$
$$= 1.104\,\text{cm}^3/\text{min}$$

$$v = \frac{K}{\mu}\frac{\Delta p}{L} = \frac{250 \times 10^{-15}}{5 \times 10^{-3}} \times \frac{3 \times 10^5}{40 \times 10^{-2}} \times 10^6 = 37.5\,\mu\text{m/s}$$

Or:

$$v = \frac{Q}{A} = \frac{1.104 \times 10^{-6}/60}{3.14 \times (25 \div 2 \times 10^{-3})^2} \times 10^6 = 37.5\,\mu\text{m/s}$$

(2) In Darcy's experiment, if the core is placed horizontally, the diameter of core is 25 mm, the length is 40 cm, the permeability is 250 mD, the viscosity of liquid is 5 mPa · s, the pressure difference at two ends is 3 atm, the relative density of liquid is 0.9, and the porosity is 0.2. Please calculate the Reynold number.

Solution:

The real flow velocity in the pores is:

$$v_\phi = \frac{v}{\phi} = \frac{37.5}{0.2} = 187.5\,\mu\text{m/s}$$

$$R_e = \frac{v_\phi \sqrt{K}\rho}{1750\mu\phi^{\frac{3}{2}}} = \frac{187.5 \times 10^{-4} \times \sqrt{250 \times 10^{-3}} \times 0.9 \times 1}{1750 \times 5 \times 0.2^{1.5}}$$
$$= 1.08 \times 10^{-5}$$

Re < 0.1, the flow is laminar flow, which satisfies Darcy's law.

2.2 Basic mathematical models

2.2.1 *Introduction*

In the process of oil-gas flow, what we are most concerned about are the magnitude of energy, the distribution of pressure and the amount of production. The mathematical model of flow is a crucial way to conduct theoretical research on how to calculate the change characteristics and rules of these physical quantities, which is also the theoretical basis of other subjects, such as production engineering, reservoir engineering, reservoir numerical simulation, and modern well testing.

A complete mathematical model of flow generally consists of two parts: basic differential equations of flow and boundary (and initial) conditions. The detailed compositions of a mathematical model are shown in Figure 2.3.

The governing equation describes the flow behaviors in the differential form, and it is the combination of the equation of motion, the equation of state and the equation of continuity. The equation of

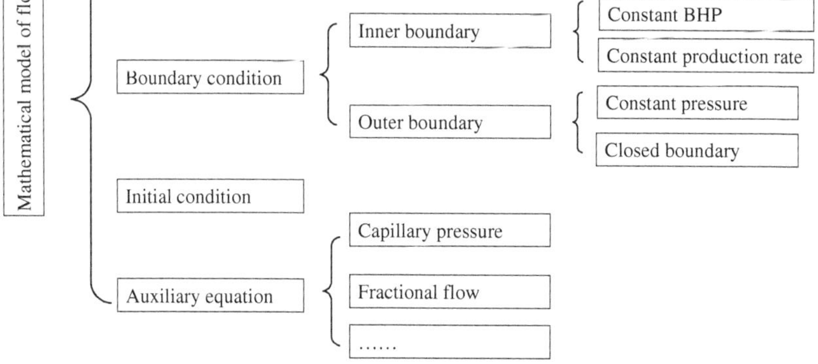

Figure 2.3. Compositions of a mathematical model of flow

state indicates the relationship of pressure, volume and temperature (PVT) for rock, water, oil and gas. The continuity equation or mass conservation equation is the mathematical expression connecting the physical quantities in the reservoir. Usually, the continuity equation is written in terms of pressure and saturation. The boundary condition and initial condition are used to obtain the solution of the differential equation governing the flow system. The auxiliary equations are not always in the mathematical model. They are used for some special occasions to help find the solutions of the flow model, such as capillary pressure equation and fractional flow equation for multiphase flow, which will be introduced in later chapters.

2.2.2 *Concepts*

Mathematical model of flow: *the mathematical model of flow is the equation or equation set used to describe all the mechanical phenomena and motion laws in the processes of reservoir fluid flow.*

Equation of state: *the equation of state is the equation that describes the changes of fluid (oil, gas and water) density or rock porosity with pressure for the petroleum reservoir system.*

Representative elementary volume (REV): *REV is also called the representative volume element (RVE) or the unit cell. It is the smallest volume that can be representative of the entire system, and is important to study the physical properties of complex media.*

Equation of continuity: *The equation of continuity or the mass conservation equation describes a relationship between the amount of reservoir fluids entering and leaving the REV and the accumulation of the reservoir fluids remaining in the REV.*

Mass flow velocity: *mass flow velocity is the product of flow velocity and density of fluid in the porous medium.*

Ordinary differential equation: *ordinary differential equation is a differential equation that involves one independent variable, its functions and its ordinary derivatives. Such as* $\frac{d^2p}{dr^2} + \frac{1}{r}\frac{dp}{dr} = 0$.

Partial differential equation: *partial differential equation is an equation that involves several independent variables, their functions and partial derivatives. Such as* $\frac{\partial^2 p}{\partial x^2} + \frac{\partial^2 p}{\partial y^2} = \frac{1}{\eta}\frac{\partial p}{\partial t}$.

2.2.3 *Analysis*

The analytical solution of a differential equation refers to a solution that can be expressed by a mathematical expression directly. This book focuses on analytical solutions to analyze the law of flow in porous media quantitatively. Analytical solutions can be classified into general solution and specific solution. There are arbitrary constants in the general solution, and the number of arbitrary constants is the same as the order of the differential equation. Specific solution is one special case of the general solution. There is no arbitrary constant in the special solution. The constants can be determined by a set of initial and boundary conditions.

Numerical solution is the solution obtained by finite difference method, finite element method, or numerical interpolation method. When an analytical solution cannot be obtained directly by using calculus, the numerical solution can be helpful to obtain the solution to simulate the flow system by means of numerical analysis. Reservoir numerical simulation mainly demands the procedures to replace the differential equation with a difference equation, discretize the petroleum reservoir by using a numerical grid, build the equation for each grid point, solve the equation set of difference equations by using linear algebra and obtain the solution for pressure and saturation on each grid point at different times.

2.2.3.1 *Establishment of the governing differential equations*

Here, we take the single-phase flow as an example to derive the differential equations that govern the flow in porous media.

(1) Equation of continuity

As shown in Figure 2.4, we take out a cubic shape REV from the reservoir. The mass flow velocity at the center point M of the REV is:

$$\rho v \tag{2.17}$$

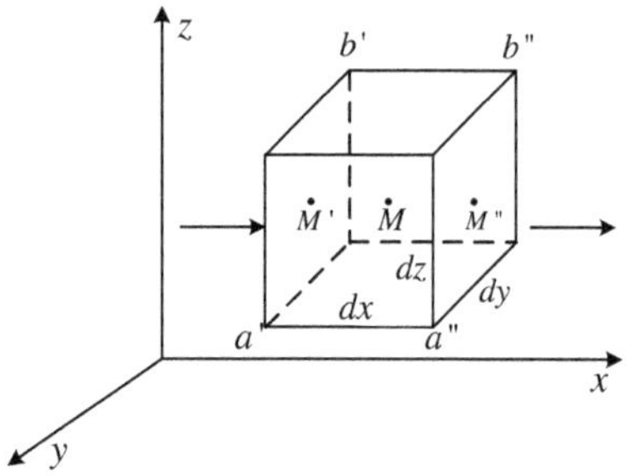

Figure 2.4. Representative elementary volume (REV) from the reservoir

The component of mass flow velocity along different directions are:

$$\rho v_x, \rho v_y, \rho v_z \tag{2.18}$$

The change rate of mass flow velocity along x, y and z directions are:

$$\frac{\partial(\rho v_x)}{\partial x}, \frac{\partial(\rho v_y)}{\partial y}, \frac{\partial(\rho v_z)}{\partial z} \tag{2.19}$$

The component of mass flow velocity along x direction at the center point M' on the plane $a'b'$ is:

$$\rho v_x - \frac{\partial(\rho v_x)}{\partial x}\frac{dx}{2} \tag{2.20}$$

The component of mass flow velocity along x direction at the center point M'' on the plane $a''b''$ is:

$$\rho v_x + \frac{\partial(\rho v_x)}{\partial x}\frac{dx}{2} \tag{2.21}$$

The mass inflow of REV along x direction for a period of time dt is:

$$\left[\rho v_x - \frac{\partial(\rho v_x)}{\partial x}\frac{dx}{2}\right] dydzdt \tag{2.22}$$

The mass outflow of REV along x direction for a period of time dt is:

$$\left[\rho v_x + \frac{\partial(\rho v_x)}{\partial z}\frac{dx}{2}\right] dydzdt \tag{2.23}$$

The mass difference of inflow and outflow of the REV along x direction for a period of time dt is:

$$-\frac{\partial(\rho v_x)}{\partial x}dxdydzdt \tag{2.24}$$

For the same reason, the mass difference of inflow and outflow of the REV along directions y and z for a period of time dt are:

$$-\frac{\partial(\rho v_y)}{\partial y}dxdydzdt, -\frac{\partial(\rho v_z)}{\partial z}dxdydzdt \tag{2.25}$$

The mass difference of inflow and outflow of the entire REV is:

$$-\left[\frac{\partial(\rho v_x)}{\partial x}+\frac{\partial(\rho v_y)}{\partial y}+\frac{\partial(\rho v_z)}{\partial z}\right]dxdydzdt \tag{2.26}$$

The difference in the mass of inflow and outflow through the REV is due to the effect of elastic energy of rocks and fluids in the REV, which results in the accumulation of mass in the REV. The mass accumulation because of this process is shown as follows.

The pore volume in the REV is:

$$\varphi dxdydz \tag{2.27}$$

The fluid mass in the REV is:

$$\rho\varphi dxdydz \tag{2.28}$$

The accumulation rate of fluid mass per unit time is:

$$\frac{\partial(\rho\phi)}{\partial t}dxdydz \tag{2.29}$$

The total accumulation of fluid mass in dt time is:

$$\frac{\partial(\rho\phi)}{\partial t}dxdydzdt \tag{2.30}$$

According to the law of mass conservation:

$$-\left[\frac{\partial(\rho v_x)}{\partial x}+\frac{\partial(\rho v_y)}{\partial y}+\frac{\partial(\rho v_z)}{\partial z}\right]dxdydzdt=\frac{\partial(\rho\phi)}{\partial t}dxdydzdt \tag{2.31}$$

It can be simplified to:

$$-\left[\frac{\partial(\rho v_x)}{\partial x}+\frac{\partial(\rho v_y)}{\partial y}+\frac{\partial(\rho v_z)}{\partial z}\right]=\frac{\partial(\rho\phi)}{\partial t} \tag{2.32}$$

This equation is the continuity equation of single-phase fluid, which is also named as the mass conservation equation.

It can be also expressed as:

$$-div(\rho\vec{v})=\frac{\partial(\rho\varphi)}{\partial t} \tag{2.33}$$

Or:

$$-\nabla\bullet(\rho\vec{v})=\frac{\partial(\rho\varphi)}{\partial t} \tag{2.34}$$

where div( ) is the divergence symbol; and ∇ is the Hamiltonian operator.

Because the flow velocity is a vector with direction, the product of flow velocity and Hamiltonian operator in Equation (2.34) is $\cdot$ (dot product). If the parameters in brackets are scalars, such as the pressure in each direction, it is recorded as ∇p.

(2) Equation of motion

In the flow mechanics of porous media, the equation of motion is generally expressed by the flow velocity in the differential form of Darcy's law. The differential forms of the equation of motion in three directions are:

$$v_x=-\frac{K}{\mu}\frac{\partial p}{\partial x},\ v_y=-\frac{K}{\mu}\frac{\partial p}{\partial y},\ v_z=-\frac{K}{\mu}\frac{\partial p}{\partial z} \tag{2.35}$$

It can also be written as:

$$\vec{v}=-\frac{K}{\mu}grad(p)\quad \text{or}\ \vec{v}=-\frac{K}{\mu}\nabla p \tag{2.36}$$

where *grad* represents gradient.

(3) Equation of state for fluid

The definition of the elastic compressibility coefficient of liquid is:

$$C_L=-\frac{1}{V_L}\frac{dV_L}{dp} \tag{2.37}$$

According to the equation of mass conservation, the liquid mass M doesn't change when the fluid is shrinking or swelling, that is:

$$M = \rho V_L \tag{2.38}$$

Take the derivative to the above equation:

$$dV_L = -\frac{M}{\rho^2} d\rho \tag{2.39}$$

Substitute the equation into the elastic compression coefficient equation:

$$C_L = \frac{1}{\rho} \frac{d\rho}{dp} \tag{2.40}$$

By using the integration method of separated variables, the following results are obtained:

$$\int_{p_0}^{p} C_L dp = \int_{\rho_0}^{\rho} \frac{1}{\rho} d\rho \tag{2.41}$$

$$C_L(p - p_0) = \ln \frac{\rho}{\rho_0} \tag{2.42}$$

$$\rho = \rho_0 e^{C_L(p - p_0)} \tag{2.43}$$

By using the Maclaurin Series expansion and only keep the first two items (with enough accuracy), the equation of state for liquid can be obtained as:

$$\rho = \rho_0[1 + C_L(p - p_0)] \tag{2.44}$$

where ρ_0 is the initial density, kg/m^3; p is the current pressure, MPa; and p_0 is the initial pressure, MPa.

It can be indicated from the equation of state for liquid that the larger the pressure, the larger the liquid density, and vice versa.

(4) Equation of state for rock

The rock elastic compressibility coefficient is defined as:

$$C_f = \frac{1}{V_f} \frac{dV_p}{dp} \tag{2.45}$$

where V_f is the apparent volume (containing pores and grains), and V_p is the pore volume.

Due to the definition of porosity $\phi = \frac{V_p}{V_f}$, it can be obtained by taking the derivatives for both sides:

$$d\varphi = \frac{dV_p}{V_f} \tag{2.46}$$

Substituting it into the equation of the rock elastic compressibility coefficient:

$$C_f = \frac{d\varphi}{dp} \tag{2.47}$$

$$d\varphi = C_f dp \tag{2.48}$$

By using the integration method of separated variables, it can be obtained that:

$$\int_{\varphi_0}^{\varphi} d\varphi = \int_{p_0}^{p} C_f dp \tag{2.49}$$

$$\varphi = \varphi_0 + C_f(p - p_0) \tag{2.50}$$

where φ_0 is the initial porosity.

This porosity formula is the state equation for rocks. It is shown that the porosity decreases with decrease in pressure, and vice versa.

(5) Basic differential equation

As has been derived, the continuity equation is:

$$-\left[\frac{\partial(\rho v_x)}{\partial x} + \frac{\partial(\rho v_y)}{\partial y} + \frac{\partial(\rho v_z)}{\partial z}\right] = \frac{\partial(\rho\phi)}{\partial t} \tag{2.51}$$

Substitute the equation of motion into the equation of continuity:

$$\frac{\partial}{\partial x}\left(\rho\frac{K}{\mu}\frac{\partial p}{\partial x}\right) + \frac{\partial}{\partial y}\left(\rho\frac{K}{\mu}\frac{\partial p}{\partial y}\right) + \frac{\partial}{\partial z}\left(\rho\frac{K}{\mu}\frac{\partial p}{\partial z}\right) = \frac{\partial(\rho\phi)}{\partial t} \tag{2.52}$$

This is another important expression of continuity equation.

In the flow mechanics of porous media, there are two types of basic differential equations for single-phase flow.

The first is rigid steady state flow. In the steady state flow of the rigid reservoir, $C_L = 0$, $C_f = 0$, and the equation of state indicates that $\rho = \rho_0$, $\phi = \phi_0$. Then the basic differential equation for steady state flow in the rigid reservoir is:

$$\frac{\partial^2 p}{\partial x^2} + \frac{\partial^2 p}{\partial y^2} + \frac{\partial^2 p}{\partial z^2} = 0 \tag{2.53}$$

It is also expressed as:

$$\nabla^2(p) = 0 \tag{2.54}$$

where ∇^2 is the Laplacian operator, sometimes it is also expressed by Δ.

This equation is named as Laplacian equation, it is also called harmonic equation and elliptic equation.

For two-dimensional flow, the basic differential equations in Cartesian coordinates and polar coordinates are:

$$\frac{\partial^2 p}{\partial x^2} + \frac{\partial^2 p}{\partial y^2} = 0, \qquad \frac{d^2 p}{dr^2} + \frac{1}{r}\frac{dp}{dr} = 0 \tag{2.55}$$

The second is elastic transient flow. In the transient flow of the elastic reservoir, the basic differential equation can be obtained:

$$\frac{\partial^2 p}{\partial x^2} + \frac{\partial^2 p}{\partial y^2} + \frac{\partial^2 p}{\partial z^2} = \frac{\mu C_t}{K}\frac{\partial p}{\partial t} \tag{2.56}$$

It can also be written as:

$$\nabla^2(p) = \frac{\mu C_t}{K}\frac{\partial p}{\partial t} \tag{2.57}$$

where C_t is the total compressibility coefficient, MPa^{-1}. $C_t = \varphi C_L + C_f$.

This equation is named as Fourier equation, and it is also called as diffusion equation, heat conduction equation and parabolic equation.

Sometimes for convenience, the basic differential equation is often expressed in the form of cylindrical coordinates (r, θ, z):

$$\frac{1}{r}\frac{\partial}{\partial r}\left(r\frac{\rho K}{\mu}\frac{\partial p}{\partial r}\right)+\frac{\partial}{\partial z}\left(\frac{\rho K}{\mu}\frac{\partial p}{\partial z}\right)+\frac{1}{r^2}\frac{\partial}{\partial \theta}\left(\frac{\rho K}{\mu}\frac{\partial p}{\partial \theta}\right)=\frac{\partial(\varphi\rho)}{\partial t} \tag{2.58}$$

In two-dimensional radial flow, the basic differential equation in Cartesian coordinates $(x,\ y)$ and polar coordinates $(r,\ \theta)$ is:

$$\frac{\partial^2 p}{\partial x^2}+\frac{\partial^2 p}{\partial y^2}=\frac{\mu C_t}{K}\frac{\partial p}{\partial t},\quad \frac{\partial^2 p}{\partial r^2}+\frac{1}{r}\frac{\partial p}{\partial r}=\frac{\mu C_t}{K}\frac{\partial p}{\partial t} \tag{2.59}$$

2.2.3.2 *Definite conditions*

(1) Boundary conditions

There are two types of outer boundary conditions: supply boundary and closed boundary.

There are two types of inner boundary conditions: constant bottom hole pressure (BHP) and constant production rate.

The above classifications are the boundary conditions commonly used in this book. Strictly speaking, there is a third boundary condition: the mixed boundary of the first and the second.

(2) Initial condition

The initial condition is the original state of physical parameters before reservoir production, which usually refers to the distribution of pressure and saturation.

2.2.3.3 *Basic mathematical model*

The establishment of the flow mathematical model includes the basic governing differential equations, boundary conditions and initial conditions. Here, let us take an actual reservoir shown in Figure 2.5 as an example to explain the establishment of a basic flow mathematical model.

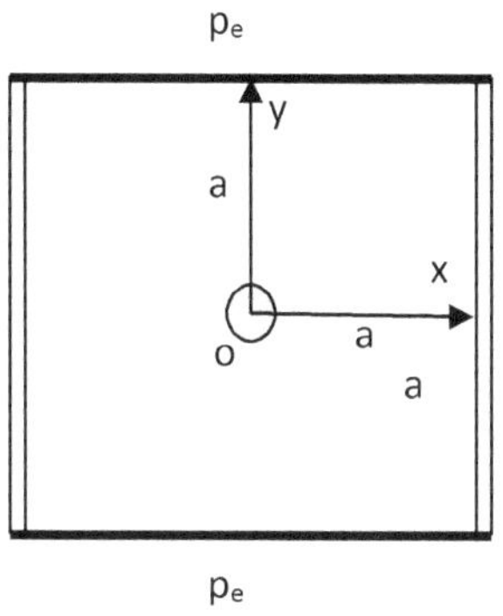

Figure 2.5. Schematic diagram of a rectangular reservoir

There is a production well in the center of the square reservoir and the length of the boundary is a. The fluid in the reservoir is the single-phase Newtonian fluid. The horizontal outer boundaries are supply boundaries, and the pressure of supply boundary is p_e. The vertical outer boundaries are the closed boundaries, the initial formation pressure is p_i, the permeability is K, the fluid viscosity is μ, the porosity is φ, the total elastic compressibility coefficient of rock is C_t, the thickness of reservoir is h, the wellbore radius is R_w, and the production rate is Q. Take the center of well as the origin of the Cartesian coordinate system. The flow mathematical model of this problem is:

$$\begin{cases} \dfrac{\partial^2 p}{\partial x^2} + \dfrac{\partial^2 p}{\partial y^2} = \dfrac{\mu C_t}{K}\dfrac{\partial p}{\partial t} & \text{Governing equation} \\ p|_{y=\pm a} = p_e & \text{Supply outer boundary} \\ \left.\dfrac{\partial \mathrm{p}}{\partial \mathrm{x}}\right|_{x=\pm a} = 0 & \text{Closed outer boundary} \\ r\left.\dfrac{\partial p}{\partial r}\right|_{\mathrm{r}=\sqrt{x^2+y^2}=R_w} = \dfrac{Q\mu}{2\pi K h} & \text{Constant production rate inner boundary} \\ p|_{\mathrm{t}=0} = p_i & \text{Initial boundary} \end{cases} \tag{2.60}$$

2.2.3.4 *Example*

1. There is a production well at the center of the circular supply boundary, which produces with the constant production rate Q. The known parameters are R_e, R_w, K, h, μ, p_e. Write out the mathematical model of the single-phase steady state flow.

$$\begin{cases} \dfrac{d^2p}{dr^2} + \dfrac{1}{r}\dfrac{dp}{dr} = 0 & \text{governing equation} \\ p|_{r=R_e} = p_e & \text{supply outer boundary condition} \\ r\left.\dfrac{\partial p}{\partial r}\right|_{r=R_w} = \dfrac{Q\mu}{2\pi Kh} & \text{constant production rate inner boundary condition} \end{cases}$$

Because the steady state flow has nothing to do with time, there is no initial condition in the flow mathematical model.

2. There is an injection well at the center of the circular closed boundary with constant injection rate Q. The known parameters are R_e, R_w, K, h, μ, p_i. Write out the mathematical model of the single-phase steady state flow.

$$\begin{cases} \dfrac{\partial^2p}{\partial r^2} + \dfrac{1}{r}\dfrac{\partial p}{\partial r} = \dfrac{\mu C_t}{K}\dfrac{\partial p}{\partial t} & \text{governing equation} \\ \left.\dfrac{\partial p}{\partial r}\right|_{r=R_e} = 0 & \text{closed outer boundary condition} \\ r\left.\dfrac{\partial p}{\partial r}\right|_{r=R_w} = \dfrac{-Q\mu}{2\pi Kh} & \text{constant production rate inner boundary condition} \\ p|_{t=0} = p_i & \text{initial condition} \end{cases}$$

For the constant rate of the injection well, the negative sign with Q is used in the inner boundary condition of the flow mathematical model.

3. There is a production well at the center of the circular closed boundary. The change of outer boundary pressure with time is $p_e(t)$, and the change of the production rate of the well with time is $Q(t)$. The known parameters are R_e, R_w, K, h, μ, p_i. Write out

the mathematical model of the single-phase transient flow.

$$\begin{cases} \dfrac{\partial^2 p}{\partial r^2} + \dfrac{1}{r}\dfrac{\partial p}{\partial r} = \dfrac{\mu C_t}{K}\dfrac{\partial p}{\partial t} & \text{governing equation} \\ \mathrm{p}|_{\mathrm{r}=R_e} = p_e(t) & \text{outer boundary condition} \\ r\dfrac{\partial p}{\partial r}\Big|_{\mathrm{r}=R_w,t} = \dfrac{-Q(t)\mu}{2\pi K h} & \text{inner boundary condition} \\ p|_{t=0} = p_i & \text{initial condition} \end{cases}$$

This shows that the flow mathematical model can take account of the variable boundary conditions.

Chapter 3

Classical flow theory in porous media

In classical flow mechanics in porous media, the influencing factors considered in the flow process are relatively simple, such as it is assumed that the medium is homogeneous and isotropic with intergranular pore structure, the fluid is a Newtonian fluid, the reservoir is under isothermal condition, the well is vertical well, etc. The classical flow problems in porous media mainly include the single-phase steady state flow, the multi-well interference, the elastic energy driven flow, the oil-water two-phase flow, the natural gas flow, the oil-gas two-phase flow and the gas-water two-phase flow. The first four problems are the key problems in this book.

3.1 Single-phase steady state flow theory

3.1.1 *Introduction*

Assumptions associated with this flow problem are: homogeneous, isotropic, uniformly thick, isothermal, rigid reservoir, single phase and Newtonian fluid.

In general, oil and water co-exist in the reservoir, and the elastic effect cannot be ignored under high pressure. The steady state flow of single-phase liquid is an ideal case, and it does not actually exist in the reservoir, but it can be approximately regarded as steady state flow in two cases: first, it can happen when the rock core experiment is conducted in the laboratory; second, it can happen when the oil production rate or water injection rate in the oil field reaches a steady state so that formation pressure and flow rate remain unchanged.

For the single-phase steady state case, the flow in rock core experiment in the lab and the flow between well lines in the field can be described as linear flow. If the vertical well penetrates the reservoir completely, the flow can be described as radial flow. If the well only produces oil on the top of a thick reservoir, the flow can be described as spherical flow.

3.1.2 *Concepts*

Steady state flow: *when the fluid flow factors do not change with time, it is called steady state flow. These factors mainly include pressure, flow rate, flow velocity, pressure gradient, etc.*

Transient flow: *when the fluid flow factors change with time, it is called transient flow.*

Pressure gradient: *the pressure gradient is the rate of formation pressure change along a certain direction, such as* $\frac{dp}{dx}$ *along the* x *direction.*

Flow resistance: *flow resistance is the energy consumed by unit volume flow when fluid flows through porous media. The flow resistance of linear flow is* $\frac{\mu L}{KBh}$, *and that of radial flow is* $\frac{\mu}{2\pi Kh} \ln \frac{R_e}{R_w}$.

Flow net diagram: *the flow net diagram is a diagram composed of a series of equipotential lines and streamlines.*

Equipotential lines: *equipotential lines connect all the points that have the same pressure in a flow net diagram.*

Streamline: *the velocity of a fluid is continuously distributed in space and time. A curve tangent to the velocity vector at any point is called a streamline. The streamline has a direction and it is always perpendicular to the equipotential line.*

Skin factor: *the skin factor is used to account for the additional flow resistance near the wellbore caused by sudden changes in reservoir properties. The symbol is* S.

Effective well radius: *The effective well radius is the imaginary well radius used to account for the effect of formation damage or well stimulation, symbol* R_{we}.

Well test: *well test is a special test design to determine well productivity and reservoir properties. It is performed by changing the work schedule, such as the production rate, and monitoring the corresponding pressure change.*

Productivity index curve: *the productivity index curve is drawn with flow rate on* y *axis and bottom hole pressure or production pressure difference on* x *axis. It can indicate the productivity of the well.*

Mobility: *mobility is the ratio of reservoir permeability to fluid viscosity, namely* $\frac{K}{\mu}$.

3.1.3 *Analysis*

3.1.3.1 *Linear flow*

In linear flow, the supply pressure is p_e, the bottom hole pressure of the production well is p_w, reservoir thickness is h, the width is B, the permeability is K, the single-phase liquid viscosity is μ, and flow distance is L as shown in Figure 3.1. Please give the pressure distribution, pressure gradient, flow velocity v, flow Q, etc.

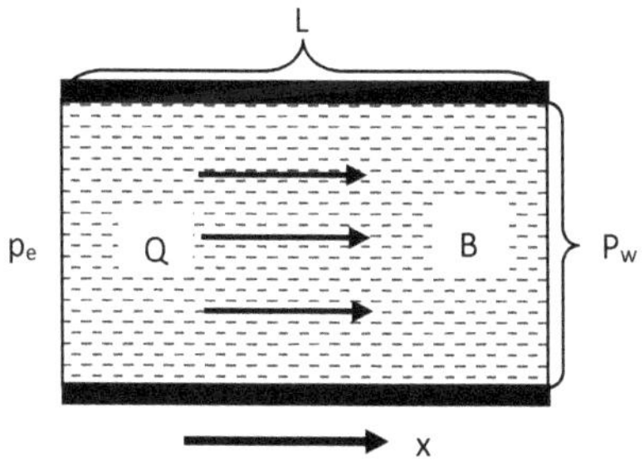

Figure 3.1. Linear flow model for steady state problem

(1) Mathematical model

$$\begin{cases} \dfrac{d^2p}{dx^2} = 0 & \text{governing equation} \\ p|_{x=0} = p_e & \text{outer boundary} \\ p|_{x=L} = p_w & \text{inner boundary} \end{cases} \tag{3.1}$$

Since it is steady state flow, the initial conditions can be ignored.

(2) Pressure distribution

The general solution of the governing equation is solved by the integration method of separation variables: $p(x) = A + Bx$, and the coefficients in the pressure distribution formula is obtained by substituting the inner and outer boundary conditions into this equation:

$$p(x) = p_e - \frac{p_e - p_w}{L}x \quad \text{or} \quad p(x) = p_w + \frac{p_e - p_w}{L}(L - x) \tag{3.2}$$

Figure 3.2 shows the distribution of formation pressure along the x direction is a straight line. It indicates that the formation pressure of linear flow changes uniformly along the flow direction and the energy loss is also uniform.

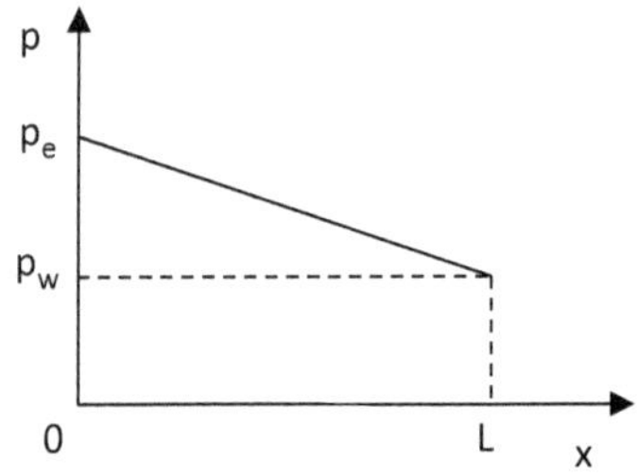

Figure 3.2. Pressure distribution of linear flow

(3) Pressure gradient

Take the derivative of the pressure equation,

$$\frac{dp}{dx} = -\frac{p_e - p_w}{L} = C_1 \tag{3.3}$$

It is known that the pressure gradient is constant, which further reflects that the pressure change is uniform.

(4) Flow velocity

$$v_x = -\frac{K}{\mu}\frac{dp}{dx} = \frac{K}{\mu}\frac{p_e - p_w}{L} = C_2 \tag{3.4}$$

The flow velocity is also constant, that is, the flow velocity in linear flow is the same everywhere.

(5) Production rate

$$Q = Av_x = Bhv_x = \frac{KBh}{\mu}\frac{(p_e - p_w)}{L} \tag{3.5}$$

Thus, Darcy's law can be derived by means of mathematical modelling.

(6) Flow resistance

$$R = \frac{\Delta p}{Q} = \frac{\mu L}{KBh} \tag{3.6}$$

(7) Flow net diagram

There is a flow net for fluid flow in reservoirs. The key points of drawing the flow net are as follows: the pressure difference between any two adjacent equipotential lines is equal; the flow rate between any two adjacent streamlines is equal; the streamline is always perpendicular to the equipotential line and has a flow direction.

Figure 3.3 shows the flow field net diagram of linear flow. It can be seen that the equipotential lines are evenly distributed and the

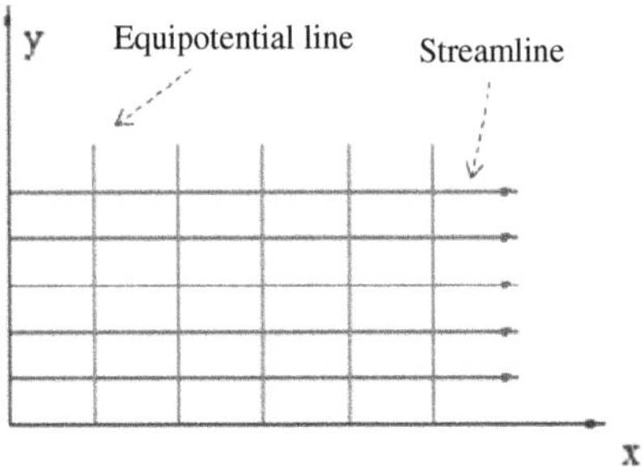

Figure 3.3. Flow net of linear flow

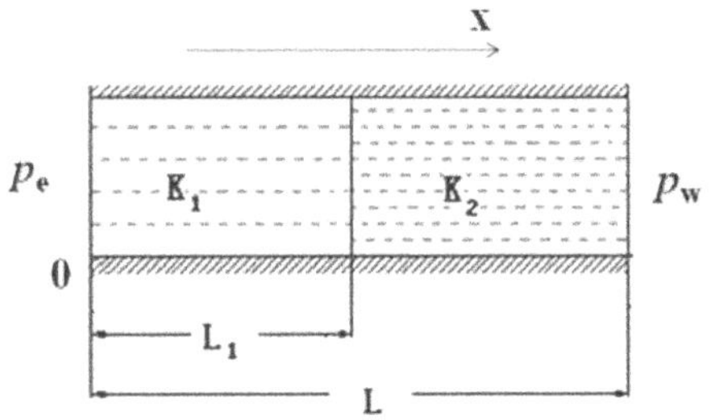

Figure 3.4. Schematic diagram of two-region permeability model for linear flow

energy loss is uniform. The streamlines are parallel, and the flow velocity is equal everywhere on the streamline.

(8) Two-region permeability model

As shown in Figure 3.4, the permeability of the linear flow model changes suddenly at L_1 along the x direction from K_1 to K_2, and the flow rate Q and pressure distribution are calculated.

If the two regions with different permeabilities are connected in the way shown in Figure 3.4 (just like two electric resistances in a series circuit) and the total pressure difference remains unchanged, then the flow rate Q can be calculated.

The flow resistance in K_1 region is:

$$R_1 = \frac{\mu L_1}{K_1 B h} \tag{3.7}$$

The flow resistance in K_2 region is:

$$R_2 = \frac{\mu(L - L_1)}{K_2 B h} \tag{3.8}$$

Total flow resistance is:

$$R = R_1 + R_2 = \frac{\mu L_1}{K_1 B h} + \frac{\mu(L - L_1)}{K_2 B h} \tag{3.9}$$

The flow rate can be obtained as:

$$Q = \frac{p_e - p_w}{R} \tag{3.10}$$

Let the pressure at L_1 be p_d, and the flow rates in K_1 and K_2 regions be equal:

$$\frac{p_e - p_d}{R_1} = \frac{p_d - p_w}{R_2} = Q \tag{3.11}$$

Then the pressure distribution can be obtained from the pressure distribution equation.

In K_1 region, the pressure distribution is:

$$p(x) = p_e - \frac{p_e - p_d}{L_1} x = p_e - \frac{QR_1}{L_1} x \tag{3.12}$$

At the L_1 location, the pressure is:

$$p_d = p_e - QR_1 \tag{3.13}$$

In K_2 region, the pressure distribution is:

$$p(x) = p_d - \frac{p_d - p_w}{L - L_1}(x - L_1) = p_e - QR_1 - \frac{QR_2}{L - L_1}(x - L_1) \tag{3.14}$$

3.1.3.2 *Radial flow*

As shown in Figure 3.5, the radial flow is in the planar reservoir with the outer supply boundary, the permeability is K, the single-phase liquid viscosity is μ, the thickness is h, the outer boundary radius is R_e, the supply pressure is p_e, a well in the center of the reservoir fully penetrates the reservoir, the wellbore radius is R_w, and the bottom hole pressure is p_w. Please derive the pressure distribution, pressure gradient, flow velocity v, flow Q, etc.

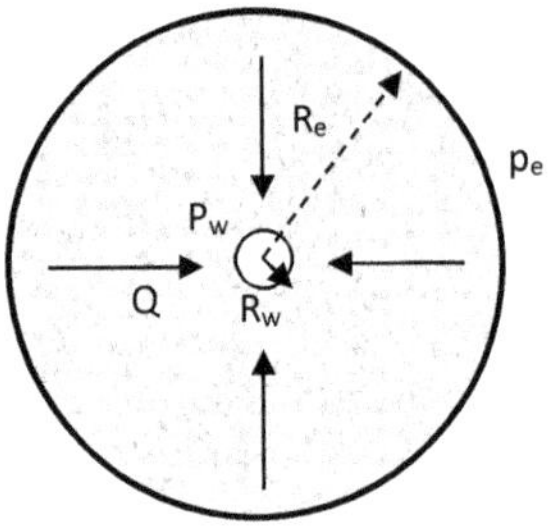

Figure 3.5. Radial flow model for steady state problem

(1) Mathematical model

The mathematical model in polar coordinates is established as follows:

$$\begin{cases} \dfrac{d^2p}{dr^2} + \dfrac{1}{r}\dfrac{\partial p}{\partial r} = 0 & \text{governing equation} \\ p|_{r=R_e} = p_e & \text{outer boundary} \\ p|_{r=R_w} = p_w & \text{inner boundary} \end{cases} \tag{3.15}$$

Since it is steady state flow, the initial conditions can be ignored.

(2) Pressure distribution

The general solution of the basic differential equation is solved by the integration method of separation variables:

$$p(r) = A + B \ln x \tag{3.16}$$

Substitute the inner and outer boundary conditions into the pressure distribution equation:

$$p(r) = p_e - \frac{p_e - p_w}{\ln \frac{R_e}{R_w}} \ln \frac{R_e}{r} \quad \text{or}$$

$$p(r) = p_w + \frac{p_e - p_w}{\ln \frac{R_e}{R_w}} \ln \frac{r}{R_w} \tag{3.17}$$

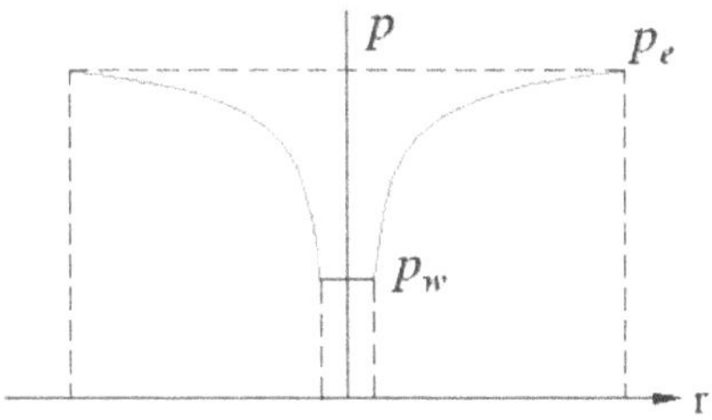

Figure 3.6. Pressure distribution of radial flow

Figure 3.6 shows the radial flow pressure distribution along the radial direction. The near-wellbore pressure drawdown is much larger than that for the wells near the outer boundary. If the pressure distribution curve surface is conical in shape, it is often called "cone of depression".

The cone of depression indicates that the closer to the bottom hole, the faster the energy loss, i.e., the energy loss mainly occurs in the near-wellbore area.

(3) Pressure gradient

Take the derivative of the pressure equation,

$$\frac{dp}{dr} = \frac{p_e - p_w}{\ln \frac{R_e}{R_w}} \frac{1}{r} \tag{3.18}$$

It can be seen that the pressure gradient varies. The closer to the bottom hole of the well, the smaller the r, and the larger the pressure gradient, the more dramatic the pressure change.

(4) Flow velocity

$$v_r = \frac{K}{\mu} \frac{dp}{dr} = \frac{K}{\mu} \frac{p_e - p_w}{\ln \frac{R_e}{R_w}} \frac{1}{r} \tag{3.19}$$

The flow velocity also varies. The closer to the bottom hole of the well, the smaller the r, and the higher the flow velocity.

(5) Flow rate

$$Q = Av_r = 2\pi rhv_r = \frac{2\pi Kh}{\mu}\frac{(p_e - p_w)}{\ln \frac{R_e}{R_w}} \tag{3.20}$$

This formula is the flow rate of radial flow, also known as Dupuit equation.

(6) Flow resistance

$$R = \frac{\Delta p}{Q} = \frac{\mu}{2\pi Kh}\ln\frac{R_e}{R_w} \tag{3.21}$$

(7) Flow net diagram

As can be seen from Figure 3.7, the closer the equipotential line to the bottom hole of wellbore, the smaller the spacing of the equipotential line. It indicates that the energy loss mainly occurs near the wellbore. The streamline points to the bottom hole of the well along the radial direction.

(8) Two-region permeability model

As shown in Figure 3.8, the permeability of the radial flow changes along r direction at R_d from K_1 to K_2, and the flow rate Q and pressure distribution are calculated.

Similar to plane linear flow, two regions with different permeabilities are connected in the way like a series circuit, and the total pressure difference remains unchanged, so the flow rate Q can be calculated.

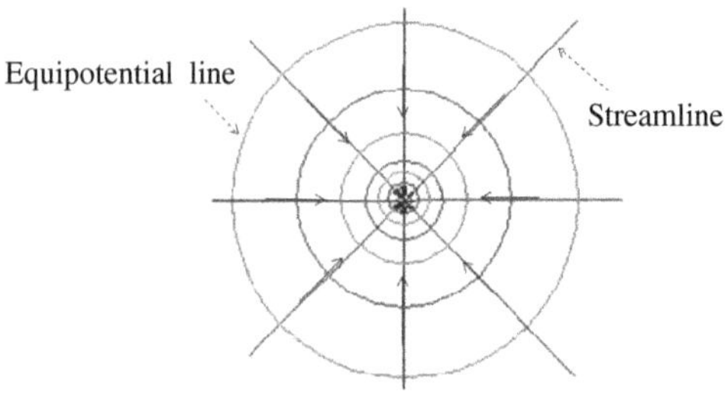

Figure 3.7. Flow net of radial flow

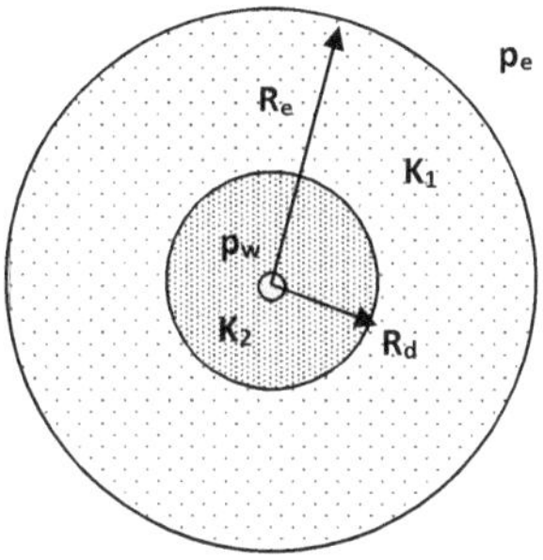

Figure 3.8. Schematic diagram of two-region permeability model for radial flow

The flow resistance in K_1 region is:

$$R_1 = \frac{\mu}{2\pi K_1 h} \ln \frac{R_e}{R_d} \tag{3.22}$$

The flow resistance in K_2 region is:

$$R_2 = \frac{\mu}{2\pi K_2 h} \ln \frac{R_d}{R_w} \tag{3.23}$$

The total flow resistance is:

$$R = R_1 + R_2 = \frac{\mu}{2\pi K_1 h} \ln \frac{R_e}{R_d} + \frac{\mu}{2\pi K_2 h} \ln \frac{R_d}{R_w} \tag{3.24}$$

The flow rate can be obtained as:

$$Q = \frac{p_e - p_w}{R} \tag{3.25}$$

Let the pressure at R_d be p_d, and the flow rates in K_1 and K_2 regions be equal:

$$\frac{p_e - p_d}{R_1} = \frac{p_d - p_w}{R_2} = Q \tag{3.26}$$

Then the pressure distribution can be obtained from the pressure distribution equation.

In the K_1 region, the pressure distribution is:

$$p(r) = p_e - \frac{p_e - p_d}{\ln \frac{R_e}{R_d}} \ln \frac{R_e}{r} = p_e - \frac{QR_1}{\ln \frac{R_e}{R_d}} \ln \frac{R_e}{r} \tag{3.27}$$

The permeability at R_d, the pressure distribution is:

$$p_d = p_e - QR_1 \tag{3.28}$$

In K_2 region, the pressure distribution is:

$$p(r) = p_d - \frac{p_d - p_w}{\ln \frac{R_d}{R_w}} \ln \frac{R_d}{r} = p_e - QR_1 - \frac{QR_2}{\ln \frac{R_d}{R_w}} \ln \frac{R_d}{r} \tag{3.29}$$

(9) Average reservoir pressure

The average reservoir pressure is calculated by the area-weighted average method. If a small ring element is taken, as shown in the shaded area in Figure 3.9, the average reservoir pressure is:

$$\overline{P} = \frac{\int_s p dA}{A} = p_e - \frac{p_e - p_w}{2 \ln \frac{R_e}{R_w}} \approx p_e \tag{3.30}$$

The average reservoir pressure is often used to replace the supply pressure in the actual oil field.

(10) Injection well problem

In the reservoir model, the properties of the reservoir and fluid are unchanged. The well is an injection well, and the injection rate of

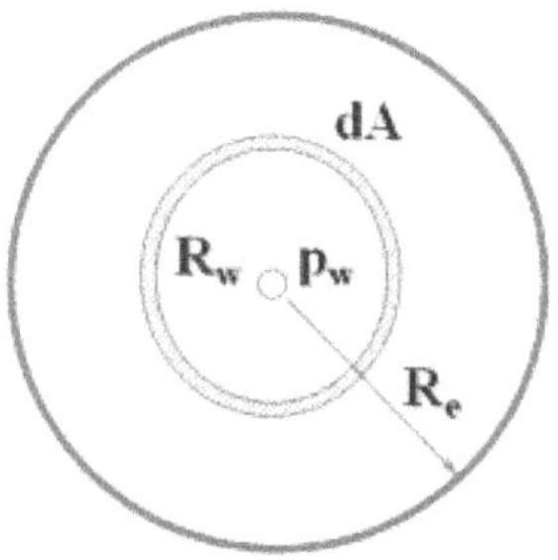

Figure 3.9. Schematic diagram of the small ring-shape element

the injection well is known as Q_{in}, hence the pressure distribution can be obtained.

It is worthy to note that the injection rate of the injection well is indicated by "–" (negative sign) when the mathematical model is being established.

Therefore, the mathematical model of the injection problem can be established as follows:

$$\begin{cases} \dfrac{d^2p}{dr^2} + \dfrac{1}{r}\dfrac{\partial p}{\partial r} = 0 & \text{governing equation} \\ p|_{r=R_e} = p_e & \text{outer boundary} \\ r\dfrac{dp}{dr}\bigg|_{r=R_w} = \dfrac{-Q_{in}\mu}{2\pi Kh} & \text{inner boundary} \end{cases} \tag{3.31}$$

The general solution of the basic differential equation is still $p = A + B\ln r$. The coefficients in the general solution can be obtained by substituting the boundary conditions into the general solution:

$$p(r) = p_e + \frac{Q_{in}\mu}{2\pi Kh}\ln\frac{R_e}{r} \tag{3.32}$$

$$p_{win} = p_e + \frac{Q_{in}\mu}{2\pi Kh}\ln\frac{R_e}{R_w} \tag{3.33}$$

The pressure distribution is shown in Figure 3.10, which is the inverted cone of depression for the injection well.

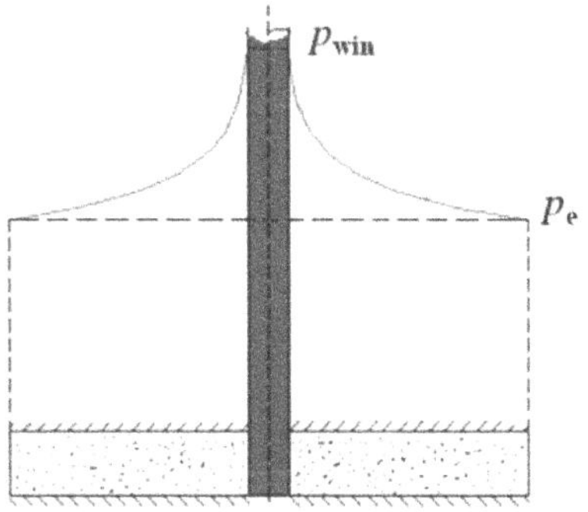

Figure 3.10. Cone of depression for injection well

3.1.3.3 *Spherical flow*

There are few practical applications of spherical flow in practice, and the general procedures to derive the solution is the same as linear flow and radial flow.

The general solution of the basic differential equation is:

$$p(r) = A + B/r \tag{3.34}$$

The pressure distribution is:

$$p(r) = p_e - \frac{p_e - p_w}{\frac{1}{R_w} - \frac{1}{R_e}} \left(\frac{1}{r} - \frac{1}{R_e} \right) \tag{3.35}$$

The flow rate is:

$$Q = \frac{4\pi K}{\mu} \frac{(p_e - p_w)}{\frac{1}{R_w} - \frac{1}{R_e}} \tag{3.36}$$

3.1.3.4 *Skin factor and effective well radius*

The pressure distribution derived in radial flow is the ideal case, where fluid flowing from the reservoir into the bottom hole does not increase flow resistance. However, most of the wells in the field are affected by the drilling and completion process, so the flow resistance will change, and it results in non-ideal wells. Non-ideal wells affected by the flow resistance change can be classified into three types, as shown in Figure 3.11: partially penetrated well, perforated well and both partially penetrated and perforated well. These factors will increase flow resistance.

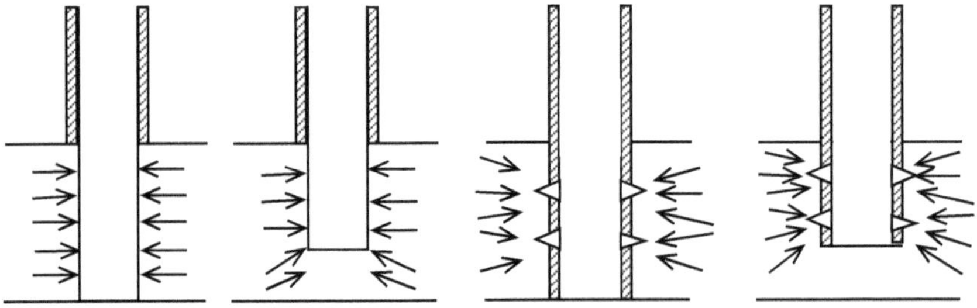

Figure 3.11. Ideal radial flow and non-ideal radial flow

In addition to the above three types of wells, in order to reduce flow resistance near the wellbore, hydraulic fracturing or acidizing are used to increase permeability in the near-wellbore area, which can reduce flow resistance.

There are generally two quantities used to account for the effect of the non-ideal well, namely the skin coefficient S and the effective well radius R_{we}.

The relationship between the two quantities is:

$$R_{we} = R_w e^{-S} \tag{3.37}$$

where R_{we} is the effective well radius of the non-ideal well, m; R_w is the radius of the ideal well, m; and S is skin factor.

When applied to the production rate equation, it is:

$$Q = \frac{2\pi Kh}{\mu} \frac{(p_e - p_w)}{\ln \frac{R_e}{R_{we}}} \tag{3.38}$$

or

$$Q = \frac{2\pi Kh}{\mu} \frac{(p_e - p_w)}{\ln \frac{R_e}{R_w} + S} \tag{3.39}$$

3.1.3.5 *Steady state well test*

For wells that conform to Darcy's law, the production formula is as follows:

$$Q = \frac{2\pi Kh}{\mu} \frac{(p_e - p_w)}{\ln \frac{R_e}{R_w}} \tag{3.40}$$

Let

$$J = \frac{2\pi Kh}{\mu \ln \frac{R_e}{R_w}} \tag{3.41}$$

$$\Delta \mathrm{p} = p_e - p_w \tag{3.42}$$

Then

$$Q = J\Delta p \tag{3.43}$$

where J is defined as the productivity index.

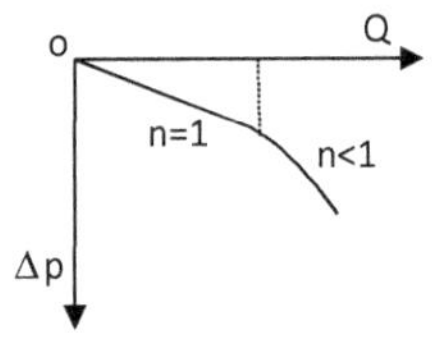

Figure 3.12. Productivity index curve from steady state well test

Plot the production rate Q and production pressure differential Δp in the same figure to obtain the productivity index curve, as shown in Figure 3.12. The actual curve obtained from the well test is not always linear.

The Darcy flow is in the straight line segment, and the production rate is in a linear relationship with the production pressure difference.

The non-Darcy flow is in the late production stage. The increase in production rate decreases with the increase of the production pressure difference. An exponential form can be used to account for the non-Darcy flow phenomenon in this case:

$$Q = J(\Delta p)^n \tag{3.44}$$

When $n = 1$, it is Darcy flow.

A steady state well test is applied in the straight line segment with $n = 1$, where the slope of the line is J. The production pressure difference can be obtained by changing the well production rate Q and monitoring the corresponding change of the bottom hole pressure p_w. By connecting the production rate and the pressure difference point by point, the straight line section can be found, which can be used for reservoir parameter estimation and well productivity analysis.

The steady state well test can be used for several different purposes:

(1) Determine a reasonable working schedule

A reasonable working schedule means that the well produces maximal oil with minimal energy consumption. It can be seen from the productivity index curve that a reasonable working schedule should be selected at the point where the straight line starts to curve.

(2) Determine the well productivity

J is the production rate per unit pressure difference, which can be used as an indicator to compare different wells. In other words, the larger J is, the stronger the well's productivity will be.

(3) Judge the effect of stimulation

When comparing the index curve before and after well stimulation, the increase of J indicates that well stimulation has a positive effect on productivity, and vice versa.

(4) Calculate the reservoir properties

In the index curve, J can be obtained. The principle of steady state well test shows:

$$J = \frac{2\pi Kh}{\mu \ln \frac{R_e}{R_w}} \tag{3.45}$$

If R_e and R_w have been determined, $\frac{Kh}{\mu}$ can be calculated.

If the viscosity of the fluid has been determined, Kh can be obtained.

If the effective reservoir thickness h is determined in the flow coefficient, the mobility $\frac{K}{\mu}$ can be calculated.

If μ and h have both been determined, the average permeability K can be calculated.

If $\frac{Kh}{\mu}$ and R_e have both been determined, the effective well radius or skin factor can be calculated.

3.1.3.6 *Example*

1. In the linear flow regime, there are two zones with different permeabilities along the flow direction. Please derive the pressure distribution along the x direction and draw the flow net diagram.

Solution:

K_1 in L_1 region is less than K_2 in L_2 region. According to the pressure gradient equation, both segments should be straight lines.

Flow resistance per unit length is $R_L = \frac{\mu}{KBh}$, pressure gradient in terms of flow is $\frac{dp}{dx} = \frac{Q\mu}{KBh}$, so $R_{L_1} > R_{L_2}$, $\left.\frac{dp}{dx}\right|_{L1} > \left.\frac{dp}{dx}\right|_{L2}$. Compared with L_2 segment, liquid in L_1 segment consumes more pressure differences per unit length and it has a large pressure gradient and a large pressure distribution slope, thus it has dense pressure lines in the flow net diagram.

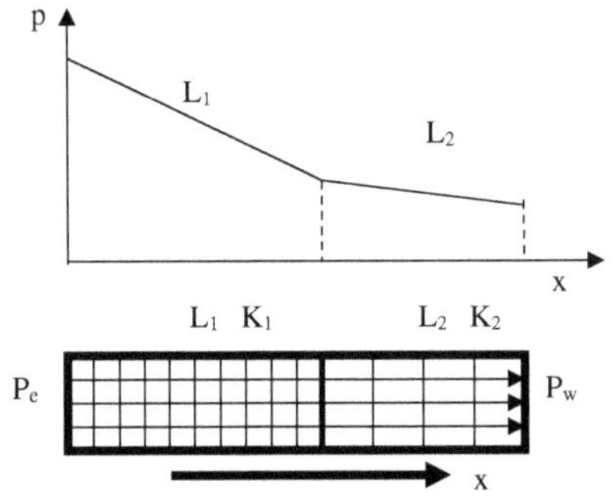

2. For the single-phase steady state radial flow, it is given that $R_e = 300\,\text{m}$, $P_e = 20\,\text{MPa}$, $K = 50\,\text{mD}$, $h = 5\,\text{m}$, $\mu = 5\,\text{mPa}\cdot\text{s}$, $\varphi = 0.2$, a central well, $R_w = 0.1\,\text{m}$ and $p_w = 15\,\text{MPa}$. Please give the results for the flow rate Q (m^3/d) and the flow velocity at the wall of the wellbore $v_{wh}(\mu\text{m/s})$.

Solution:

$$Q = \frac{2\pi Kh}{\mu}\frac{(p_e - p_w)}{\ln\frac{R_e}{R_w}}$$

$$= \frac{2 \times 3.14 \times 50 \times 10^{-15} \times 5}{5 \times 10^{-3}} \times \frac{(20 - 15) \times 10^6}{\ln\frac{300}{0.1}} \times 86400$$

$$= 16.95\, m^3/d$$

$$v_{wh} = \frac{Q}{2\pi R_w h\phi} = \frac{16.95}{2 \times 3.14 \times 0.1 \times 5 \times 0.2} \div 86400 \times 10^6$$

$$= 312.4\, \mu m/s$$

The velocity of flow at the wall of the wellbore is the largest.

3. For the single-phase steady state radial flow, it is known that $R_e = 300\,\text{m}$, $p_e = 20\,\text{MPa}$, $K = 50\,\text{mD}$, $h = 5\,\text{m}$, $\mu = 5\,\text{mPa}\cdot\text{s}$,

a central producing well, $R_w = 0.1\,\text{m}$, $p_w = 15\,\text{MPa}$. There are four ring shape areas with the distance from the center as 0.1–1 m, 1–10 m, 10–100 m and 100–300 m, respectively. Please find the ratio of pressure loss to total pressure difference in the four areas, and analyze the characteristics of energy loss.

Solution:

Let's first find the pressures at $r = 1\,\text{m}$, $10\,\text{m}$ and $100\,\text{m}$.

$$p(r = 1) = p_e - \frac{p_e - p_w}{\ln \frac{R_e}{R_w}} \ln \frac{R_e}{r}$$

$$= 20 - \frac{20 - 15}{\ln \frac{300}{0.1}} \ln \frac{300}{1} = 16.44\,MPa$$

$$p(r = 10) = p_e - \frac{p_e - p_w}{\ln \frac{R_e}{R_w}} \ln \frac{R_e}{r}$$

$$= 20 - \frac{20 - 15}{\ln \frac{300}{0.1}} \ln \frac{300}{10} = 17.88\,MPa$$

$$p(r = 100) = p_e - \frac{p_e - p_w}{\ln \frac{R_e}{R_w}} \ln \frac{R_e}{r}$$

$$= 20 - \frac{20 - 15}{\ln \frac{300}{0.1}} \ln \frac{300}{100} = 19.31\,MPa$$

Then, the pressure difference consumed by the four segments are respectively:

$$\Delta p(100\text{–}300\,m) = p_e - p(r = 100)$$

$$= 20 - 19.31 = 0.69\,MPa$$

$$\Delta p(10\text{–}100\,m) = p(r = 100) - p(r = 10)$$

$$= 19.31 - 17.88 = 1.43\,MPa$$

$$\Delta p(1\text{–}10\,m) = p(r = 10) - p(r = 1)$$

$$= 17.88 - 16.44 = 1.44\,MPa$$

$$\Delta p(0.1\text{–}1\,m) = p(r = 1) - p_w$$

$$= 16.44 - 15 = 1.44\,MPa$$

The ratio of pressure loss to the total pressure difference consumed in the four stages are:

$$R_{\Delta p(100\text{–}300\,m)} = \frac{\Delta p(100\text{–}300\,m)}{p_e - p_w} = \frac{0.69}{20-15} = 0.138$$

$$R_{\Delta p(10\text{–}100\,m)} = \frac{\Delta p(10\text{–}100\,m)}{p_e - p_w} = \frac{1.43}{20-15} = 0.286$$

$$R_{\Delta p(1\text{–}10\,m)} = \frac{\Delta p(1\text{–}10\,m)}{p_e - p_w} = \frac{1.44}{20-15} = 0.288$$

$$R_{\Delta p(0.1\text{–}1\,m)} = \frac{\Delta p(0.1\text{–}1m)}{p_e - p_w} = \frac{1.44}{20-15} = 0.288$$

The energy loss within 1 m accounts for 28.8% of the total loss, the energy loss within 10 m accounts for 57.6%, and the large area beyond 100 m only accounts for 13.8%. Therefore, the energy loss near the well accounts for a large proportion, and stimulation near the well has an important impact in improving the development effect.

4. Provided that the bottom hole pressure is given during the establishment of the Dupuit equation model, the flow rate in the radial flow is Q, and all other conditions remain unchanged. Please find the bottom hole pressure and reservoir pressure distribution.

Solution:

If the production rate is Q, the mathematical model for bottom hole pressure is:

$$\begin{cases} \dfrac{d^2p}{dr^2} + \dfrac{1}{r}\dfrac{\partial p}{\partial r} = 0 & \text{governing equation} \\ p|_{r=R_e} = p_e & \text{outer boundary} \\ r\dfrac{dp}{dr}\bigg|_{r=R_w} = \dfrac{Q\mu}{2\pi Kh} & \text{inner boundary} \end{cases}$$

By applying the separation variable integration method, the general solution can be obtained as follows: $p(r) = A + B \ln x$, and the bottom hole pressure and pressure distribution can be obtained by substituting the inner and outer boundary conditions into the general solution.

$$p_w = p_e - \frac{Q\mu}{2\pi Kh} \ln \frac{R_e}{R_w}$$

$$p(r) = p_e - \frac{Q\mu}{2\pi Kh} \ln \frac{R_e}{r}$$

or

$$p(r) = p_w + \frac{Q\mu}{2\pi Kh} \ln \frac{r}{R_w}$$

5. A 198 mm drill bit was used to drill the oil layer in a certain well. The depth of the oil layer ranged from 2646.5 m to 2660.5 m. The steady state well test result is listed in Table 3.1. It is considered that as productivity can still be improved, acidification is carried out near the wellbore of the reservoir, and the steady state well test is carried out after acidification, with the results shown in Table 3.2. It is known that oil supply boundary radius $R_e = 300$ m, the wellbore bottom hole radius $R_w = 0.1$ m, original formation pressure $p_i = 29.0\,MPa$, formation volume factor for oil $B_o = 1.12$, and oil relative density $\gamma_o = 0.85$. It is assumed that the reservoir is still homogeneous after acidification.

① Draw the productivity index curves for the two steady state well tests on the same coordinate;

② Find the coefficient Kh/μ before and after acidification;

③ Analyze the effectiveness of the well stimulation.

Table 3.1. Steady state well test results before well stimulation

Choke size	Production rate			Wellhead pressure		Bottom hole pressure
	Oil	Gas	Gas-oil ratio	Tubing	Casing	Flowing
mm	t/d	m^3/d	m^3/t	MPa	MPa	MPa
6	97.2	24.3	250	10.3	11.7	26.6
5	80.0	20.0	253	11.2	12.2	27.2
3	40.0	4.9	127	12.3	13.2	28.0

Table 3.2. Steady state well test results after well stimulation

Choke size	Production rate			Wellhead pressure		Bottom hole pressure
	Oil	Gas	Gas-oil ratio	Tubing	Casing	Flowing
mm	t/d	m^3/d	m^3/t	MPa	MPa	MPa
3	55.1	6.1	110	12.6	13.4	28.2
4	90	13.2	144	12.2	13.1	27.7
5	115.7	19.7	170	11.9	12.8	27.5
6	150.2	36.8	245	11.2	12.4	26.8
7	162.1	58.7	362	10.4	12.1	26.55

Table 3.3. Computed results for steady state well test before and after acidification

Before	Δp	MPa	2.40	1.80	1.00		
	Q_v	m^3/d	128.08	105.41	52.71		
After	Δp	MPa	0.80	1.30	1.50	2.20	2.45
	Q_v	m^3/d	72.60	118.59	152.45	197.91	213.59

Solution:

First, the oil production at the wellhead should be converted to the volume flow at the bottom of the wellbore, that is, $Q_v = Q_m B_o/\gamma_o$, and the pressure difference between the original formation pressure and the bottom of the wellbore should be calculated. Note that not all the data are useful.

① Draw the productivity index curves for the two steady state well tests on the same coordinate.

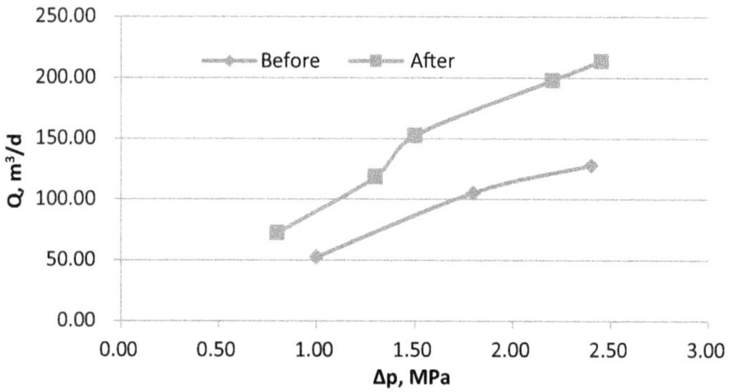

The slope of the two lines are $m_{\text{Before}} = 54.48$ before the acidification and $m_{\text{After}} = 84.88$ after the acidification.

To obtain the coefficient Kh/μ before and after acidification,

$$J = \frac{2\pi Kh}{\mu \ln \frac{R_e}{R_w}} = m$$

The coefficient Kh/μ before and after acidification are respectively:

$$\left.\frac{Kh}{\mu}\right|_{\text{Before}} = \frac{\ln \frac{R_e}{R_w}}{2\pi} m = \frac{\ln \frac{300}{0.1}}{2 \times 3.14} \times 54.48 = 69.46\, m^3/(d \cdot MPa)$$

$$\left.\frac{Kh}{\mu}\right|_{\text{After}} = \frac{\ln \frac{R_e}{R_w}}{2\pi} m = \frac{\ln \frac{300}{0.1}}{2 \times 3.14} \times 84.88 = 108.21\, m^3/(d \cdot MPa)$$

Since $m_{\text{Before}} = 54.48 < m_{\text{After}} = 84.88$, the stimulation is effective.

3.2 Multi-well interference theory

3.2.1 *Introduction*

The radial flow is a well in the center of circular formation. However, in many oil fields, there are actually several wells producing or injecting in the same reservoir. The reservoir is connected. This part will discuss on how the production wells and injection wells interfere with each other.

The essence of multi-well interference is energy rebalancing, and the characteristics of energy rebalancing follow the principle of superposition. This chapter mainly introduces four principles of superpositions of steady state flow of single-phase liquid: superposition of pressure drawdown, superposition of pressure function, superposition of potential function and superposition of flow velocity.

In the analysis of multi-well interference theory, the first is to analyze the flow of the simplest case with two wells, then to analyze the case with many wells, and finally to analyze the case with the well array. The methods used here are the principle of superposition, mirror image method and electric analog method respectively.

3.2.2 Concepts

Multi-well interference: *when multiple wells in the same formation work simultaneously, changes in the work schedule of any one well will affect other wells. This phenomenon is called multi-well interference.*

Principle of superposition for pressure drawdown: *when multiple wells in the same formation work simultaneously, the pressure drawdown at any point in the formation is equal to the algebraic sum of the pressure drawdown generated at this point when each well works independently.*

Potential function: *the potential function is also known as velocity potential function. The definition of the potential is:*

$$\Phi = \frac{K}{\mu} p + C \tag{3.46}$$

where Φ *is the potential function;* K *is the permeability;* μ *is the viscosity;* p *is the pressure; and* C *is the integration constant determined by the boundary conditions.*

Principle of superposition for potential: *when multiple wells in the same formation work simultaneously, the potential (or potential function) at any point in the formation is equal to the algebraic sum of potential (or potential function) generated at this point when each well works independently.*

Principle of superposition for flow velocity: *when multiple wells in the same formation work simultaneously, the flow velocity at any point in the formation is equal to the vector sum of the flow velocity at this point when each well works independently.*

Point source: *the point source is the point from which the flow comes out in the 2D radial flow case and it creates the streamlines along the radial direction. In the context of petroleum engineering, it usually refers to an injection well and the production rate is associated with a negative sign by convention.*

Point sink: *the point sink is the point at which the flow converges and it creates the streamlines along the radial direction. In the context of petroleum engineering, it generally refers to a production well and the production rate is associated with a positive sign by convention.*

Stagnation point: *the stagnation point is where the flow velocity is 0 due to the superposition of the flow velocity when two or more wells are producing at the same time.*

Diverging streamline: *the diverging streamline is the streamline where the streamlines diverge apart when the fluid flows through it. The streamline located in the center of two point sinks with equal production rates is the diverging streamline. It can generally be regarded as a fault.*

Main streamline: *the main streamline is the streamline that has the highest flow velocity. The straight streamline connecting a point sink and a point source is the main streamline.*

Tonguing phenomenon: *the streamlines connecting the point source and point sink have the highest at the main streamline and get smaller when they move away from the main streamline. If a fluid is injected through point source, the shape of the fluid front line looks like a tongue, hence it is called tonguing phenomenon.*

Mirror image method: *when there is a fault or supply boundary, the boundary can be regarded as a mirror. A virtual well of the actual well can be reflected by the mirror and account for the effect of fault or supply boundary on the flow. This method is used to solve the flow problem with boundary and convert it to a multi-well interference problem.*

Electric analog method: *when the wells located on multiple lines are produced at the same time, a circuit diagram composed of resistance is used to represent the flow resistance and describe the flow process, and the equation can be built based on the circuit diagram to solve the problem of well array flow.*

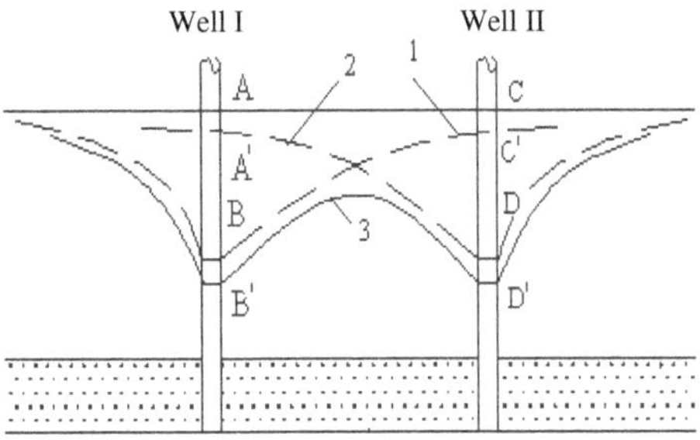

(a) Pressure distribution curve of two sinks

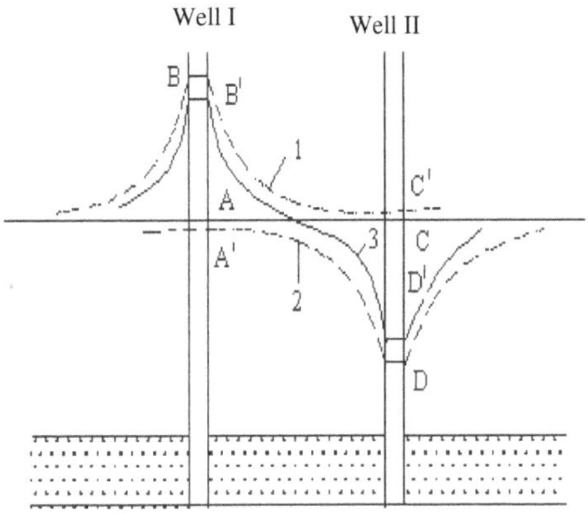

(b) Pressure distribution curve of one source and one sink

Figure 3.13. Principle of superposition of pressure drawdown

3.2.3 *Analysis*

3.2.3.1 *Principle of superposition for pressure drawdown*

As shown in Figure 3.13, the dotted lines 1 and 2 in these two figures are respectively the pressure distributions when only well I and well II are working separately. The solid line 3 is the superposed pressure distribution of two wells working at the same time.

If the original formation pressure of reservoir is p_i (in a horizontal stratum, it can be regarded to be the same as the supply pressure p_e), the pressure drawdowns when well I and well II work separately are:

$$\Delta p_1 = p_i - p_1 \quad \text{and} \quad \Delta p_2 = p_i - p_2 \tag{3.47}$$

According to the principle of superposition for pressure drawdown, the pressure drawdown (superposition of pressure drawdown value) generated when two wells work simultaneously $\Delta p = \Delta p_1 + \Delta p_2$ are:

$$p = p_i - \Delta p \tag{3.48}$$

For the single well bottom-hole pressure, according to the superposition of the pressure drawdown, the bottom hole pressure drawdown for well I and well II are respectively:

$$\Delta p_{1w} = AA' + AB = AB' \quad \text{and} \quad \Delta p_{2w} = CC' + CD = CD' \tag{3.49}$$

As shown in Figure 1.13(b), well I is an injection well, its pressure drawdown is negative, the bottom hole pressure drawdown for well I and well II are respectively:

$$\Delta p_{1w} = AA' - AB = AB' \quad \text{and} \quad \Delta p_{2w} = CD - CC' = CD' \tag{3.50}$$

The bottom hole pressure drawdown of well I is generally negative, which means that its pressure is greater than the original formation pressure, and only in this way can the injection fluid be more easily injected into the formation.

The pressure drawdown generated when the well works alone can be obtained by applying the equation of radial flow pressure distribution. If $p_i = p_e$, then the pressure drawdown at any point M in the formation is:

$$\begin{aligned}\Delta p_M &= p_i - p_M \\ &= p_i - \left(p_e - \frac{p_e - p_w}{\ln \frac{R_e}{R_w}} \ln \frac{R_e}{r}\right) = \frac{p_e - p_w}{\ln \frac{R_e}{R_w}} \ln \frac{R_e}{r}\end{aligned} \tag{3.51}$$

or

$$\begin{aligned}\Delta p_M &= p_i - p_M \\ &= p_i - \left(p_e - \frac{Q\mu}{2\pi Kh} \ln \frac{R_e}{r}\right) = \frac{Q\mu}{2\pi Kh} \ln \frac{R_e}{r}\end{aligned} \tag{3.52}$$

The principle of superposition of the multi-well pressure drawdown is shown as follows:

$$\Delta p_M = \sum_{i=1}^{n} \Delta p_{Mi} \tag{3.53}$$

where Δp_M is the pressure drawdown at any point M; N is total number of wells; Δp_{Mi} is pressure drawdown generated at point M in well I when working alone, the production well has "+" sign and the injection well has "−" sign.

For two sinks with equal production rates, the pressure distribution is symmetric around the central line between the two wells. For one source and one sink case with equal flow rates, the pressure at the center of the two wells is equal to the original formation pressure.

3.2.3.2 *Principle of superposition for pressure function*

When it is single-phase steady state flow, the mathematical model of flow in the infinite formation with constant production rate is as follows:

$$\begin{cases} \dfrac{d^2P}{dr^2} + \dfrac{1}{r}\dfrac{dP}{dr} = 0 \\ r\dfrac{dp}{dr}\bigg|_{r=R_w} = \dfrac{Q\mu}{2\pi kh} \\ P|_{r=\infty} = P_e \end{cases} \tag{3.54}$$

The pressure function is obtained as follows:

$$p = \frac{Q\mu}{2\pi Kh}\ln r + C \tag{3.55}$$

where r is the distance from the point to the center of the well, namely the polar radius.

If several wells in the same reservoir work simultaneously, according to the principle of superposition of pressure function, the superposed pressure is:

$$p_M = \sum_{i=1}^{n} p_{Mi} \tag{3.56}$$

where p_M is the pressure function at any point M in the infinite reservoir; and p_{Mi} is pressure function at M point in the infinite reservoir when well I works alone.

As shown in Figure 3.13, when the two wells work separately, the basic differential equations are as follows:

$$\frac{\partial^2 p_1}{\partial x^2} + \frac{\partial^2 p_1}{\partial y^2} = 0 \quad \text{and} \quad \frac{\partial^2 p_2}{\partial x^2} + \frac{\partial^2 p_2}{\partial y^2} = 0 \tag{3.57}$$

When the two wells work simultaneously, according to the principle of superposition of Laplace equation, the following equation can be obtained:

$$\frac{\partial^2 (p_1 + p_2)}{\partial x^2} + \frac{\partial^2 (p_1 + p_2)}{\partial y^2} = 0 \tag{3.58}$$

This superposition is essentially the superposition of pressure functions, not the superposition of pressure values.

3.2.3.3 *Principle of superposition for potential*

(1) Introduction to the potential

Just like the commonly known gravitational field, electrical field, magnetic field and electromagnetic field, there are flow fields in the fluid mechanics of porous media.

The field shows the potential of energy transformation and the movement of matter under the action of energy. Pressure in flow field is energy, and pressure difference driving fluid flow is motion under the action of energy.

The potential can be used to represent capability of one type of energy converted into another type of energy. For example, the gravitational potential is used to indicate the amount of potential energy to be converted into mechanical energy. The "potential" in the flow field can be understood as the conversion of pressure energy into kinetic energy.

The potential gradient is defined as the flow velocity, and the expression is:

$$v = -\frac{d\Phi}{dx} \tag{3.59}$$

By using Darcy's law:

$$v = -\frac{K}{\mu}\frac{dp}{dx} \tag{3.60}$$

The expression of potential is obtained as follows:

$$\Phi = \frac{K}{\mu}p + C \tag{3.61}$$

In radial flow, the flow rate on the flow area with radial radius r is:

$$Q = 2\pi r h \frac{K}{\mu}\frac{dp}{dr} \tag{3.62}$$

According to the definition of potential, there is:

$$\frac{Q}{2\pi r h} = \frac{d\Phi}{dr} \tag{3.63}$$

By using the separation variable integration method, the potential of a point on the plane is:

$$\Phi = \frac{Q}{2\pi h}\ln r + C \tag{3.64}$$

or

$$\Phi = \frac{q}{2\pi}\ln r + C \tag{3.65}$$

where q is the flow rate per unit reservoir thickness, which is often used when solving problems with potential function; and C is a constant determined by the boundary.

(2) Principle of superposition for potential

If the production rate of each well is known, according to the principle of superposition of potential, it can be expressed as:

$$\Phi_M = \sum_i^n \pm\frac{q_i}{2\pi}\ln r_i + C \tag{3.66}$$

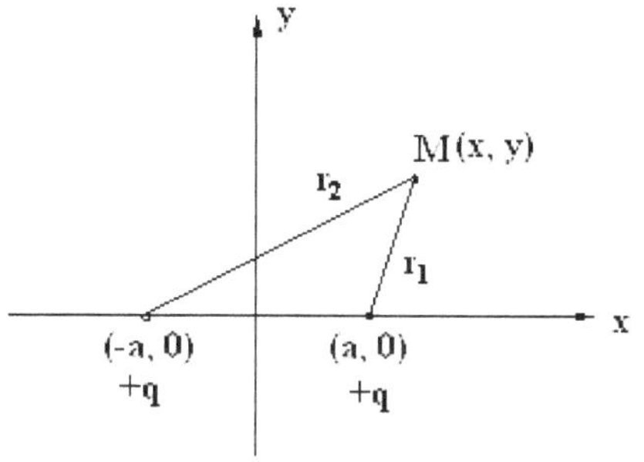

Figure 3.14. Schematic diagram of two sinks with equal production rates

where Φ_M is the potential at any point M; q_i is the production rate per unit thickness of well i; the production well has "+" sign and the injection well has "−" sign; r_i is distance from well i to point M; and C is a constant related to the boundary condition.

(3) Problem of the two sinks with equal production rates

As shown in Figure 3.14, the distance between the two wells is $2a$, the well radius is R_w, the radius of the circular supply boundary is R_e ($R_e \gg 2a$ and $2a \gg R_w$), the supply pressure is p_e, the two wells have the same production rate q, the permeability is K, the reservoir thickness is h, the liquid viscosity is μ, and the bottom hole pressure is p_w. Use the principle of superposition of potential to calculate the production rate and pressure distribution.

According to the principle of superposition of potential, the potential of any point M is:

$$\Phi_M = \frac{q}{2\pi}\ln r_1 + \frac{q}{2\pi}\ln r_2 + C = \frac{q}{2\pi}\ln(r_1 \cdot r_2) + C \tag{3.67}$$

Take a special point (generally on the supply boundary or on the wall of the wellbore).

For the point taken on the supply boundary, $r_1 = r_2 = R_e$, and there is:

$$\Phi_e = \frac{q}{2\pi}\ln(R_e \cdot R_e) + C \tag{3.68}$$

For the point taken on the wall of well 1, $r_1 = R_w$, $r_2 = 2a - R_w \approx 2a$ (a is much larger than R_w), it can be obtained that:

$$\Phi_{w1} = \frac{q}{2\pi} \ln(R_w \cdot 2a) + C \tag{3.69}$$

Subtracting Φ_{w1} from Φ_e, and cancelling the constant, it can be obtained that:

$$\Phi_e - \Phi_{w1} = \frac{q}{2\pi} \ln \frac{R_e^2}{R_w \cdot 2a} \tag{3.70}$$

By using the definition of potential, it can be obtained that:

$$\Phi_e - \Phi_{w1} = \frac{K}{\mu}(p_e - p_w) \tag{3.71}$$

Combining the above two equations, it can be obtained that:

$$\frac{q}{2\pi} \ln \frac{R_e^2}{R_w \cdot 2a} = \frac{K}{\mu}(p_e - p_w) \tag{3.72}$$

The production rate is:

$$q = \frac{2\pi K}{\mu} \frac{p_e - p_w}{\ln \frac{R_e^2}{R_w \cdot 2a}} \tag{3.73}$$

When obtaining the pressure distribution, subtract the potential at a special point from the potential at any point M, and cancel the constant C.

If the point on the wall of well 1 is used as the special point, then:

$$\Phi_{w1} - \Phi_M = \frac{q}{2\pi} \ln \frac{R_w \cdot 2a}{r_1 \cdot r_2} \tag{3.74}$$

According to the definition of potential, we can get:

$$\Phi_{w1} - \Phi_M = \frac{K}{\mu}(p_w - p_M) \tag{3.75}$$

Combining the right sides of the above two equations, it can be obtained that:

$$\frac{q}{2\pi}\ln\frac{R_w\cdot 2a}{r_1\cdot r_2}=\frac{K}{\mu}(p_w-p_M) \tag{3.76}$$

The pressure distribution of any point M is obtained as:

$$p_M=p_w-\frac{q\mu}{2\pi K}\ln\frac{R_w\cdot 2a}{r_1\cdot r_2} \tag{3.77}$$

or

$$p_M=p_w+\frac{q\mu}{2\pi K}\ln\frac{r_1\cdot r_2}{R_w\cdot 2a} \tag{3.78}$$

If the point on the supply boundary is selected as the special point, the pressure distribution is:

$$p_M=p_e-\frac{q\mu}{2\pi K}\ln\frac{R_e^2}{r_1\cdot r_2} \tag{3.79}$$

or

$$p_M=p_e+\frac{q\mu}{2\pi K}\ln\frac{r_1\cdot r_2}{R_e^2} \tag{3.80}$$

According to the expression of potential at any point M, the equipotential lines are expressed as:

$$r_1\cdot r_2=C_1 \tag{3.81}$$

Draw the flow net diagram as shown in Figure 3.15. These curves with arrows are streamlines, which are perpendicular to the equipotential lines. The flow net diagram shows a “dumbbell shape”.

As can be seen from the flow net diagram in Figure 3.15, the streamline flows symmetrically to the two wells, and the y-axis acts as a diverging streamline to separate the fluid flow.

According to the principle of superposition of flow velocity, the flow velocity at any M point is the vector superposition of the flow velocity when the two wells are working independently, as shown in Figure 3.16.

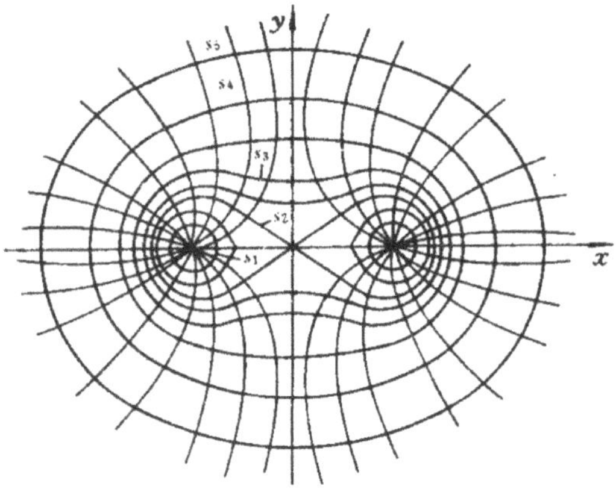

Figure 3.15. Flow net of two sinks with equal production rates

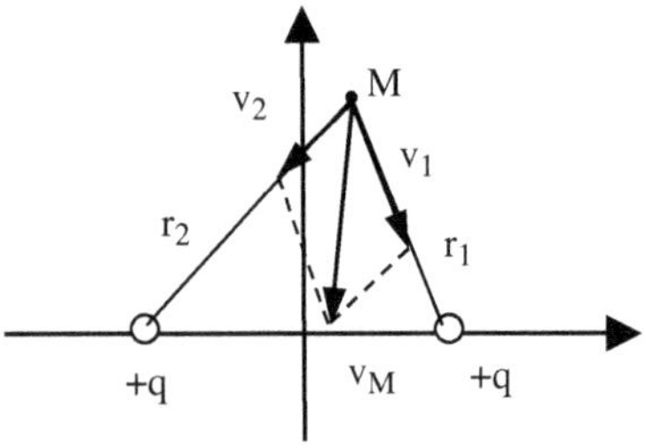

Figure 3.16. Superposition of flow velocity of two sinks with equal production rates

According to the definition of flow velocity, the following equation can be obtained:

$$v_1 = \frac{q}{2\pi \cdot r_1}, \quad v_2 = \frac{q}{2\pi \cdot r_2} \tag{3.82}$$

When the point M is on the x-axis, v_1 and v_2 are in opposite directions. If it is at the midpoint, v_1 and v_2 are equal and in opposite directions, the flow velocity is 0, and the flow velocity in the area near this point is also very small. The stagnation point may exist in the production process of several wells, and this area around the stagnation point is the area for further exploitation.

(4) One source and one sink with equal production rates

As shown in Figure 3.17, the distance between the two wells is $2a$, the well radius is R_w. Well 2 is a production well with the production

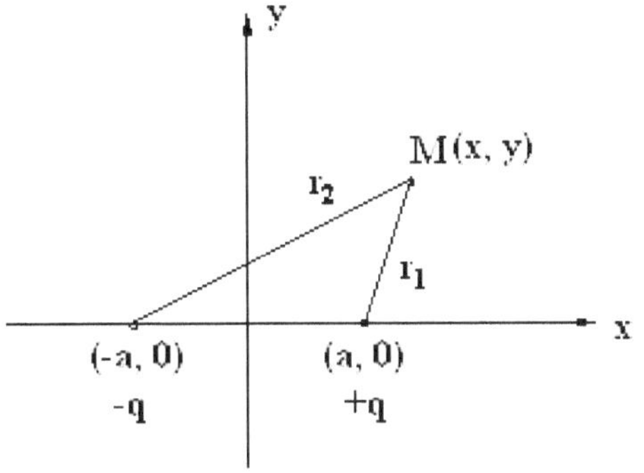

Figure 3.17. Schematic diagram of one source and one sink with equal production rates

rate q, while well 1 is an injection well with the same injection rate q. The two wells produce at the same time, the permeability is K, reservoir thickness is h, the liquid viscosity is μ, the bottom hole pressure of well 2 is p_w and the bottom hole pressure of well 1 is p_{win}. Use the principle of superposition of potential to calculate the production rate and pressure distribution.

According to the principle of superposition of potential, the potential of any point M is:

$$\Phi_M = \frac{q}{2\pi}\ln r_1 + \frac{-q}{2\pi}\ln r_2 + C = \frac{q}{2\pi}\ln\frac{r_1}{r_2} + C \tag{3.83}$$

When the point on wall 2 of the wellbore is taken as the special point, where $r_1 = 2a - R_w \approx 2a$, $r_2 = R_w$, it can be obtained that:

$$\Phi_2 = \frac{q}{2\pi}\ln\frac{2a}{R_w} + C \tag{3.84}$$

When the point on the wall of well 2 is taken as the special point, $r_1 = R_w$, $r_2 = 2a - R_w \approx 2a$, there is:

$$\Phi_1 = \frac{q}{2\pi}\ln\frac{R_w}{2a} + C \tag{3.85}$$

The boundary constant C is eliminated by subtracting the point source potential from the point sink potential, and the following equation is obtained:

$$\Phi_1 - \Phi_2 = \frac{q}{\pi} \ln \frac{R_w}{2a} \tag{3.86}$$

By using the definition of potential, it can be obtained that:

$$\Phi_1 - \Phi_2 = \frac{K}{\mu}(p_{win} - p_w) \tag{3.87}$$

Combining the above two equations, it can be obtained that:

$$\frac{q}{\pi} \ln \frac{2a}{R_w} = \frac{K}{\mu}(p_{win} - p_w) \tag{3.88}$$

The production rate is:

$$q = \frac{\pi K}{\mu} \frac{p_{win} - p_w}{\ln \frac{2a}{R_w}} \tag{3.89}$$

When obtaining the pressure distribution, subtract the potential at a special point from the potential at any point M, and cancel the constant C.

If the potential on well wall 1 is used for the special point, then:

$$\Phi_1 - \Phi_M = \frac{q}{2\pi} \ln \frac{r_2 \cdot 2a}{r_1 \cdot R_w} \tag{3.90}$$

By using the definition of potential, it can be obtained that:

$$\Phi_B - \Phi_M = \frac{K}{\mu}(p_{win} - p_M) \tag{3.91}$$

Then:

$$\frac{q}{2\pi} \ln \frac{r_2 \cdot 2a}{r_1 \cdot R_w} = \frac{K}{\mu}(p_{win} - p_M) \tag{3.92}$$

The pressure distribution of any point M is obtained as:

$$p_M = p_{win} - \frac{q\mu}{2\pi K} \ln \frac{r_2 \cdot 2a}{r_1 \cdot R_w} \tag{3.93}$$

If the special point is selected on the wall of well 2, the pressure distribution is:

$$p_M = p_w + \frac{q\mu}{2\pi K} \ln \frac{r_1 \cdot 2a}{r_2 \cdot R_w} \tag{3.94}$$

According to the expression of M potential at any point, equipotential lines can be expressed as:

$$\frac{r_1}{r_2} = C_2 \tag{3.95}$$

Draw the flow net diagram as shown in Figure 3.18. Those curves with arrows are streamlines, which are perpendicular to equipotential lines. The flow net diagram is "spindle-shaped".

According to the principle of superposition of flow velocity, the flow velocity at any M point in the formation is the vector superposition of the flow velocity when the two wells are working independently, as shown in Figure 3.19.

According to the definition of flow velocity, the following equation can be obtained:

$$v_1 = \frac{q}{2\pi \cdot r_1}, \quad v_2 = \frac{q}{2\pi \cdot r_2} \tag{3.96}$$

When M is on the y-axis, v_1 and v_2 are equal. After superposition, the direction of flow velocity v_M is perpendicular to the y-axis and points to the positive x direction. Moreover, the y-axis is an equipotential line.

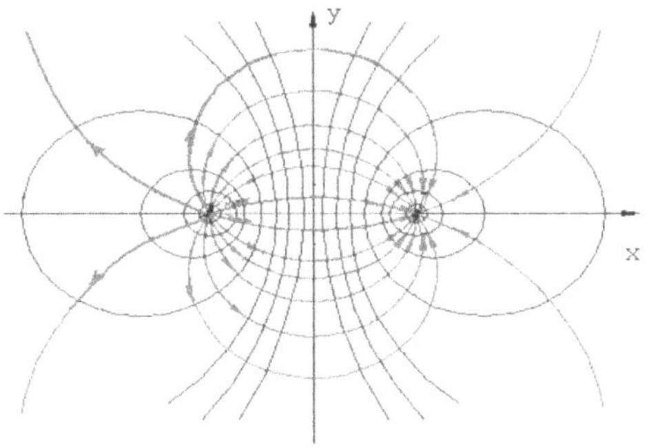

Figure 3.18. Flow net of one source and one sink with equal production rates

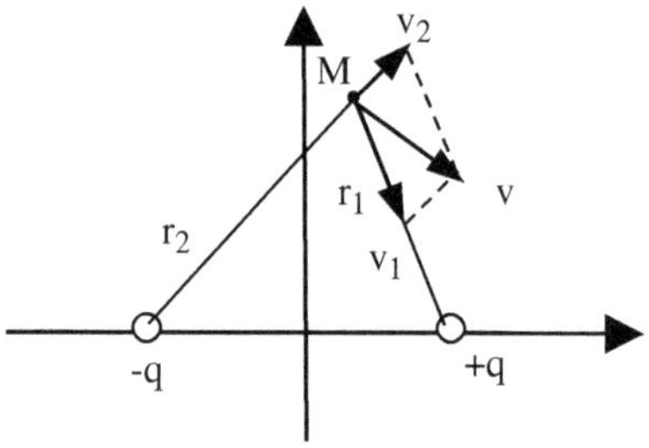

Figure 3.19. Superposition diagram of flow velocity of one source and one sink

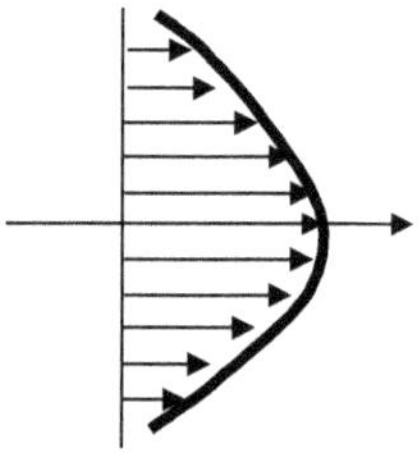

Figure 3.20. Schematic diagram of tonguing phenomenon

When point M is on the line perpendicular to the x-axis, r_1 and r_2 will be smaller as they get closer to the x-axis, and the included angle between v_1 and v_2 will be smaller, so v_M will be larger. Thus, the "tonguing phenomenon" in Figure 3.20 is formed, which can also be described as the shape of the point source fluid displacing the point sink fluid between the two wells.

When M is on the x-axis, v_M is the largest in the direction perpendicular to the x-axis and the flow of fluid is the fastest on the x-axis. Therefore, the x-axis is called the "main streamline". In the actual production of oil fields, the water cut degree is generally higher in the area between the injection well and the production well, the water breakthrough time of the oil well is related to the distance between the two wells, and it is easy to form an invalid water injection channel in this area, which is often unfavorable to water flooding.

3.2.3.4 *Mirror image method*

So far we have introduced the method to deal with the flow problem in the infinite reservoir with the ideal circular outer boundary.

That is, the outer boundary is sufficiently far away from the wells. In reality, the boundary may be close to the wells, and it can be with a line shape. Furthermore, the fault can form a closed boundary to the reservoir, and also have a line shape. In this section, we will introduce the mirror image method to deal with the flow problem in the finite reservoir.

(1) Fault boundary problem (the linear closed boundary)

As shown in Figure 3.21, there is a production well located near a linear fault boundary, and the fluid cannot flow across this boundary. The production rate of the well is $+q$, and the distance between the well and the fault is a. The fault acts like a diverging streamline in the problem of two point sinks with equal flow rates, and the flow fields in these two problems are consistent. Therefore, the problem of a well near the linear fault can be converted into the problem of two point sinks with equal flow rates by adding a virtual production well.

According to the analysis of the problem of two sinks with equal production rates, the production rate of this well is:

$$q = \frac{2\pi K}{\mu} \frac{p_e - p_w}{\ln \frac{R_e^2}{R_w \cdot 2a}} \tag{3.97}$$

The pressure distribution near the well is:

$$p_M = p_w + \frac{q\mu}{2\pi K} \ln \frac{r_1 \cdot r_2}{R_w \cdot 2a} \tag{3.98}$$

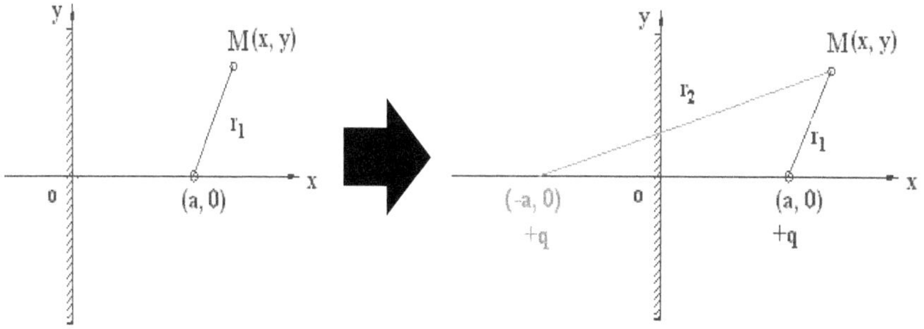

Figure 3.21. Mirror image of a well located near a linear closed boundary

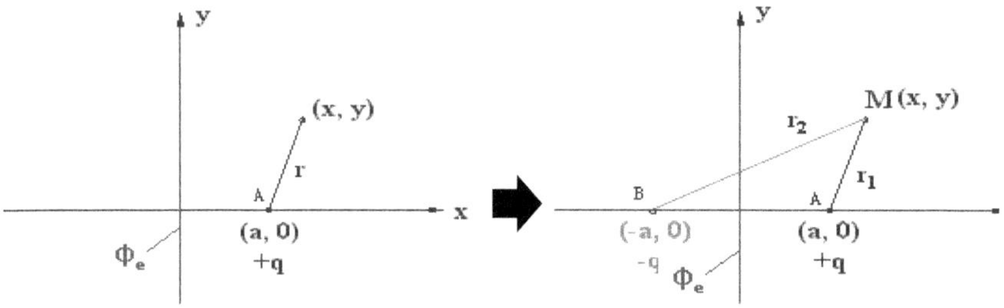

Figure 3.22. Mirror image of a well located near a linear supply boundary

(2) Linear supply boundary

As shown in Figure 3.22, there is a production well located near the linear supply boundary (y-axis) with production rate $+q$, and the distance between the well and the boundary is a. The linear supply boundary acts like an equipotential line in the problem of one point source and one point sink with equal flow rates, and the flow fields of these two problems are consistent. Therefore, the problem of a well near the linear supply boundary can be converted into the problem of one point source and one point sink with equal flow rates by adding a virtual injection well.

According to the analysis of the problem of one source and one sink with equal production rates, when calculating the production rate of the well, the special point can be selected on the line supply boundary. The potential is:

$$\Phi_e = C \tag{3.99}$$

When selecting the special point on the actual production well, the potential is:

$$\Phi_w = \frac{q}{2\pi} \ln \frac{R_w}{2a} + C \tag{3.100}$$

The constant C is eliminated by subtracting the potential on the actual production well from the one on the line supply boundary, then:

$$\Phi_e - \Phi_w = \frac{q}{2\pi} \ln \frac{2a}{R_w} \tag{3.101}$$

According to the definition of potential, there is:

$$\Phi_e - \Phi_w = \frac{K}{\mu}(p_e - p_w) \tag{3.102}$$

Combining the above two equations, we can get:

$$\frac{q}{2\pi} \ln \frac{2a}{R_w} = \frac{K}{\mu}(p_e - p_w) \tag{3.103}$$

The production rate for one point sink and one point source is obtained as:

$$q = \frac{2\pi K}{\mu} \frac{p_e - p_w}{\ln \frac{2a}{R_w}} \tag{3.104}$$

Take the point on the wall of the actual wellbore as the special point, and obtain the pressure distribution as:

$$p_M = p_w + \frac{q\mu}{2\pi K} \ln \frac{r_1 \cdot 2a}{r_2 \cdot R_w} \tag{3.105}$$

The flow rate of the straight line supply boundary is compared with that of the circular supply boundary.

For the straight line supply boundary:

$$q = \frac{2\pi K}{\mu} \frac{p_e - p_w}{\ln \frac{2a}{R_w}} \tag{3.106}$$

For the circular supply boundary:

$$q = \frac{2\pi K}{\mu} \frac{p_e - p_w}{\ln \frac{R_e}{R_w}} \tag{3.107}$$

Since the key difference between the production rates with the straight line supply boundary and the circular supply boundary is $\ln 2$, $\ln 2 = 0.693$ is a relatively small value, so:

$$\ln \frac{2a}{R_w} \approx \ln \frac{R_e}{R_w} \tag{3.108}$$

When R_e/R_w is large (it is generally the case), the production rates with the line supply boundary and the circular supply boundary are very close. Even if the supply boundary is irregular, it still can be

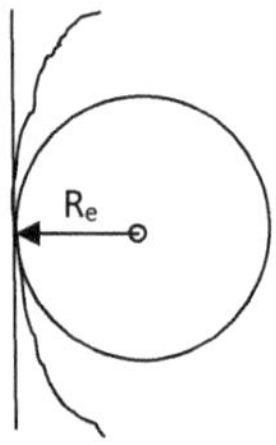

Figure 3.23. Schematic diagram of a reservoir with different outer boundaries

approximated to either the linear supply boundary or the circular supply boundary with little difference, as shown in Figure 3.23.

(3) Mirror image of complex boundary

For a complex boundary with more than two boundaries, the general principle to deal with the boundary effect is:

① For the linear fault boundary, the actual well can be imaged as a virtual well with the same production rate sign;

② For the line supply boundary, the actual well can be imaged as a virtual well with the opposite production rate sign;

③ The boundary must also be imaged with the boundary type kept the same, i.e., the image of the fault is still the fault boundary, and the same for supply boundary.

④ If the virtual well image encounters other boundaries, it still needs to be further imaged.

⑤ If the virtual boundary of the image encounters other boundaries, the virtual boundary still needs to be further imaged.

As shown in Figure 3.24, there is a production well in the reservoir with a straight line fault and a straight line supply boundary that are perpendicular to each other. The supply pressure is p_e and bottom hole pressure is p_w. The distances between the well and the boundaries are both a. Please find the flow rate q, the pressure distribution p, and the flow velocity v_A at the intersection point of the two boundaries.

Draw the imaged wells as shown in Figure 3.24. There are three imaged wells and one actual well. A total of four wells work at the same time.

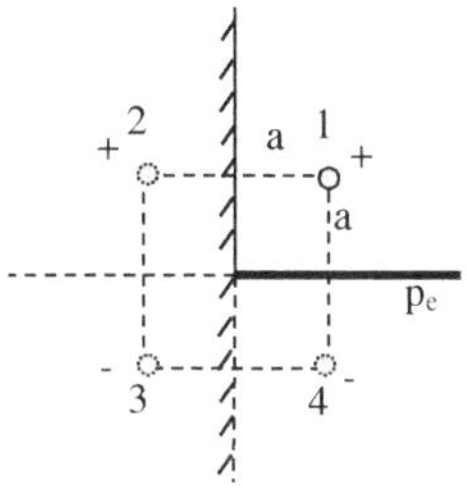

Figure 3.24. Mirror image of a well with one supply boundary and one supply boundary that are perpendicular

Take a point M in the actual flow zone, and according to the principle of superposition of potential, the potential of point M is:

$$\Phi_M = \frac{q}{2\pi} \ln \frac{r_1 r_2}{r_3 r_4} + C \tag{3.109}$$

Take the special point on the line supply boundary, then:

$$\Phi_e = C \tag{3.110}$$

Take the special point on the actual well wall, then:

$$\Phi_w = \frac{q}{2\pi} \ln \frac{R_w \cdot 2a}{2a \cdot 2\sqrt{2}a} + C \tag{3.111}$$

Cancel constant C by subtracting the potential on the wall of the production well from that on the supply boundary:

$$\Phi_e - \Phi_w = \frac{q}{2\pi} \ln \frac{2a \cdot 2\sqrt{2}a}{R_w \cdot 2a} \tag{3.112}$$

From the definition of potential, we can get:

$$\Phi_e - \Phi_w = \frac{K}{\mu}(p_e - p_w) \tag{3.113}$$

Then:

$$\frac{q}{2\pi} \ln \frac{2a \cdot 2\sqrt{2}a}{R_w \cdot 2a} = \frac{K}{\mu}(p_e - p_w) \tag{3.114}$$

The production rate is:

$$q = \frac{2\pi K(p_e - p_w)}{\mu \ln \frac{2\sqrt{2}a}{R_w}} \tag{3.115}$$

To find the pressure distribution:

$$\Phi_M - \Phi_w = \frac{q}{2\pi} \ln \frac{r_1 r_2 2\sqrt{2}a}{r_3 r_4 R_w} \tag{3.116}$$

According to the definition of potential:

$$\Phi_M - \Phi_w = \frac{K}{\mu}(p_M - p_w) \tag{3.117}$$

Then:

$$\frac{q}{2\pi} \ln \frac{r_1 r_2 2\sqrt{2}a}{r_3 r_4 R_w} = \frac{K}{\mu}(p_M - p_w) \tag{3.118}$$

The pressure distribution is obtained as:

$$p_M = p_w + \frac{q\mu}{2\pi K} \ln \frac{r_1 r_2 2\sqrt{2}a}{r_3 r_4 R_w} \tag{3.119}$$

To find v_A, the distance from point A to the four wells is:

$$v = \frac{q}{2\pi\sqrt{2}a} \tag{3.120}$$

According to the principle of superposition of flow velocity, see Figure 3.25.

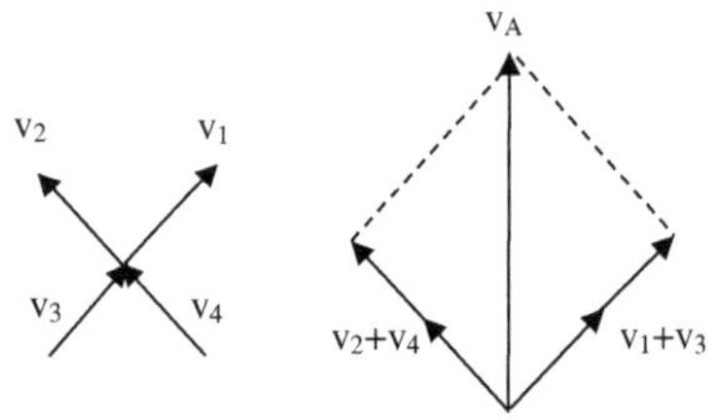

Figure 3.25. Superposition of flow velocity at the boundaries' intersection

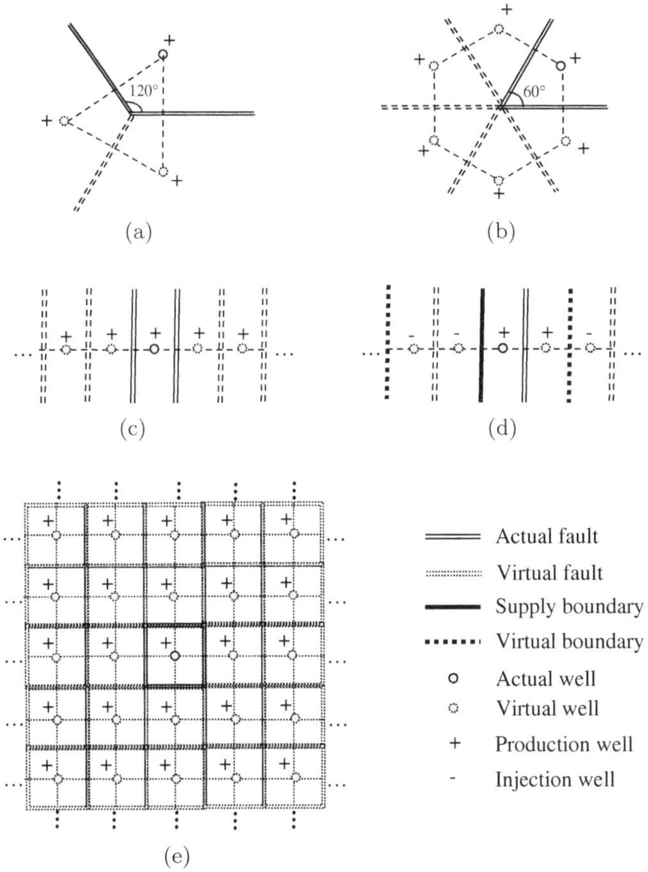

Figure 3.26. Mirror image of the complex boundaries

The flow velocity at point A is:

$$v_A = 2\sqrt{2}v = \frac{q}{\pi a} \tag{3.121}$$

The direction of the velocity is along the positive direction of y-axis.

(4) Mirror images of other complex boundaries

Figure 3.26 lists several different cases for more complex boundaries.

3.2.3.5 *Electric analog method*

The electric analog method can be applied when dealing with well line or well array problem.

(1) Straight well line problem

As shown in Figure 3.27(a), there is a straight well line (i.e., wells located along a straight line). The distance between the well line and the supply boundary is L, the well spacing in the well line is $2a$, there are n wells on the well line, and the bottom hole pressure of each well is equal to p_w.

A single well is taken from the well line as a basic unit to be studied, as shown in Figure 3.27(b). The real flow can be regarded as consisting of two parts, as shown in Figure 3.27(c). The first part is the linear flow from the line supply boundary to the area around the well, and the flow resistance is the outer resistance. The second

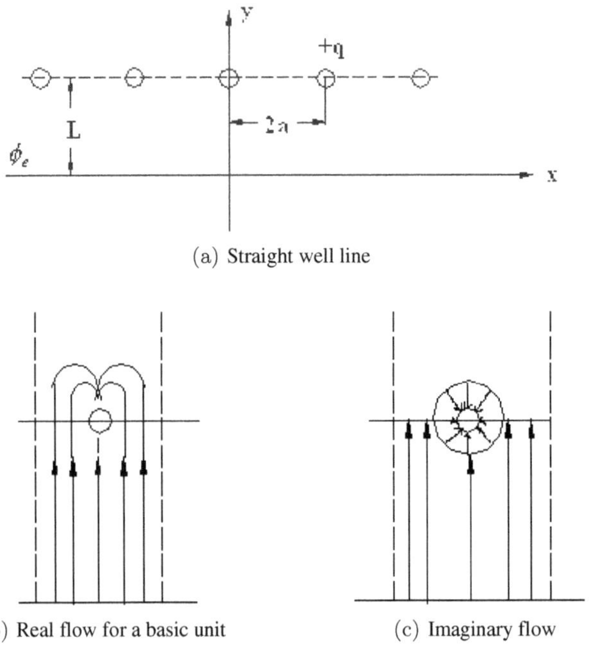

Figure 3.27. Straight well line problem solved through electric analog method

part is the radial flow from the area around the well (imaginary supply radius) to the bottom hole, and the flow resistance is the outer resistance. The two flow resistances are connected as a series circuit.

The outer resistance is:

$$R_{ou} = \frac{\mu L}{n \cdot 2a \cdot Kh} \tag{3.122}$$

The inner resistance is:

$$R_{in} = \frac{1}{n} \cdot \frac{\mu}{2\pi Kh} \ln \frac{a}{\pi Rw} \tag{3.123}$$

where R_{ou} is the outer flow external resistance; R_{in} is the inner flow resistance; n is the number of wells; L is the distance between well line and the supply boundary; and a is the half well spacing.

In electronics, Ohm's law is similar to Darcy's law here, and that is the reason why it is called the electric analog method.

Ohm's law is

$$I = \frac{\Delta U}{R_{\text{electric}}} \tag{3.124}$$

where I is the electric current, ΔU is the voltage difference, and R_E is the electric resistance.

It is similar to Darcy's law:

$$Q = \frac{\Delta \text{p}}{R_{\text{flow}}} \tag{3.125}$$

Therefore, the flow process can be represented by the circuit diagram, as shown in Figure 3.28. The flow equation is:

$$Q = \frac{p_e - p_w}{R_{ou} + R_{in}} \tag{3.126}$$

It is also possible to draw the circuit diagram, build the equations and solve the well array problem when multiple wells are arranged.

In the flow process shown in Figure 3.29(a), there are three rows of wells in the U-shape formation (closed boundary on the three sides) with constant supply pressure on the upper boundary, and each parameter is shown in the figure.

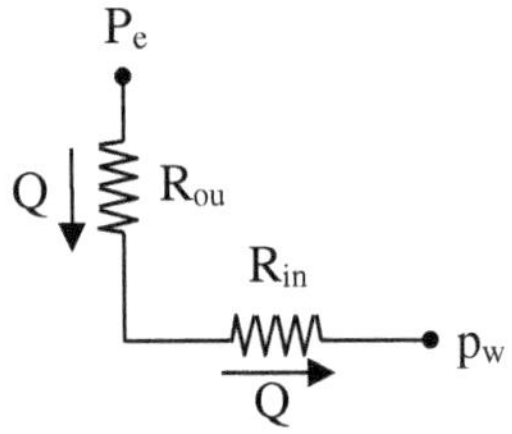

Figure 3.28. Circuit diagram of the well line problem

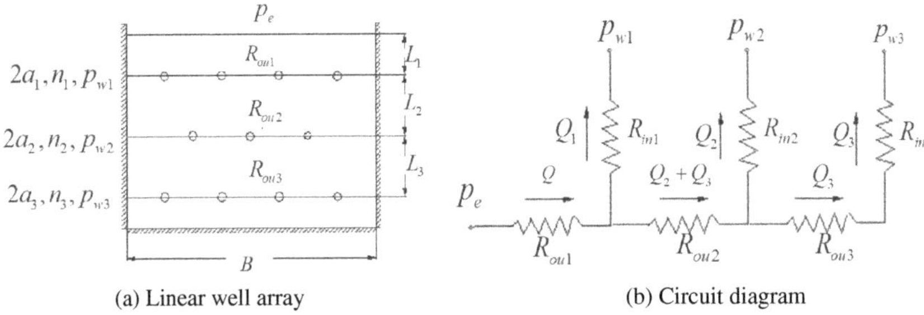

Figure 3.29. Circuit diagram of linear well array problem

The assumptions are as follows:

① The spacing of wells on the same row is $2a$, but can be different on different rows;

② The bottom hole pressure p_w of each well in the same row is the same, but can be different on different rows;

③ The production rate is the same for the wells in the same row, but can be different on different rows;

④ The well diameter is same for the wells in the same row, but can be different on different rows.

If each row of wells meets the above assumptions, the electric analog method can be applied to solve the problem. The steps are as follows:

Step 1: draw the circuit diagram

The circuit diagram is shown in Figure 3.29(b). It can be seen from the diagram that the total flow rate Q is originated from supply boundary and overcomes the outer resistance on the L_1 well line.

After that, part of Q_1 overcomes the inner resistance and flows to the bottom hole of the wells on the L_1 well line. The rest $Q_2 + Q_3$ overcome the outer resistance of L_2 well line, and then Q_2 flows through the inner resistance of to the bottom hole on the L_2 well line. The remaining Q_3 overcomes the outer and inner resistances and flows into the bottom hole on the L_3 well line.

Step 2: calculate the outer and inner flow resistances of each well line.

Outer resistance:

$$\begin{aligned} R_{ou\,1} &= \frac{\mu L_1}{n_1 \cdot 2a_1 \cdot Kh} \\ R_{ou\,2} &= \frac{\mu L_2}{n_2 \cdot 2a_2 \cdot Kh} \\ R_{ou\,3} &= \frac{\mu L_3}{n_3 \cdot 2a_3 \cdot Kh} \end{aligned} \tag{3.127}$$

Inner resistance:

$$\begin{aligned} R_{in\,1} &= \frac{1}{n_1} \cdot \frac{\mu}{2\pi Kh} \ln \frac{a_1}{\pi Rw_1} \\ R_{in\,2} &= \frac{1}{n_2} \cdot \frac{\mu}{2\pi Kh} \ln \frac{a_2}{\pi Rw_2} \\ R_{in3} &= \frac{1}{n_3} \cdot \frac{\mu}{2\pi Kh} \ln \frac{a_3}{\pi Rw_3} \end{aligned} \tag{3.128}$$

Step 3: build the equation based on the circuit diagram.

$$\begin{aligned} p_e - p_{w1} &= QR_{ou\,1} + Q_1 R_{in\,1} \\ p_{w1} - p_{w2} &= -Q_1 R_{in\,1} + (Q_2 + Q_3) R_{ou\,2} + Q_2 R_{in\,2} \\ p_{w2} - p_{w3} &= -Q_2 R_{in\,2} + Q_3 R_{ou\,3} + Q_3 R_{in\,3} \\ Q &= Q_1 + Q_2 + Q_3 \end{aligned} \tag{3.129}$$

There are four equations and four unknowns: Q, Q_1, Q_2 and Q_3, and the equations can be solved.

(2) Annular well line problem

In addition to straight well line, the annular well row is also common in large reservoirs.

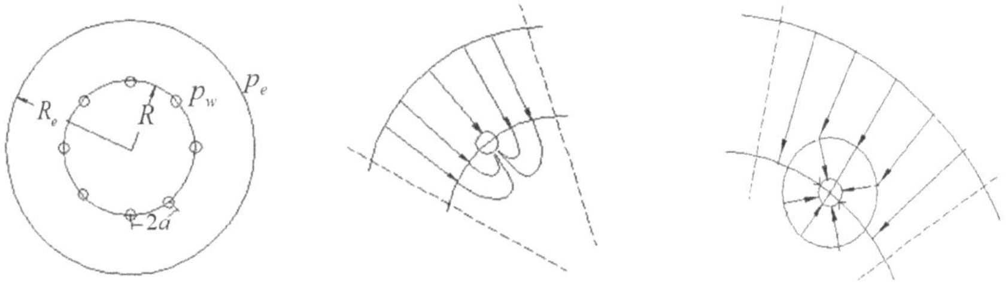

(a) Annular well line (b) Real (c) Imaginary

Figure 3.30. Annular well line problem solved through electric analog method

Figure 3.30(a) shows the annular well line problem, where the circular supply boundary radius is R_e, the supply pressure is P_e, the annular well row radius is R, the well spacing is $2a$, the number of wells is n, and the bottom hole pressure of each well is equal to p_w.

In the annular well line, a single well is taken as the basic unit, and the real flow in Figure 3.30(b) can be regarded as consisting of two parts: the radial flow from the annular supply boundary to the area near the well with the outer resistance and the radial flow from the area around the well (imaginary supply radius) to the bottom hole of wells with the inner resistance. These two resistances are connected like a series circuit, as shown in Figure 3.30(c).

The outer and inner resistances are:

$$R_{ou} = \frac{\mu}{2\pi Kh} \ln \frac{R_e}{R} \tag{3.130}$$

$$R_{in} = \frac{1}{n} \cdot \frac{\mu}{2\pi Kh} \ln \frac{a}{\pi R_w} \tag{3.131}$$

The circuit diagram for the annular well line problem can be the same as the straight well line case as shown in Figure 3.28.

Circuit diagrams can also be used when multiple annular wells are presented.

In the flow process shown in Figure 3.31, there are four lines of wells in the circular formation. The outermost row is the injection well, and the bottom hole pressure is p_{win} which provides the supply pressure to drive the flow.

The electric analog method is applied to solve the production rate of each well.

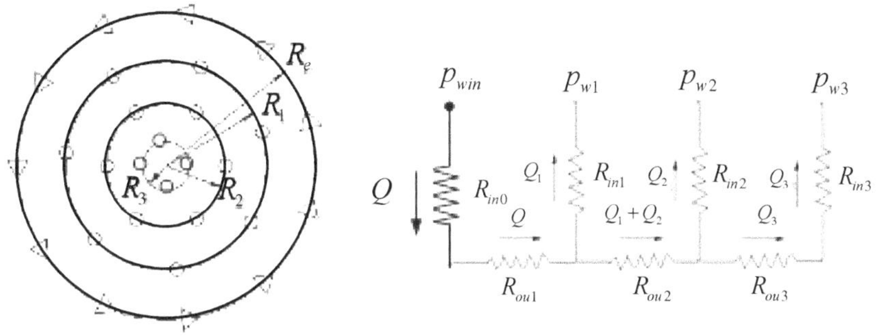

(a) Multiple annular well lines (b) Circuit diagram

Figure 3.31. Multiple annular well lines problem and its circuit diagram

Step 1: draw the circuit diagram.

As shown in Figure 3.31(b), it is different from the circuit diagram in Figure 3.29(b) due to the fact that the outermost circle is the injection well, not a supply boundary. Therefore, fluid flow Q first passes through the inner resistance of the injection well, and the rest of the parts are the same as the well array problem.

Step 2: calculate the inner and outer resistances

The outer resistances are:

$$
\begin{aligned}
R_{ou1} &= \frac{\mu}{2\pi Kh} \ln \frac{R_e}{R_1} \\
R_{ou2} &= \frac{\mu}{2\pi Kh} \ln \frac{R_1}{R_2} \\
R_{ou3} &= \frac{\mu}{2\pi Kh} \ln \frac{R_2}{R_3}
\end{aligned}
\tag{3.132}
$$

The inner resistances are:

$$
\begin{aligned}
R_{in0} &= \frac{1}{n_0} \cdot \frac{\mu}{2\pi Kh} \ln \frac{a_0}{\pi R_{w0}} \\
R_{in1} &= \frac{1}{n_1} \cdot \frac{\mu}{2\pi Kh} \ln \frac{a_1}{\pi Rw1} \\
R_{in2} &= \frac{1}{n_2} \cdot \frac{\mu}{2\pi Kh} \ln \frac{a_2}{\pi Rw2} \\
R_{in3} &= \frac{1}{n_3} \cdot \frac{\mu}{2\pi Kh} \ln \frac{a_3}{\pi Rw3}
\end{aligned}
\tag{3.133}
$$

Step 3: build the circuit equation.

$$\begin{aligned} p_{win} - p_{w1} &= QR_{in0} + QR_{ou1} + Q_1R_{in1} \\ p_{w1} - p_{w2} &= -Q_1R_{in1} + (Q_2 + Q_3)R_{ou2} + Q_2R_{in2} \\ p_{w2} - p_{w3} &= -Q_2R_{in2} + Q_3R_{ou3} + Q_3R_{in3} \\ Q &= Q_1 + Q_2 + Q_3 \end{aligned} \tag{3.134}$$

3.2.4 *Example*

1. There are stagnation oil zones between the two sinks with equal production rates. Try to figure out how to extract more oil from this zone.

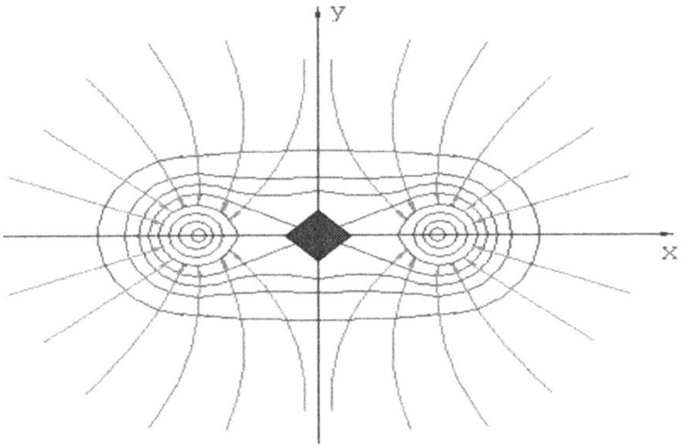

Solution:

(1) Drill an infill well near the stagnation point;
(2) Change one of the production wells to an injection well;
(3) Increase (decrease) the production rate of one well and decrease (increase) the production rate of the other well.
(4) Shut in one well and keep the other open.

2. Compute the production rate q of the well in the reservoir with the circular supply boundary, the linear supply boundary and the linear fault boundary well. Let $R_e = 300\,\text{m}$, $p_e = 20\,\text{MPa}$, $K = 50\,\text{mD}$, $h = 5\,\text{m}$, $\mu = 5\,\text{MPa}\cdot\text{s}$, $R_w = 0.1\,\text{m}$, $p_w = 15\,\text{MPa}$, and $a = 50\,\text{m}$ away from the linear fault.

Solution:

Circular supply boundary:

$$q = \frac{2\pi K}{\mu} \frac{(p_e - p_w)}{\ln \frac{R_e}{R_w}}$$

$$= \frac{2 \times 3.14 \times 50 \times 10^{-15}}{5 \times 10^{-3}} \times \frac{(20 - 15) \times 10^{-6}}{\ln \frac{300}{0.1}} \times 86400 = 3.39\,\mathrm{m}^2/\mathrm{d}$$

Linear supply boundary:

$$q = \frac{2\pi K}{\mu} \frac{(p_e - p_w)}{\ln \frac{2a}{R_w}}$$

$$= \frac{2 \times 3.14 \times 50 \times 10^{-15}}{5 \times 10^{-3}} \times \frac{(20 - 15) \times 10^{-6}}{\ln \frac{2 \times 300}{0.1}} \times 86400 = 3.12\,\mathrm{m}^2/\mathrm{d}$$

Linear fault boundary:

$$q = \frac{2\pi K}{\mu} \frac{(p_e - p_w)}{\ln \frac{R_e^2}{R_w \cdot 2a}}$$

$$= \frac{2 \times 3.14 \times 50 \times 10^{-15}}{5 \times 10^{-3}} \times \frac{(20 - 15) \times 10^{-6}}{\ln \frac{300^2}{0.1 \times 2 \times 50}} \times 86400 = 2.98\,\mathrm{m}^2/\mathrm{d}$$

3. There are two supply boundaries at upper and lower sides of the reservoir. Try to analyze how to use the electric analog method to solve the production rate.

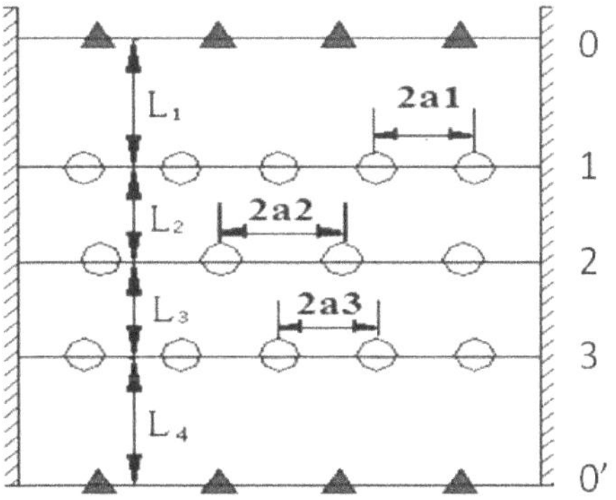

Solution:

Since the upper and lower sides are both supply boundaries, let us assume the fluids from both supply boundaries are flowing into the third row of production wells, and the circuit diagram can be shown as follows:

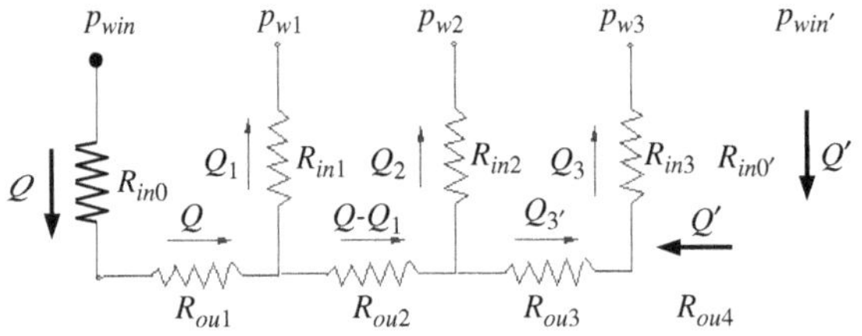

The equation can be obtained from the circuit diagram as:

$$
\begin{aligned}
p_{win} - p_{w1} &= QR_{in0} + QR_{ou1} + Q_1R_{in1} \\
p_{w1} - p_{w2} &= -Q_1R_{in1} + (Q - Q_1)R_{ou2} + Q_2R_{in2} \\
p_{w2} - p_{w3} &= -Q_2R_{in2} + Q_3'R_{ou3} + Q_3R_{in3} \\
p_{w3} - p_{win}' &= -Q_3R_{in3} - Q'R_{ou4} - Q'R_{in0}' \\
Q &= Q_1 + Q_2 + Q_3' \\
Q_3 &= Q_3' + Q'
\end{aligned}
$$

If some of the obtained flow rates are associated with negative values, that means the assumption about the third row is not correct, and we need to make other assumptions, such as we need to assume that the second or fourth row is the one that the fluids from both sides flow into, and rebuild the circuit diagram and the corresponding equation.

3.3 Single-phase transient flow theory

3.3.1 *Introduction*

Under high pressure, it is impossible for rocks and fluids not to have elastic effect. Rigid reservoir or rigid fluid is only in ideal or experimental state, and the elastic effect of actual reservoir is very strong.

For unsaturated reservoirs, the natural elastic energy is the important driving force in the early stage of development. Therefore, this section can also be called elastic driving flow theory.

In the actual reservoir, if a well is opened, the initial reservoir energy is sufficient and the oil well production is high, but the production will gradually decrease with decrease in formation pressure. Under certain conditions, the pressure and production rates may reach a stable state, thus this process rebalances the change of energy or pressure. In this process, the flow factors are related to time, which is called "transient flow".

3.3.2 *Concepts*

Incompressible liquid: *an incompressible liquid is a liquid that has no compressibility and is an ideal liquid.*

Slightly compressible liquid: *when the pressure changes, the density of liquid affected by compression changes very little. Generally its elastic compression coefficient is* $10^{-2} \sim 10^{-4}\ MPa^{-1}$, *and also refers to the oil and water in the formation. If its compressibility is ignored, it can be called incompressible liquid.*

Compressible fluid: *a compressible fluid is a fluid that has considerable compressibility, such as gas.*

Pseudo-steady state flow: *it is called pseudo-stable flow when the change of pressure at every point in the formation is the same.*

Total elastic compressibility of rock: *the volume of fluid discharged due to pore shrinkage and fluid volume expansion from the unit volume rock when the pressure is reduced by one unit. The equation is* $C_t = \phi C_L + C_f$, *with common unit* MPa^{-1}.

Hydraulic diffusivity: *the hydraulic diffusivity is the area of pressure drop spreading in the reservoir per unit time in the elastic driving process, and the equation is:*

$$\eta = \frac{K}{\mu C_t} \tag{3.135}$$

where η is the hydraulic diffusivity, m^2/s; K is the permeability, m^2; μ is the viscosity, $Pa \cdot s$; and C_t is the total compressibility, Pa^{-1}.

Exponential integral function: *it is a commonly used integral form, and its expression is:*

$$-Ei(-x) = \int_x^{+\infty} \frac{e^{-u}}{u} du \tag{3.136}$$

where $-Ei()$ is the exponential integral function; and x is the independent variable in the function.

Degree of elastic recovery: *it refers to the degree of recovery obtained by utilizing elastic drive in an unsaturated reservoir. Under the constant production rate condition, when the bottom hole pressure reaches saturation pressure, the elastic drive ends. Under the constant bottom hole pressure condition, if the bottom hole pressure is kept larger than the saturation pressure, the end of elastic flooding occurs when the well production rate is close to 0.*

Elastic reserves: *it refers to the amount of fluid discharged from pores in a reservoir when the pressure at all points is above the saturation pressure.*

Transient well test: *the measured bottom hole pressure changes with time when the well is being produced at a certain production rate (or when the well is being shut in after a period of production at a certain production rate), then the reservoir parameters can be obtained. It is generally divided into constant production rate drawdown test and pressure buildup test.*

Constant production rate drawdown test: *when the well is produced at a constant rate, the test is carried out by measuring the pressure drawdown with time.*

Shut-in pressure buildup test: *when a well is shut-in after a period of production at a certain production rate, the test is carried out by measuring the increase in pressure.*

Wellbore storage effect: *when a well is just opened or shut-in, surface production and underground production are not equal due to* *the compressibility of the fluid in the wellbore. This phenomenon is called wellbore storage effect.*

Wellbore storage coefficient: *it refers to the change of fluid volume caused by unit pressure change in wellbore, symbol C, commonly used unit is* m^3/MPa.

Afterflow: *due to wellbore storage effect, when the well is just shut-in (generally the surface shut-in), some fluid in the formation continues to flow into the bottom hole of the well.*

3.3.3 *Analysis*

3.3.3.1 *Physical process of elastic transient flow*

In an elastic energy driving flow process, the elastic energy comes from two parts: the liquid and the rock.

During the process of reservoir production, the formation pressure is often reduced. Due to the compressibility of the liquid, the elastic energy release causes its volume to expand, and due to the compressibility of the rock, the elastic energy release causes the pore volume to shrink. The combination of the two parts of elastic energy is used to overcome the flow resistance and make the fluid flow.

Pressure drop induces the release of elastic energy, and the release of elastic energy delays the process of pressure drop. If both the fluid and the rock are rigid, the pressure drop can instantly reach an equilibrium state. Therefore, the propagation process of pressure drop in formation is the physical process of transient flow.

After oil well production, pressure drop first occurs in the bottom of the wellbore, therefore, the release of elastic energy starts from the bottom of a well. If the pressure drop is defined to be the propagation of "pressure waves" to a distance far away, the pressure wave propagation starts from the bottom of the wellbore and gradually extends outward, such that the overall process is the extending and deepening process of the cone of depression. Moreover, the fluid flows only in the region affected by the cone of depression, and the fluid cannot flow in the region not affected by it.

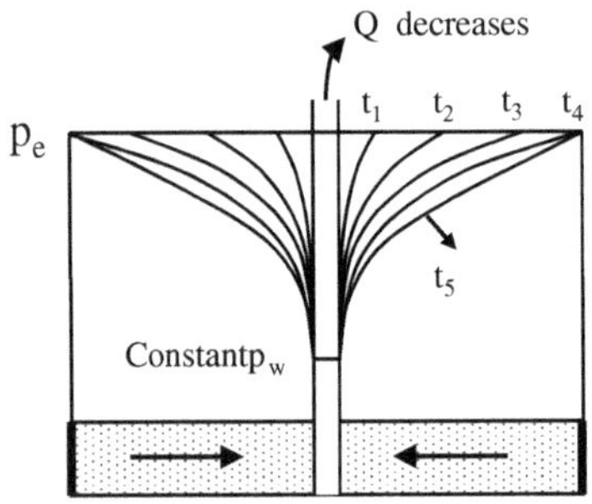

Figure 3.32. Pressure propagation with outer supply boundary and constant BHP

The flow process is introduced in detail with the combination of the different outer and inner boundaries.

(1) Supply outer boundary and constant bottom hole pressure

As shown in Figure 3.32, the horizontal formation has a circular supply boundary and a production well is located in the center with constant bottom hole pressure p_w. The pressure distribution diagram at different times is drawn with the well axis as the center line. As can be seen from the figure, the propagation process of pressure wave is divided into three stages:

① Infinite formation flow stage. At the early flow stage, before the fluid flow approaches to outer boundary, the production time t is less than t_4. The bottom hole pressure remains unchanged, and the pressure drawdown near the well quickly reaches the maximum. Since the energy loss is mainly concentrated near the well, the production rate Q is also large, and then gradually decreases. Pressure wave spreads outward. Within the region of the cone of depression, the pressure decreases; and out of the region of the cone of depression, the pressure remains the original formation pressure (i.e., supply pressure P_e) and the production Q is completely from the elastic energy drive Q_{elastic}, i.e. the supply boundary has no contribution to the flow, $Q_{\text{supply}} = 0$, $Q = Q_{\text{elastic}}$;

② Transitional stage. Between the arrival time to the boundary t_4 and the time to approach steady state flow t_5, the pressure wave has spread to the supply boundary, and both bottom-hole pressure and boundary pressure remain unchanged. Elastic energy in the formation continues to release, and the pressure gradually decreases. At this time, $Q = Q_{\text{elastic}} + Q_{\text{supply}}$.

③ Steady state flow stage. The elastic energy is completely released, the pressure in the formation remains unchanged, and the steady state flow regime is achieved. The flow follows the law of steady state flow described before. The well production is completely from the supply boundary, that is, $Q = Q_{\text{supply}}$.

(2) Supply outer boundary and constant production rate

As shown in Figure 3.33, the horizontal formation has a circular supply boundary, with a production well in the center. The production rate is Q and kept constant, and the pressure distribution at different times is drawn with the well axis as the central line. As can be seen from the figure, the propagation process of pressure wave is divided into three stages:

① Infinite formation flow stage. At the early flow stage, before the fluid flow approaches the outer boundary, the production time t is less than t_4. Well production is constant, and in order to guarantee the constant production rate, the bottom hole pressure decreases with pressure wave spreading outward. Within the region of the cone of depression, the pressure decreases; and out of the region of the cone of depression, the pressure remains the original formation pressure

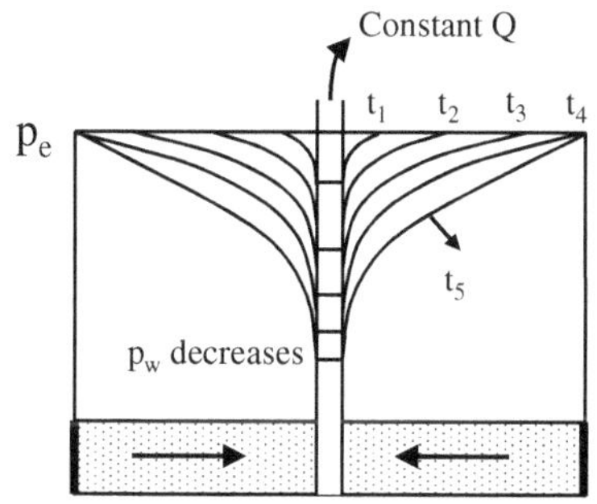

Figure 3.33. Pressure propagation with outer supply boundary and constant Q

and the production Q is completely from the elastic energy drive Q_{elastic}, i.e. the supply boundary has no contribution to the flow, $Q = Q_{\text{elastic}}$;

② Transitional stage. Between the arrival time to the boundary t_4 and the time to approach steady state flow t_5, the pressure wave has spread to the supply boundary, the boundary pressure remains unchanged, the bottom hole pressure and the pressure in the formation continue to drop, and the elastic energy is also released, $Q = Q_{\text{elastic}} + Q_{\text{supply}}$.

③ Steady state flow stage. The elastic energy is completely released, the bottom hole pressure and the formation pressure remain unchanged, and the steady state flow is achieved. The flow follows the law of steady state flow described before. The well production is completely from the supply boundary, that is, $Q = Q_{\text{supply}}$.

(3) Closed outer boundary and constant bottom hole pressure

As shown in Figure 3.34, the horizontal formation has a circular closed boundary with a producing well in the center of the reservoir. The pressure distribution diagram at different times is drawn with the well axis as the central line to produce with the constant bottom hole pressure p_w. As can be seen from the figure, the propagation process of pressure wave can be divided into three stages:

① Infinite formation flow stage. At the early flow stage, before the fluid flow approaches the outer boundary the production time t is

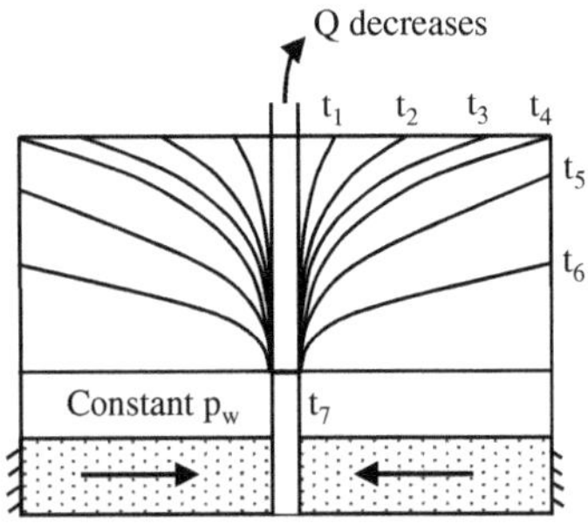

Figure 3.34. Pressure propagation with outer closed boundary and constant BHP

less than t_4. The flow characteristics is the same as the case of supply outer boundary and constant bottom hole pressure. The pressure at the bottom of the well remains unchanged, and the pressure drawdown near the well quickly reaches the maximum. Out of the region of the cone of depression at this stage, the closed boundary has no contribution to the flow, $Q = Q_{\text{elastic}}$.

② Transitional stage. Between the arrival time to the boundary t_4 and the time to approach steady state flow t_5, the pressure wave has spread to the closed boundary, the bottom hole pressure remains unchanged, the boundary pressure gradually decreases with the formation pressure. The elastic energy in the formation continues to release, and there is no external energy supplement, $Q = Q_{\text{elastic}}$, and the well production Q continues to decrease.

③ Steady state flow stage. The elastic energy is completely released, pressure in the formation and boundary pressure are consistent with bottom hole pressure, the well production $Q = 0$, and the steady state flow is achieved.

(4) Closed outer boundary and constant production rate

As shown in Figure 3.35, the horizontal formation has a circular closed boundary with a production well in the center. The production rate is Q and kept constant, and the pressure drop distribution diagram at different times is drawn with the well axis as the central

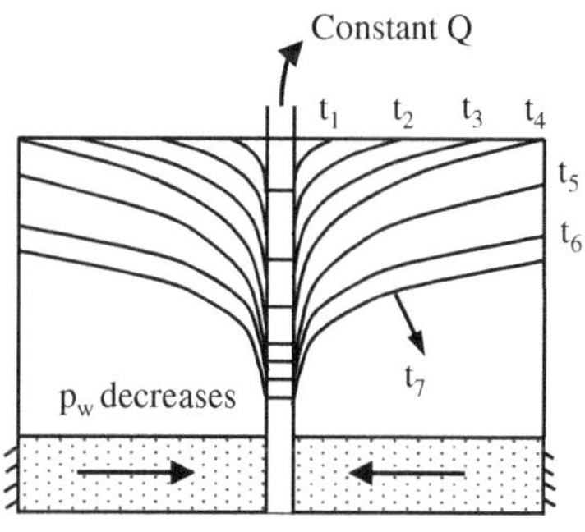

Figure 3.35. Pressure propagation with outer closed boundary and constant Q

line. As can be seen from the figure, the propagation process of pressure wave can be divided into three stages:

① Infinite formation flow stage. At the early flow stage, before the fluid flow approaches the outer boundary, the production time t is less than t_4. Similar to the flow characteristics in the case of supply outer boundary and constant production rate, the pressure at the bottom of the well keeps decreasing, and the pressure wave spreads outwards and deepens. Out of the region of the cone of depression, the fluid does not flow, and the boundary has no influence on the flow, $Q = Q_{\text{elastic}}$.

② Transitional stage. Between the arrival time to the boundary t_4 and the time to approach steady state flow t_5, the pressure wave has spread to the closed boundary. The boundary pressure, bottom hole pressure and formation pressure are all decreasing, the elastic energy is continuously released, and there is no external energy supplement. $Q = Q_{\text{elastic}}$.

③ Pseudo-steady state flow stage. In order to ensure production, the pressure of the whole formation keeps falling, and keeps the same decreasing rate so as to achieve the pseudo-steady state, which is different from the previous three modes. $Q = Q_{\text{elastic}}$. Before the bottom hole pressure drops to saturation pressure, it is always under the pseudo-steady state flow stage.

3.3.3.2 *Establishment of the governing equation*

Let us assume that the reservoir is homogeneous, and the permeability and viscosity of the liquid remain unchanged.

Substituting the equation of motion and the equation of state into the continuity equation gives:

$$\frac{\partial}{\partial x}\left(\rho\frac{K}{\mu}\frac{\partial p}{\partial x}\right)+\frac{\partial}{\partial x}\left(\rho\frac{K}{\mu}\frac{\partial p}{\partial x}\right)+\frac{\partial}{\partial x}\left(\rho\frac{K}{\mu}\frac{\partial p}{\partial x}\right)=\frac{\partial(\rho\phi)}{\partial t} \tag{3.137}$$

For the left side of the equation, the exponential form of the equation of state for liquid is:

$$\rho = \rho_0[1 + C_L(p - p_0)] \tag{3.138}$$

Substitute it into the differential equation:

$$\frac{\partial}{\partial x}\left(\rho\frac{K}{\mu}\frac{\partial p}{\partial x}\right) = \rho_0\frac{K}{\mu}\frac{\partial}{\partial x}\left[[1 + C_L(p - p_0)]\frac{\partial p}{\partial x}\right]$$

$$= \rho_0\frac{K}{\mu}\frac{\partial^2 p}{\partial x^2} \tag{3.139}$$

Similarly, the y and z directions can also be solved, then:

$$\frac{\partial}{\partial x}\left(\rho\frac{K}{\mu}\frac{\partial p}{\partial x}\right) + \frac{\partial}{\partial x}\left(\rho\frac{K}{\mu}\frac{\partial p}{\partial x}\right) + \frac{\partial}{\partial x}\left(\rho\frac{K}{\mu}\frac{\partial p}{\partial x}\right)$$

$$= \rho_0\frac{K}{\mu}\left(\frac{\partial^2 p}{\partial x^2} + \frac{\partial^2 p}{\partial y^2} + \frac{\partial^2 p}{\partial z^2}\right) \tag{3.140}$$

For the right side of the equation, the state equation of rock is:

$$\phi = \phi_0 + C_f(p - p_0) \tag{3.141}$$

Since the rock and liquid are slightly compressible, C_L and C_f are small. There is:

$$\frac{\partial(\rho\phi)}{\partial t} = \rho_0(\phi_0 C_L + C_f)\frac{\partial p}{\partial t} \tag{3.142}$$

The total compressibility is:

$$C_t = \phi_0 C_L + C_f \tag{3.143}$$

Then the right side can be rewritten as:

$$\frac{\partial(\rho\phi)}{\partial t} = \rho_0 C_t\frac{\partial p}{\partial t} \tag{3.144}$$

The left and right equations are equal to each other, and we get:

$$\rho_0\frac{K}{\mu}\left(\frac{\partial^2 p}{\partial x^2} + \frac{\partial^2 p}{\partial y^2} + \frac{\partial^2 p}{\partial z^2}\right) = \rho_0 C_t\frac{\partial p}{\partial t} \tag{3.145}$$

It can be rewritten as:

$$\frac{\partial^2 p}{\partial x^2} + \frac{\partial^2 p}{\partial y^2} + \frac{\partial^2 p}{\partial z^2} = \frac{\mu C_t}{K}\frac{\partial p}{\partial t} \tag{3.146}$$

The hydraulic diffusivity is:

$$\eta = \frac{K}{\mu C_t} \tag{3.147}$$

The governing equation for the elastic transient flow with the single-phase slightly compressible liquid is:

$$\frac{\partial^2 p}{\partial x^2} + \frac{\partial^2 p}{\partial y^2} + \frac{\partial^2 p}{\partial z^2} = \frac{1}{\eta}\frac{\partial p}{\partial t} \tag{3.148}$$

In two-dimensional form, polar coordinates are often used, and the formula is:

$$\frac{\partial^2 p}{\partial r^2} + \frac{1}{r}\frac{\partial p}{\partial r} = \frac{1}{\eta}\frac{\partial p}{\partial t} \tag{3.149}$$

For the four types of pressure wave propagation process, if the mathematical model is established, then:

(1) supply outer boundary and constant bottom hole pressure

$$\begin{cases} \dfrac{\partial^2 p}{\partial r^2} + \dfrac{1}{r}\dfrac{\partial p}{\partial r} = \dfrac{1}{\eta}\dfrac{\partial p}{\partial t} & \text{governing equation} \\ p|_{t=o} = p_i & \text{initial condition} \\ p|_{\mathrm{r=R_e}} = p_e & \text{outer boundary condition} \\ r\dfrac{\partial p}{\partial r}\Big|_{\mathrm{r=R_w}} = \dfrac{Q\mu}{2\pi kh} & \text{inner boundary condition} \end{cases}$$

(2) supply outer boundary and constant production rate

$$\begin{cases} \dfrac{\partial^2 p}{\partial r^2} + \dfrac{1}{r}\dfrac{\partial p}{\partial r} = \dfrac{1}{\eta}\dfrac{\partial p}{\partial t} & \text{governing equation} \\ p|_{t=o} = p_i & \text{initial condition} \\ p|_{\mathrm{r=R_e}} = p_e & \text{outer boudnary condition} \\ p|_{\mathrm{r=R_w}} = p_w & \text{inner boundary condition} \end{cases}$$

(3) closed outer boundary and constant bottom hole pressure

$$
\begin{cases}
\dfrac{\partial^2 p}{\partial r^2} + \dfrac{1}{r}\dfrac{\partial p}{\partial r} = \dfrac{1}{\eta}\dfrac{\partial p}{\partial t} & \text{governing equation} \\
p|_{t=o} = p_i & \text{initial condition} \\
\left.\dfrac{\partial p}{\partial r}\right|_{\mathrm{r=R_e}} = 0 & \text{outer boudnary condition} \\
p|_{\mathrm{r=R_w}} = p_w & \text{inner boundary condition}
\end{cases}
$$

(4) closed boundary and constant production rate

$$
\begin{cases}
\dfrac{\partial^2 p}{\partial r^2} + \dfrac{1}{r}\dfrac{\partial p}{\partial r} = \dfrac{1}{\eta}\dfrac{\partial p}{\partial t} & \text{governing equation} \\
p|_{t=o} = p_i & \text{initial condition} \\
\left.\dfrac{\partial p}{\partial r}\right|_{\mathrm{r=R_e}} = 0 & \text{outer boudnary condition} \\
\left.r\dfrac{\partial p}{\partial r}\right|_{\mathrm{r=R_w}} = \dfrac{Q\mu}{2\pi kh} & \text{inner boundary condition}
\end{cases}
$$

3.3.3.3 *Solution of transient flow*

This solution is derived for transient flow in the infinite reservoir with constant production rate.

According to the propagation process of pressure wave in the formation, the flow in the first stage has nothing to do with the boundary, which is called the infinite reservoir flow stage. The flow in this stage can fully reflect the energy of the reservoir, the physical property of the reservoir and the change of the productivity of the well, and it is the basis of formulating the work schedule for the later stage of oil well production. Therefore, it is necessary to quantify the characteristics of the elastic transient flow in the infinite reservoir.

This solution is derived under these assumptions: there is a point sink in the homogeneous, uniformly thick and infinite reservoir, the slightly compressible single-phase liquid flows to the bottom of the well in the form of radial flow regime with constant production rate Q, and the original formation pressure is P_i. If the reservoir and fluid properties are known, please find the change in pressure at any point with time in the reservoir.

The mathematical model of the problem is established as follows:

$$\begin{cases} \dfrac{\partial^2 p}{\partial r^2} + \dfrac{1}{r}\dfrac{\partial p}{\partial r} = \dfrac{1}{\eta}\dfrac{\partial p}{\partial t} \\ p_{t=0} = p_i \\ p|_{r\to\infty} = p_i \\ r\dfrac{\partial p}{\partial r}\Big|_{r=R_w\to 0} = \dfrac{Q\mu}{2\pi Kh} \end{cases} \quad (3.150)$$

The solution obtained is:

$$p(r,t) = p_i - \frac{Q\mu}{4\pi Kh}\left[-E_i\left(-\frac{r^2}{4\eta t}\right)\right] \quad (3.151)$$

The pressure drawdown is:

$$\Delta p(r,t) = p_i - p(r,t) = \frac{Q\mu}{4\pi Kh}\left[-E_i\left(-\frac{r^2}{4\eta t}\right)\right] \quad (3.152)$$

The value of the exponential integral function in the formula $-E_i(-\frac{r^2}{4\eta t})$ can be found in the table provided in Appendix 4.

The exponential integral function diagram is shown in Figure 3.36.

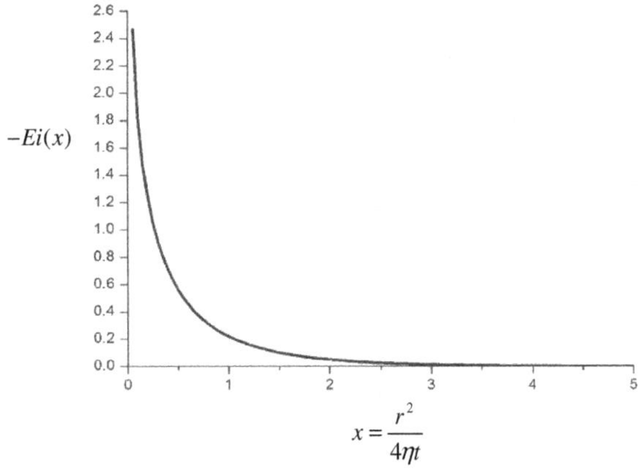

Figure 3.36. Exponential integral function diagram

It can be known that:

① At a given time, the closer to the bottom of the well, the smaller r is, the smaller x is, the larger $-Ei(x)$ is, and the greater the pressure drop, which conform to radial flow characteristics.

② At a given position, the longer the production time t, the smaller x is, the larger $-Ei(x)$ is, and the larger the pressure drawdown, which conform to the pressure wave propagation process.

③ When $x = \frac{r^2}{4\eta t} < 0.01$, the exponential integral function can be expanded as an infinite series, and taking the first two terms, it gives the approximation form as:

$$-Ei\left(-\frac{r^2}{4\eta t}\right) \approx \ln\frac{2.25\eta t}{r^2} \tag{3.153}$$

④ When $r = R_w$, because R_w is small and when $\eta = \frac{K}{\mu C_t}$ is large, it is easy to satisfy the condition $x = \frac{r^2}{4\eta t} < 0.01$, so this approximation form can be used to express the change of bottom hole pressure, then:

$$p_w(t) = p_i - \frac{Q\mu}{4\pi Kh}\ln\frac{2.25\eta t}{R_w^2} \tag{3.154}$$

The bottom hole pressure drawdown is:

$$\Delta p_w(t) = \frac{Q\mu}{4\pi Kh}\ln\frac{2.25\eta t}{R_w^2} \tag{3.155}$$

3.3.3.4 *Multi-well interference for transient flow*

In the multi-well production problem, multi-well interference also exists in elastic transient flow. The principle of multi-well interference superposition previously introduced is also valid here. In this book, the principle of superposition for pressure drawdown is introduced to deal with elastic transient flow.

Superposition principle of pressure drop can be expressed as:

$$\Delta p(r,t) = \sum_{i=1}^{n} \Delta p_i \tag{3.156}$$

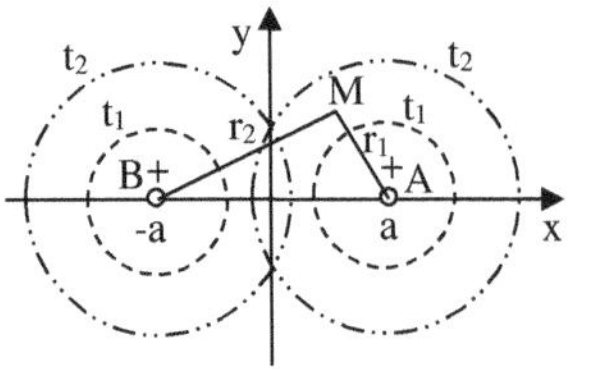

Figure 3.37. Transient flow of two point sinks with equal production rates

(1) Problem of two point sinks with equal production rates

As shown in Figure 3.37, the two production wells in the infinite formation produce oil at the same production rate Q at the same time. The distance between the two wells is $2a$, and the distances between any M point in the formation and the two wells are r_1 and r_2, respectively.

When well A is produced separately:

$$\Delta p_1(r,t) = \frac{Q\mu}{4\pi Kh}\left[-E_i\left(-\frac{r_1^2}{4\eta t}\right)\right] \tag{3.157}$$

When well B is produced separately:

$$\Delta p_2(r,t) = \frac{Q\mu}{4\pi Kh}\left[-E_i\left(-\frac{r_2^2}{4\eta t}\right)\right] \tag{3.158}$$

The superposed pressure drawdown is:

$$\Delta p(r,t) = \frac{Q\mu}{4\pi Kh}\left[-E_i\left(-\frac{r_1^2}{4\eta t}\right)\right] + \frac{Q\mu}{4\pi Kh}\left[-E_i\left(-\frac{r_2^2}{4\eta t}\right)\right] \tag{3.159}$$

The pressure at any point M changing with time is:

$$\begin{aligned} p(r,t) &= p_i - \Delta p(r,t) \\ &= p_i - \frac{Q\mu}{4\pi Kh}\left[-E_i\left(-\frac{r_1^2}{4\eta t}\right)\right] \\ &\quad - \frac{Q\mu}{4\pi Kh}\left[-E_i\left(-\frac{r_2^2}{4\eta t}\right)\right] \end{aligned} \tag{3.160}$$

It can be implied that:

$$\left.\frac{\partial p}{\partial x}\right|_{x=0} = 0 \tag{3.161}$$

That means no fluid passes through the y-axis, the y-axis is the diverging streamline, which can be regarded as a fault.

Even when the cones of depression for these two wells are not overlapped, such as at t_1, and there is no well interference, the superposed solution is still valid to obtain the pressure distribution.

(2) One point sink and one source with equal flow rates

As shown in Figure 3.38, one source and one sink with the same flow rate work simultaneously in the infinite reservoir. The distance between the two wells is $2a$, and the distances between any M point in the formation and the two wells are r_1 and r_2, respectively.

When well A is produced separately:

$$\Delta p_1(r,t) = \frac{Q\mu}{4\pi Kh}\left[-E_i\left(-\frac{r_1^2}{4\eta t}\right)\right] \tag{3.162}$$

When well B is produced separately (injection well or point source, Q is "–"):

$$\Delta p_2(r,t) = \frac{-Q\mu}{4\pi Kh}\left[-E_i\left(-\frac{r_2^2}{4\eta t}\right)\right] \tag{3.163}$$

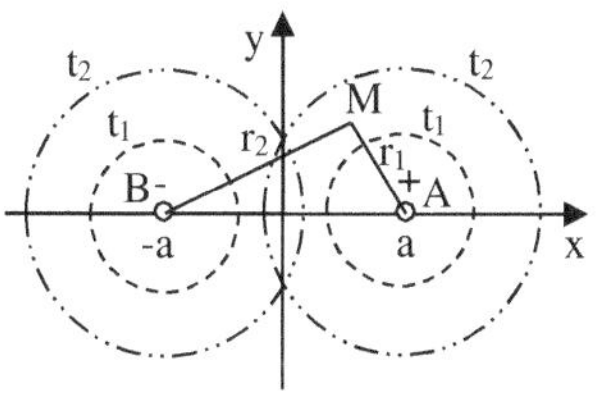

Figure 3.38. Transient flow of one point sink and one source with equal flow rates

The superposed pressure drawdown is:

$$\Delta p(r,t) = \frac{Q\mu}{4\pi Kh}\left[-E_i\left(-\frac{r_1^2}{4\eta t}\right)\right] - \frac{Q\mu}{4\pi Kh}\left[-E_i\left(-\frac{r_2^2}{4\eta t}\right)\right] \tag{3.164}$$

The pressure at any point M changing with time is:

$$\begin{aligned} p(r,t) &= p_i - \Delta p(r,t) \\ &= p_i - \frac{Q\mu}{4\pi Kh}\left[-E_i\left(-\frac{r_1^2}{4\eta t}\right)\right] \\ &\quad + \frac{Q\mu}{4\pi Kh}\left[-E_i\left(-\frac{r_2^2}{4\eta t}\right)\right] \end{aligned} \tag{3.165}$$

If point M is on the y-axis, $r_1 = r_2$, then:

$$p(r,t)|_{r_1=r_2} = p_i \tag{3.166}$$

That means the y-axis is the equipotential line, and its value is the original formation pressure P_i. Therefore, the y-axis can be regarded as the supply boundary.

(3) Mirror image of transient flow

The y-axis can be regarded as a fault for the two point sinks problem and as the supply boundary for the one point source and one point sink problem. The principle of mirror image derived for the steady state is still applicable in transient flow.

(4) Varying production rate problem

In the actual oil field production, it is not realistic to keep the production rate constant from the beginning to end, and the production rate needs to be adjusted from time to time. After changing the well production rate, the pressure changes accordingly, and the mirror image method can be helpful to deal with the varying production rate problem and obtain the bottom hole pressure.

① Step-like production rate change

As shown in Figure 3.39, a production well is located in an infinite formation, the production rate during time $0 - t_1$ is Q_1, and the production rate after time t_1 is Q_2. Please find the formation pressure distribution and bottom hole pressure distribution at different times.

When $t \leq t_1$, the production rate is:

$$p(r,t) = p_i - \frac{Q\mu}{4\pi Kh}\left[-E_i\left(-\frac{r^2}{4\eta t}\right)\right] \tag{3.167}$$

$$p_w(t) = p_i - \frac{Q\mu}{4\pi Kh}\ln\frac{2.25\eta t}{R_w^2} \tag{3.168}$$

When $t > t_1$, a special mirror image method is required.

As shown in Figure 3.39, two virtual wells are added at the well point by using the mirror image method: the virtual well 1 is the well produced with Q_1 but continues to be produced after t_1. Imaging well 2 is an injection well that starts at time t_1 with production of $-Q_1$. In this way, the production rates of one injection and one production at the same position offset each other, and the production rate at the actual position remains Q_2.

After the mirror image, the problem of varying production rate problem is converted into the problem of three-well interference problem. By applying the superposition principle of pressure drawdown, the pressure distribution can be obtained as:

$$\begin{aligned} p(r,t) = p_i &- \frac{Q_1\mu}{4\pi Kh}\left[-E_i\left(-\frac{r^2}{4\eta t}\right)\right] \\ &- \frac{(Q_2 - Q_1)\mu}{4\pi Kh}\left\{-E_i\left[-\frac{r^2}{4\eta(t-t_1)}\right]\right\} \end{aligned} \tag{3.169}$$

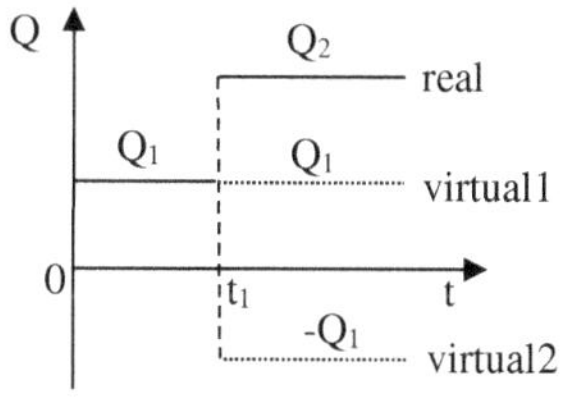

Figure 3.39. Schematic diagram of stepwise production rate change

The bottom hole pressure can be obtained as:

$$p_w(t) = p_i - \frac{Q_1\mu}{4\pi Kh}\ln\frac{2.25\eta t}{R_w^2} - \frac{(Q_2 - Q_1)\mu}{4\pi Kh}\ln\frac{2.25\eta(t-t_1)}{R_w^2} \tag{3.170}$$

② Curve-like production rate change

If the well production rate change is a curve, as shown in Figure 3.40, the production rate can be divided into n equal parts, resulting in several step-like production rate changes. By using the mirror image method, the problem of multi-well interference can be represented by the production rate changes shown in Figure 3.41.

The pressure distribution is obtained as:

$$p(r,t) = p_i - \frac{\mu}{4\pi Kh}\int_0^t \frac{Q(\tau)}{t-\tau}e^{-\frac{r^2}{4\eta(t-\tau)}}d\tau \tag{3.171}$$

③ Well shut-in problem

As shown in Figure 3.42, a well in an infinite formation is produced at a constant production rate Q_1, and the well is shut in at time T. Please derive the expression to obtain the changes

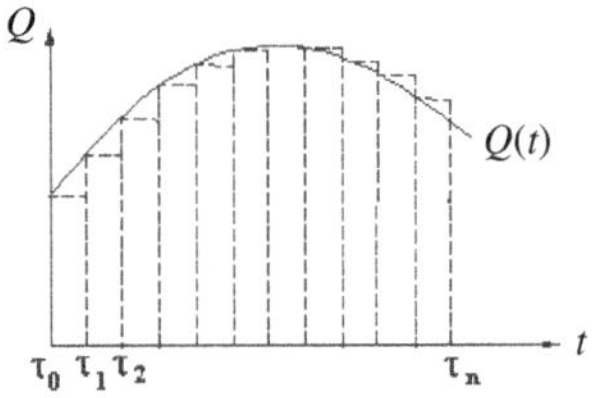

Figure 3.40. Schematic diagram of curve-like production rate change

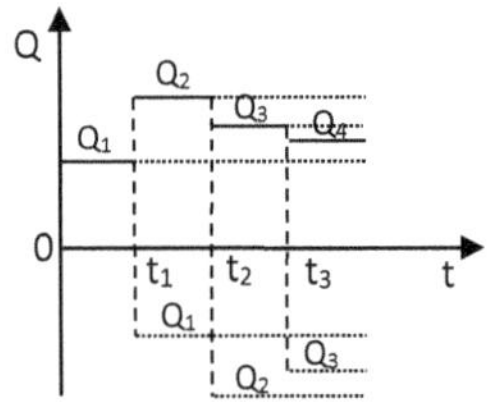

Figure 3.41. Schematic diagram of several step-like production rate changes

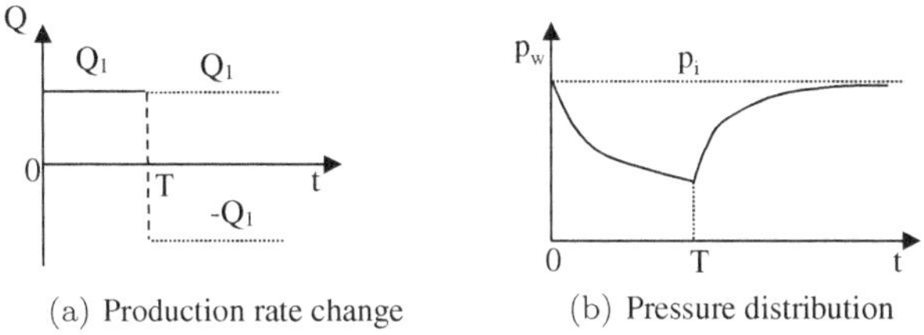

Figure 3.42. Production rate change and pressure distribution for well shut-in problem

of the formation pressure distribution and bottom hole pressure distribution.

When $t \leq T$, the pressure distribution is:

$$p(r,t) = p_i - \frac{Q\mu}{4\pi Kh}\left[-E_i\left(-\frac{r^2}{4\eta t}\right)\right] \tag{3.172}$$

And the bottom hole pressure is:

$$p_w(t) = p_i - \frac{Q\mu}{4\pi Kh}\ln\frac{2.25\eta t}{R_w^2} \tag{3.173}$$

When $t > T$, the mirror image method needs to be applied, and it is considered that the well is still producing with constant production rate Q_1, and a well at the same position at time t is injected with Q_1, but the actual production rate is 0. Two virtual wells are being produced.

The superposed pressure distribution is:

$$\begin{aligned} p(r,t) = p_i &- \frac{Q\mu}{4\pi Kh}\left[-E_i\left(-\frac{r^2}{4\eta t}\right)\right] \\ &+ \frac{Q\mu}{4\pi Kh}\left\{-E_i\left[-\frac{r^2}{4\eta(t-T)}\right]\right\} \end{aligned} \tag{3.174}$$

The bottom hole pressure is:

$$\begin{aligned} p_w(t) = p_i &- \frac{Q\mu}{4\pi Kh}\ln\frac{2.25\eta t}{R_w^2} \\ &+ \frac{Q\mu}{4\pi Kh}\ln\frac{2.25\eta(t-T)}{R_w^2} \end{aligned} \tag{3.175}$$

The bottom hole pressure can be rewritten as:

$$p_w(t) = p_i - \frac{Q\mu}{4\pi Kh} \ln \frac{t}{t-T} \tag{3.176}$$

or

$$p_w(t) = p_i + \frac{Q\mu}{4\pi Kh} \ln \frac{t-T}{t} \tag{3.177}$$

In order to facilitate recording, mark the time after the well shut-in as $\Delta t = t - T$ and the production time before well shut-in as T_p, then the bottom hole pressure is:

$$p_w(\Delta t) = p_i - \frac{Q\mu}{4\pi Kh} \ln \frac{\Delta t + T_p}{\Delta t} \tag{3.178}$$

It can also be written as:

$$p_w(\Delta t) = p_i + 0.183\frac{Q\mu}{Kh} \lg \frac{\Delta t}{\Delta t + T_p} \tag{3.179}$$

The above bottom hole pressure equation is known as Horner equation.

The change of bottom hole pressure at the time of $0 - t$ is shown in Figure 3.42(b). It can be seen that before the well shut-in, bottom hole pressure drops rapidly and gradually becomes flat. After the well shut-in, the bottom hole pressure quickly recovers. If enough time passes, it slowly rises to the original formation pressure P_i. This is a typical curve of the "well shut-in" bottom hole pressure.

3.3.3.5 *Pseudo-steady state problem*

For the pseudo-steady state flow, $\frac{\partial p}{\partial t}$ = Constant. Therefore, the governing equation is:

$$\frac{\partial^2 p}{\partial r^2} + \frac{1}{r}\frac{\partial p}{\partial r} = \text{Constant} \tag{3.180}$$

The pressure distribution of pseudo-steady state flow expressed in terms of the bottom hole pressure is:

$$p(r,t) = p_w(t) + \frac{Q\mu}{2\pi Kh}\left(\ln\frac{r}{R_w} - \frac{1}{2}\frac{r^2}{R_e^2}\right) \tag{3.181}$$

The pressure distribution of pseudo-steady state flow expressed in terms of the boundary pressure is:

$$p(r,t) = p_e(t) - \frac{Q\mu}{2\pi Kh}\left[\ln\frac{r}{R_w} - \frac{1}{2}\left(1 - \frac{r^2}{R_e^2}\right)\right] \tag{3.182}$$

The bottom hole pressure of pseudo-steady state flow expressed in terms of the boundary pressure is:

$$p_w(t) = p_e(t) - \frac{Q\mu}{2\pi Kh}\left(\ln\frac{R_e}{R_w} - \frac{1}{2}\right) \tag{3.183}$$

The bottom hole pressure of pseudo-steady state flow expressed in terms of the initial formation pressure is:

$$p_w(t) = p_i - \frac{Q\mu}{2\pi Kh}\left(\frac{2\eta t}{R_e^2} + \ln\frac{R_e}{R_w} - \frac{3}{4}\right) \tag{3.184}$$

The average formation pressure of pseudo-steady state flow expressed in terms of the bottom hole pressure is:

$$\overline{p(t)} = p_w(t) + \frac{Q\mu}{2\pi kh}\left(\ln\frac{R_e}{R_w} - \frac{3}{4}\right) \tag{3.185}$$

The average formation pressure of pseudo-steady state flow expressed in terms of the boundary pressure is:

$$\overline{p(t)} = p_e(t) - \frac{Q\mu}{8\pi Kh} \tag{3.186}$$

A well in a reservoir with closed boundary is known to have constant production rate Q and it reaches a pseudo-steady state flow. Please find the average formation pressure, bottom hole pressure, boundary pressure and the maximal degree of the elastic recovery at time t.

Method 1:

The definition of the total volume of liquid recovery is:

$$V_{tp} = Q \cdot t \tag{3.187}$$

It can be written as:

$$V_{tp} = \pi(R_e^2 - R_w^2)hC_t(p_i - \overline{p}) \tag{3.188}$$

Based on the two formulas:

$$\begin{cases} \overline{p(t)} = p_w(t) + \dfrac{Q\mu}{2\pi Kh}\left(\ln\dfrac{R_e}{R_w} - \dfrac{3}{4}\right) \\ \overline{p(t)} = p_e(t) - \dfrac{Q\mu}{8\pi kh} \end{cases} \tag{3.189}$$

The bottom hole pressure $p_w(t)$ and boundary pressure $p_e(t)$ can be obtained.

According to the formula:

$$p_w(t) = p_i - \frac{Q\mu}{2\pi Kh}\left(\frac{2\eta t}{R_e^2} + \ln\frac{R_e}{R_w} - \frac{3}{4}\right) \tag{3.190}$$

If the time T_{elastic} required for the bottom hole pressure to drop to $p_w(t) = p_b$ (p_b is the saturation pressure) is measured, the maximal degree of elastic recovery can be calculated as:

$$E_e = \frac{QT_{\text{elastic}}}{\pi R_e^2 h\varphi} \tag{3.191}$$

Method 2:

Given Q and t, the bottom hole pressure can be calculated:

$$p_w(t) = p_i - \frac{Q\mu}{2\pi Kh}\left(\frac{2\eta t}{R_e^2} + \ln\frac{R_e}{R_w} - \frac{3}{4}\right) \tag{3.192}$$

The boundary pressure and average formation pressure can be further calculated as follows:

$$p_w(t) = p_e(t) - \frac{Q\mu}{2\pi Kh}\left(\ln\frac{R_e}{R_w} - \frac{1}{2}\right) \tag{3.193}$$

$$\overline{p(t)} = p_w(t) + \frac{Q\mu}{2\pi Kh}\left(\ln\frac{R_e}{R_w} - \frac{3}{4}\right) \tag{3.194}$$

The two methods can both be applied to calculate the average formation pressure, and their results are consistent.

3.3.3.6 *Transient well test*

The relationship between pressure difference and flow rate has been used in the steady state well test, and the relationship between bottom hole pressure and time will be used in the transient flow well test.

There are two kinds of transient well test: constant production rate drawdown test and shut-in pressure buildup test.

(1) Pressure drawdown well test

When a well starts producing at a given rate, the bottom hole pressure drops. Before the pressure drawdown reaches the boundary (i.e., infinite formation), the bottom hole pressure is:

$$p_w(t) = p_i - 0.183\frac{Q\mu}{Kh}\lg\frac{2.25\eta t}{R_w^2} \tag{3.195}$$

Let

$$a = p_i - 0.183\frac{Q\mu}{Kh}\lg\frac{2.25\eta}{R_w^2} \tag{3.196}$$

$$b = 0.183\frac{Q\mu}{Kh} \tag{3.197}$$

Then

$$p_w(t) = a - b\lg t \tag{3.198}$$

The semi-log curve of $p_w(t) \sim \lg t$ is shown in Figure 3.43.

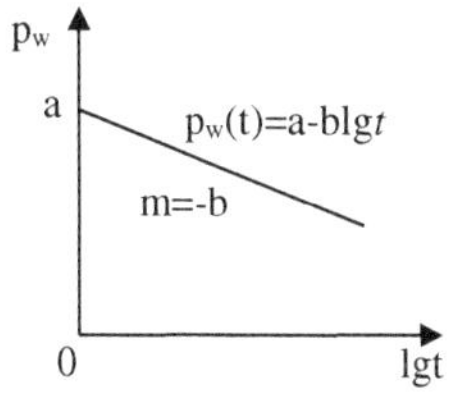

Figure 3.43. Semi-log curve of pressure and time

Then $p_w(t) \sim \lg t$ is linear, with intercept a and slope $m = -b$.

If the pressure drawdown well test curve is measured, this linear relationship can be obtained, and the formation properties can be calculated and analyzed according to intercept and slope, such as K, Kh, $\frac{Kh}{\mu}$, $\frac{K}{\mu}$, η, R_{we}, p_i, etc.

In the actual production, each well has a limited drainage area. When the pressure wave propagates rapidly to the closed boundary or the test time is long, the flow can be regarded as a pseudo-steady state flow, when:

$$p_w(t) = p_i - \frac{Q\mu}{2\pi Kh}\left(\frac{2\eta t}{R_e^2} + \ln\frac{R_e}{R_w} - \frac{3}{4}\right) \tag{3.199}$$

In combination with $\eta = \frac{K}{\mu C_t}$, and set:

$$a = p_i - \frac{Q\mu}{2\pi Kh}\left(\ln\frac{R_e}{R_w} - \frac{3}{4}\right) \tag{3.200}$$

$$b = \frac{Q}{\pi R_e^2 h C_t} \tag{3.201}$$

Then:

$$p_w(t) = a - bt \tag{3.202}$$

The curve of $p_w(t) \sim t$ is shown in Figure 3.44.

Then $p_w(t) \sim t$ forms a linear relationship, with intercept a and slope $m = -b$.

Generally, the pseudo-steady state stage is used to estimate elastic reserves.

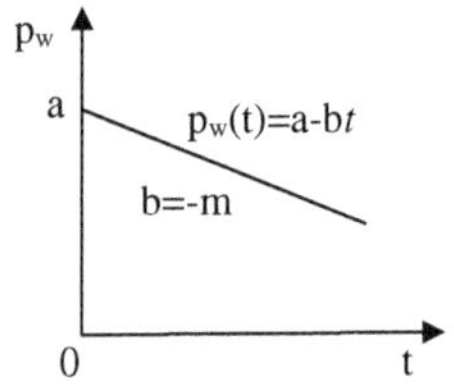

Figure 3.44. Pseudo-steady state well test

The elastic reserve N_e is:

$$N_e = \pi R_e^2 h C_t (p_i - p_b) \tag{3.203}$$

Combined with $b = \frac{Q}{\pi R_e^2 h C_t}$, the elastic reserves are obtained as:

$$N_e = \frac{Q(p_i - p_b)}{b} \tag{3.204}$$

In the actual curve, the slope m of the straight line segment can be obtained, then b can be calculated. If the original formation pressure p_i and the saturation pressure p_b are known, the elastic reserves N_e can be obtained.

(2) Pressure buildup well test

When the well is shut-in at a given rate for a period of time, the bottom hole pressure will gradually recover, and the measured pressure curve can be used to calculate formation parameters according to Horner equation.

① Horner method
Horner equation is:

$$p_w(\Delta t) = p_i + 0.183\frac{Q\mu}{Kh}\lg\frac{\Delta t}{\Delta t + T_p} \tag{3.205}$$

Let:

$$a = p_i \tag{3.206}$$

$$b = 0.183\frac{Q\mu}{Kh} \tag{3.207}$$

Then:

$$p_w(\Delta t) = a + b\lg\frac{\Delta t}{\Delta t + T_p} \tag{3.208}$$

In the semi-log coordinate, plot $p_w(\Delta t) \sim \lg\frac{\Delta t}{\Delta t + T_p}$, as shown in Figure 3.45.

It can be shown that $p_w(\Delta t) \sim \lg\frac{\Delta t}{\Delta t + T_p}$ is a straight line when $\Delta t \to \infty$, with $\lg\frac{\Delta t}{\Delta t + T_p} = 0$ as intercept, and slope of straight line as $m = b$.

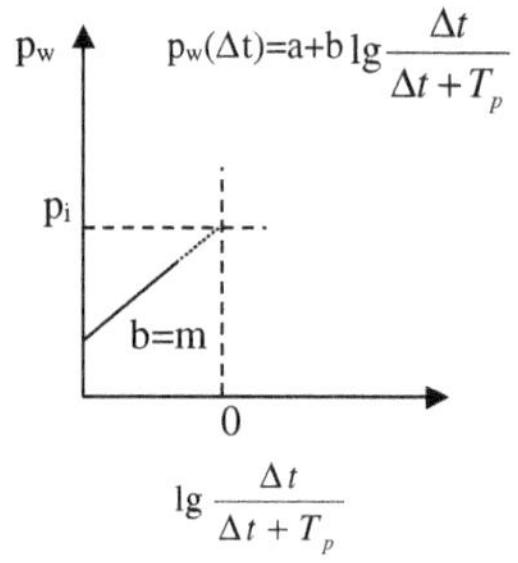

Figure 3.45. Well test curve for Horner method

If the pressure buildup curve is measured, the straight line section can be obtained. The extrapolation of the straight section can provide the original formation pressure, and the slope can provide the permeability.

② MBH method

T_p in Horner equation is not easy to be measured in practice, and MBH method simplifies it:

$$p_w(\Delta t) = p_w(\Delta t = 0) + \frac{Q\mu}{4\pi Kh} \ln \frac{2.25\eta \Delta t}{R_w^2} \tag{3.209}$$

or

$$p_w(\Delta t) = p_w(\Delta t = 0) + 0.183\frac{Q\mu}{Kh} \lg \frac{2.25\eta \Delta t}{R_w^2} \tag{3.210}$$

Let:

$$a = p_w(\Delta t = 0) + 0.183\frac{Q\mu}{Kh} \lg \frac{2.25\eta}{R_w^2} \tag{3.211}$$

$$b = 0.183\frac{Q\mu}{Kh} \tag{3.212}$$

Then:

$$p_w(\Delta t) = a + b \lg \Delta t \tag{3.213}$$

In the semi-log coordinate, plot $p_w(\Delta t) \sim \lg \Delta t$ relation curve, as shown in Figure 3.46.

It can be seen that $p_w(\Delta t) \sim \lg \Delta t$ is linear, intercept is a, and the slope is $m = b$.

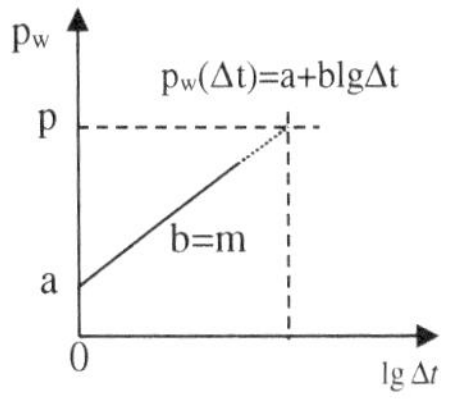

Figure 3.46. Pressure buildup curve for MBH method

If the pressure buildup curve is measured, the straight line section can be obtained. The current formation pressure can be obtained by extrapolating the straight line section, and the permeability can be obtained from the slope.

3.3.4 *Example*

1. In a circular closed reservoir, the radius R_e is 150 m, the thickness h is 5 m, the porosity is 0.2, the original oil saturation S_o is 0.7, the original water saturation S_w is 0.3, underground oil elastic compressibility $C_o = 10 \times 10^{-4}\,\text{MPa}^{-1}$, underground water elastic compressibility $C_w = 5 \times 10^{-4}\,\text{MPa}^{-1}$, rock elastic compressibility $C_f = 2 \times 10^{-4}\,\text{MPa}^{-1}$, crude oil volume coefficient $B_o = 1.12$, original formation pressure $P_i = 20\,\text{MPa}$, the saturation pressure $p_b = 8\,\text{MPa}$. If the bottom hole pressure is the saturation pressure for production until the reservoir pressure drops to the saturation pressure, please find the total produced liquid.

Solution:

The elastic compressibility of liquid is:

$$\begin{aligned} C_L &= C_o \times S_o + C_w \times S_w \\ &= (10 \times 0.7 + 5 \times 0.3) \times 10^{-4} = 8.5 \times 10^{-4}\,MPa^{-1} \end{aligned}$$

The total compressibility is:

$$C_t = \phi C_L + C_f = (0.2 \times 8.5 + 2) \times 10^{-4} = 3.7 \times 10^{-4}\,MPa^{-1}$$

The total volume of reservoir rock is:

$$V_f = \pi R_e^2 h = 3.14 \times 150^2 \times 5 = 353250\,m^3$$

The total produced fluid is:

$$Q_t = C_t V_f (p_i - p_b) = 3.7 \times 10^{-4} \times 353250 \times (20 - 8) = 1568.43\, m^3$$

2. As shown in the figure below, two producing wells near the fault in the elastic drive infinite formation were produced at constant production rates of Q_A and Q_B, respectively. Well A is shut-in at time T. Please derive the bottom hole pressure of well B.

Solution:

Draw the mirror image of well A and B as A′ and B′ as shown in the figure below.

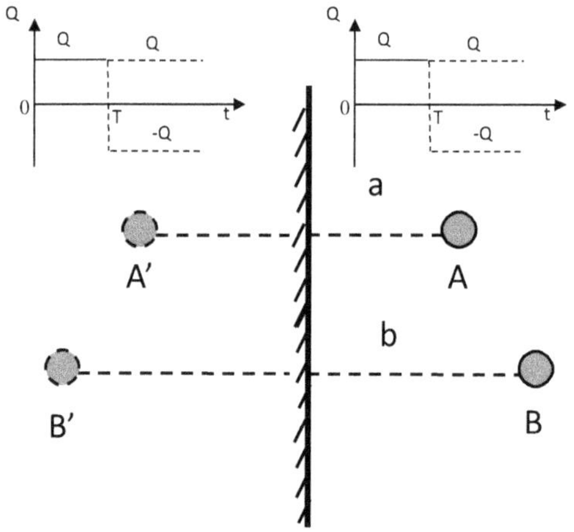

When $t \leq T$, 4 wells are produced. According to the superposition principle of pressure drop,

$$p_B(t) = p_i - \frac{Q\mu}{4\pi Kh}\left\{\ln\frac{2.25\eta t}{R_w^2} + \left[-E_i\left(-\frac{|AB|^2}{4\eta t}\right)\right] + \left[-E_i\left(-\frac{|A'B|^2}{4\eta t}\right)\right] + \left[-E_i\left(-\frac{|B'B|^2}{4\eta t}\right)\right]\right\}$$

When $t > T$, 6 wells are produced. According to the superposition principle of pressure drop,

$$p_B(t) = p_i - \frac{Q\mu}{4\pi Kh}\left\{\ln\frac{2.25\eta t}{R_w^2} + \left[-E_i\left(-\frac{|B'B|^2}{4\eta t}\right)\right] + \left[-E_i\left(-\frac{|AB|^2}{4\eta t}\right)\right]\right.$$

$$+\left[-E_i\left(-\frac{|A'B|^2}{4\eta t}\right)\right]-\left[-E_i\left(-\frac{|AB|^2}{4\eta(t-T)}\right)\right]$$
$$-\left[-E_i\left(-\frac{|A'B|^2}{4\eta(t-T)}\right)\right]\Bigg\}$$

3.4 Oil-water two-phase flow theory

3.4.1 *Introduction*

As have already been discussed, the fluid in the problems of steady state flow, multi-well interference and elastic transient flow is assumed to be the single-phase liquid, and the fluid that flows across the supply boundary has the same properties as the fluid in the reservoir. In real reservoirs, oil and water generally co-exist and are produced at the same time. Sometimes, water injection is often needed to supplement the formation energy and improve oil recovery, and more oil can be produced by water flooding. However, the properties of water and oil are quite different, especially the differences in viscosity, which makes the characteristics of the oil-water two-phase flow substantially different from that of the single-phase liquid.

In the case of oil-water two-phase flow, the relative permeability of rock is different with different saturation. At the same time, due to the effect of capillary force, the pressures of the two phases are not equal.

Under reservoir conditions, the viscosity of water is generally lower than that of oil, and water flows more easily than oil due to the difference in viscous force when liquid flows. Therefore, a common indicator to evaluate the performance of water flooding is water cut, that is, the volumetric percentage of water in the produced fluid. The higher the water cut, the lower the efficiency of water flooding.

In actual reservoir production, water cut is classified as: low water cut when it is less than 20%, medium water cut when it is between 20–60%, high water cut when it is 60–90%, and super-high water cut when it is more than 90%.

In order to control the issues of water breakthrough and high water cut in the production well during water flooding, it is necessary to study the rule of the oil-water two-phase flow during water flooding. This chapter is to provide the basic theory to solve this problem.

3.4.2 *Concepts*

Piston-like displacement: *in the process of water flooding, when oil is displaced by water, the water and oil zones have a shape contact surface, generally perpendicular to the direction of the streamline. All the oil in the zone is extracted by water, and no oil is left behind on the contact surface of water and oil.*

Non-piston displacement: *in the process of water flooding, when oil is displaced by water, in addition to the water and oil areas, there is also a mixed oil-water flow zone, that is, after water enters the oil area, the oil cannot be completely extracted, and the mixing zone is expanding gradually during the displacement process.*

Viscosity ratio of oil to water: *it refers to the ratio of oil viscosity to water viscosity under reservoir conditions. The equation is:*

$$\mu_r = \frac{\mu_o}{\mu_w} \tag{3.214}$$

where μ_r *is the oil-water viscosity ratio;* μ_o *is the viscosity of oil; and* μ_w *is the viscosity of water.*

Water cut: *it is the percentage of water in the total produced liquid volume, also known as fractional flow. The symbol is* f_w*, and the equation is:*

$$f_w = \frac{Q_w}{Q_w + Q_o} \tag{3.215}$$

where f_w *is the water cut;* Q_w *is the flow rate of water;* Q_o *is the flow rate of oil.*

Flood front water saturation: *the water saturation at the contact surface between the oil zone and the mixing zone. The contact surface is also called flood front, symbol* S_{wf}.

Average water saturation: *it is the weighted average of water saturation in terms of the length in the oil-water two-phase mixing zone, symbol* $\overline{S_w}$. *The definition of the average water saturation is:*

$$\overline{S_w} = \frac{\int_{x_1}^{x_2} S_w dx}{x_2 - x_1} \tag{3.216}$$

Before the water breakthrough, it has:

$$\overline{S_w} - S_{wc} = \frac{W(t)}{\phi A(x_f - x_0)} \tag{3.217}$$

where $W(t)$ *is the total amount of water injected into the reservoir at time* t; x_f *is the position of flood front;* $\overline{S_w}$ *is the average water saturation in the oil-water two-phase zone;* x_1 *is the starting point of oil-water two-phase zone;* x_2 *is the endpoint of oil-water two-phase zone;* S_w *is the water saturation at any point;* S_{wc} *is the connate water saturation;* ϕ *is the porosity;* A *is the drainage area;* x_f *is the location of flood front; and* x_0 *is the initial position of oil displacement by water.*

Water-free production period: *it is the period from initial production to water breakthrough.*

Water-free recovery rate: *it is the degree of oil recovery achieved at the end of the water-free production period.*

3.4.3 *Analysis*

3.4.3.1 *Piston-like displacement*

(1) Linear flow

As shown in Figure 3.47, the piston-like oil displacement by water in the reservoir has supply pressure p_e and production pressure p_w. The interface between oil zone and water zone is sharp.

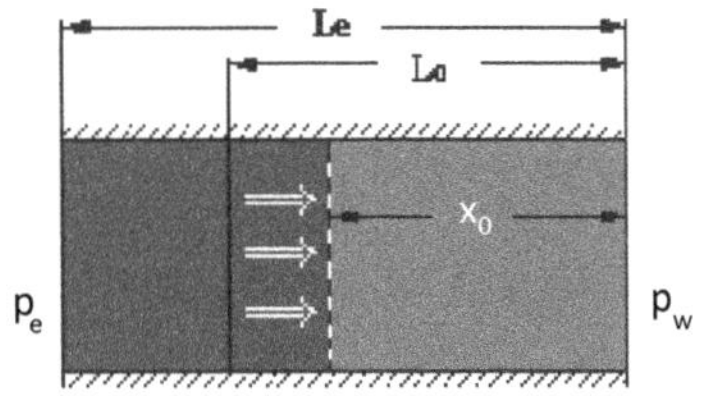

Figure 3.47. Piston-like oil displacement by water in linear flow

The flow resistances in the water and oil zones are treated as the connection of a series circuit, and they are:

$$R_{\text{fo}} = \frac{\mu_o x_o}{KBh} \tag{3.218}$$

$$R_{\text{fw}} = \frac{\mu_w (L_e - x_o)}{KBh} \tag{3.219}$$

The flow rate is:

$$Q = \frac{p_e - p_w}{R_{fw} + R_{fo}} \tag{3.220}$$

When oil zone decreases from L_o to x_o, R_{fo} decreases with the increase in R_{fw}. Generally, $\mu_w < \mu_o$, so the increased rate of water flow resistance ΔR_{fw} is less than the decreased rate of oil flow resistance ΔR_{fo}. The total flow resistance $R_{fw} + R_{fo}$ decreases, the pressure difference remains constant, and Q increases, i.e. the piston-like flow resistance and flow rate during the oil displacement process are changing, although the elastic force is not taken into account, but the flow is still transient flow.

(2) Radial flow

As shown in Figure 3.48, the piston-like displacement in the circular reservoir, the supply pressure and the bottom hole pressure are p_e and p_w, respectively. The interface between oil area and water area is sharp.

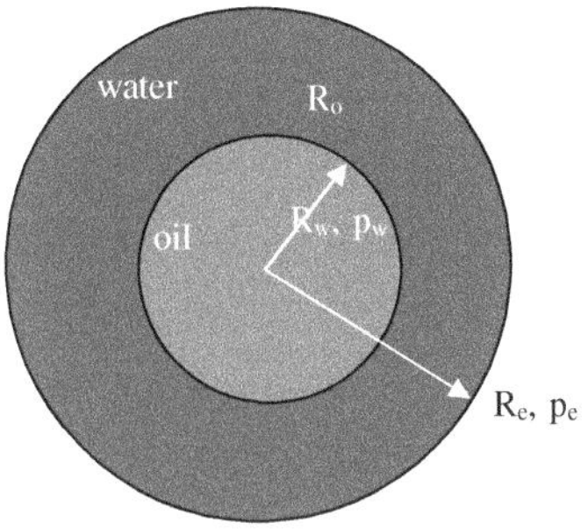

Figure 3.48. Piston-like displacement in the circular reservoir

Flow resistances of the water and oil zones are connected in the series circuit way, and they are:

$$R_{\text{fo}} = \frac{\mu_o}{2\pi Kh} \ln \frac{R_o}{R_w} \tag{3.221}$$

$$R_{\text{fw}} = \frac{\mu_w}{2\pi Kh} \ln \frac{R_e}{R_o} \tag{3.222}$$

The flow rate is:

$$Q = \frac{p_e - p_w}{R_{fw} + R_{fo}} \tag{3.223}$$

When the oil area decreases to R_o from R_e, R_{fo} decreases with the increase in R_{fw} and generally $\mu_w < \mu_o$. The increased rate of flow resistance in water zone is less than the decreased rate of flow resistance in oil zone, so the total flow resistance decreases and Q increases.

3.4.3.2 *Non-piston-like displacement*

In the process of oil displacement by water, there is only water phase flow in water zone, and the mixed oil-water flows in the zone between water and oil.

The starting position of the oil-water mixing zone is the initial contact surface of oil displacement by water, which is called the starting point of water flooding and expressed by x_0.

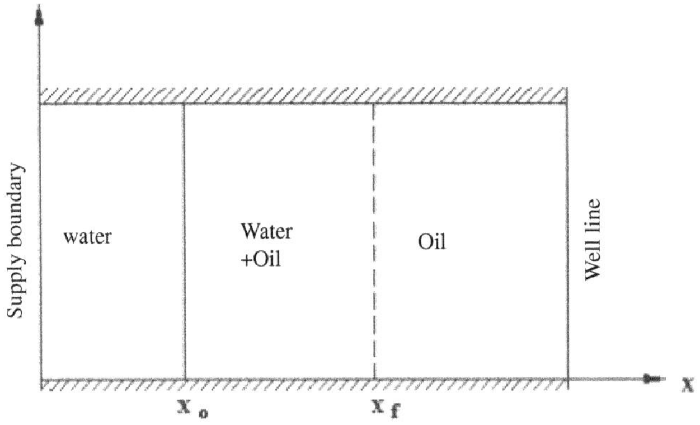

Figure 3.49. Non-piston-like oil displacement by water

The contact surface between the oil-water mixing zone and the oil zone is flood front, and its position is indicated by x_f.

As shown in Figure 3.49, only the water phase exists in the water zone, while oil and water co-exist in the oil zone, but the water does not flow and the water is present as connate water corresponding to the water saturation S_{wc}. The mixed zone has three characteristics:

① Oil and water co-existence as two phases. The water saturation S_w from the end of the water area to the end of the oil area gradually decreases, oil saturation S_o on the contrary gradually increases, which is mainly due to the viscosity difference of oil and water;

② The water does not displace all the oil in the mixing zone, and residual oil is left, whose saturation is marked as residual oil saturation S_{or};

③ At the flood front of the mixing zone, the water saturation S_{wf} is higher than the connate water saturation S_{wc}. This phenomenon is called the "shock" of water saturation, which is mainly due to the fact that capillary force and gravity are ignored (the real case is much more complicated).

The flood front water saturation has the following three characteristics:

① As shown in Figure 3.51, in the process of water flooding, the flood front position keeps advancing, but the flood front water saturation remains unchanged.

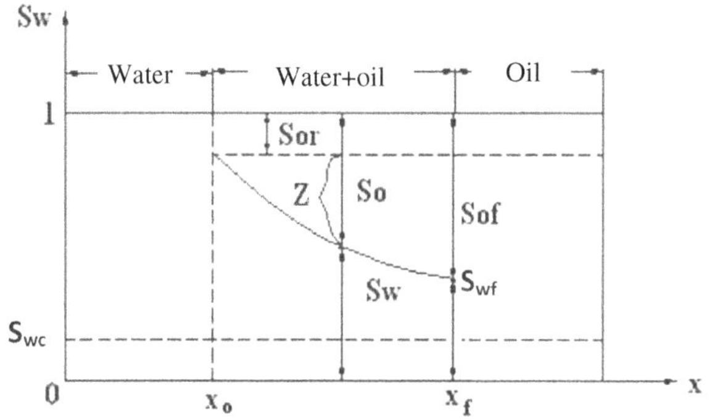

Figure 3.50. Water saturation distribution in non-piston-like displacement

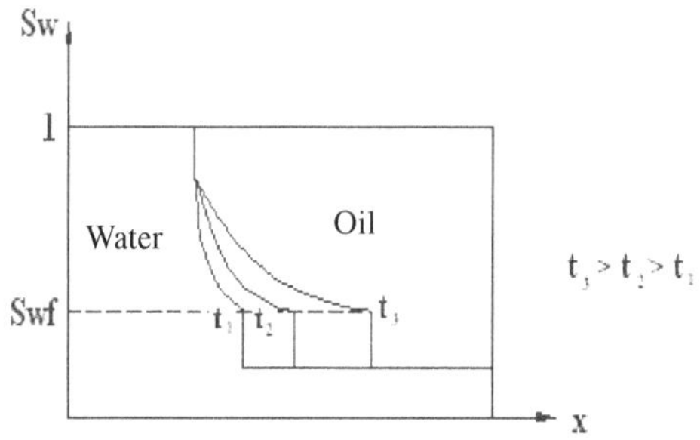

Figure 3.51. Relationship between flood front water saturation and time

② As shown in Figure 3.52, the larger the oil-water viscosity ratio, the smaller the flood front water saturation, the easier the water breakthrough occurs and the worse the water flooding performance. On the contrary, the smaller the oil-water viscosity ratio, the closer the oil-water two-phase property, the easier they co-exist, the larger the flood front water saturation, the larger the advancing surface, the slower the water breakthrough occurs, the better the water flooding performance. It is one of the important factors to enhance oil recovery by water flooding.

③ The value of flood front water saturation is affected by pore characteristics, which is reflected in the calculation of flood front water saturation by oil-water relative permeability curve. This part will be introduced in detail in the following sections.

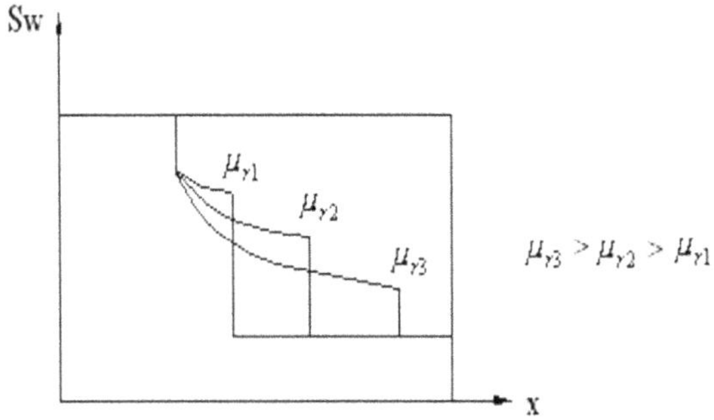

Figure 3.52. Relationship between flood front water saturation and oil-water viscosity ratio

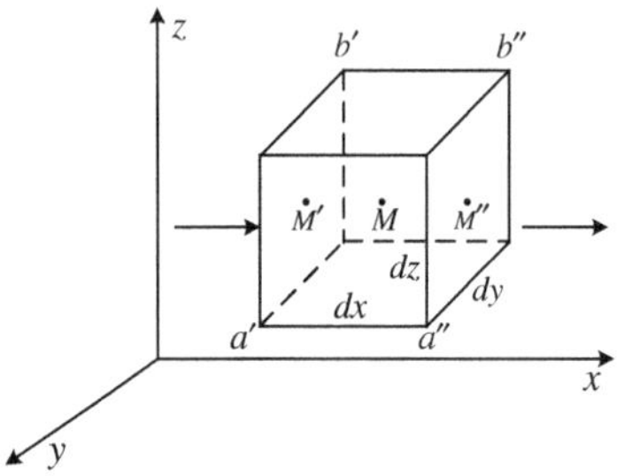

Figure 3.53. Schematic diagram of REV for oil-water two-phase flow

In the process of water flooding, when the flood front reaches the production end, that is, the mixed area covers all oil areas, it is called "water breakthrough". The period before the water breakthrough is called "water-free oil recovery period". In actual oil fields, the flooding effect is usually improved by delaying water breakthrough time and prolonging the period of water-free oil recovery, which is usually achieved by increasing water viscosity (by injecting polymer).

3.4.3.3 *Continuity equation of oil displacement by water*

The mass conservation law of the REV is still applied, but the difference is that the liquid flowing in and out of the REV has oil and water phases, and the two phases should be analyzed respectively, as shown in Figure 3.53.

In the x direction, the mass flow velocity of oil and water flowing into the REV is:

$$\rho_o v_{ox} - \frac{\partial(\rho_o v_{ox})}{\partial x}\frac{dx}{2} \tag{3.224}$$

$$\rho_w v_{wx} - \frac{\partial(\rho_w v_{wx})}{\partial x}\frac{dx}{2} \tag{3.225}$$

In the x direction, the mass flow velocity of the two phases of oil and water flowing out of the REV is:

$$\rho_o v_{ox} + \frac{\partial(\rho_o v_{ox})}{\partial x}\frac{dx}{2} \tag{3.226}$$

$$\rho_w v_{wx} + \frac{\partial(\rho_w v_{wx})}{\partial x}\frac{dx}{2} \tag{3.227}$$

Then, in the direction of x, the difference in the inflow and outflow masses of oil and water in dt time is:

$$-\frac{\partial(\rho_o v_{ox})}{\partial x} dxdydzdt \tag{3.228}$$

$$-\frac{\partial(\rho_w v_{wx})}{\partial x} dxdydzdt \tag{3.229}$$

Similarly, in the directions of y and z, the difference in the inflow and outflow masses of oil and water in dt time is:

$$-\frac{\partial(\rho_o v_{oy})}{\partial y} dxdydzdt - \frac{\partial(\rho_o v_{oz})}{\partial z} dxdydzdt \tag{3.230}$$

$$-\frac{\partial(\rho_w v_{wy})}{\partial y} dxdydzdt - \frac{\partial(\rho_w v_{wz})}{\partial z} dxdydzdt \tag{3.231}$$

The difference of oil and water inflow and outflow in the whole REV is:

$$-\left[\frac{\partial(\rho_o v_{ox})}{\partial x} + \frac{\partial(\rho_o v_{oy})}{\partial y} + \frac{\partial(\rho_o v_{oz})}{\partial z}\right] dxdydzdt \tag{3.232}$$

$$-\left[\frac{\partial(\rho_w v_{wx})}{\partial x} + \frac{\partial(\rho_w v_{wy})}{\partial y} + \frac{\partial(\rho_w v_{wz})}{\partial z}\right] dxdydzdt \tag{3.233}$$

Due to the change of saturation in REV caused by the inflow and outflow of oil and water, the mass changes of oil and water in the REV are as follows:

$$\frac{\partial(\phi\rho_o S_o)}{\partial t}dxdydzdt \tag{3.234}$$

$$\frac{\partial(\phi\rho_w S_w)}{\partial t}dxdydzdt \tag{3.235}$$

According to the mass conservation law, the continuity equation of oil and water two-phase flow can be obtained:

$$-\left[\frac{\partial(\rho_o v_{ox})}{\partial x}+\frac{\partial(\rho_o v_{oy})}{\partial y}+\frac{\partial(\rho_o v_{oz})}{\partial z}\right]=\frac{\partial(\phi\rho_o S_o)}{\partial t} \tag{3.236}$$

$$-\left[\frac{\partial(\rho_w v_{wx})}{\partial x}+\frac{\partial(\rho_w v_{wy})}{\partial y}+\frac{\partial(\rho_w v_{wz})}{\partial z}\right]=\frac{\partial(\phi\rho_w S_w)}{\partial t} \tag{3.237}$$

There are two auxiliary equations: $S_o + S_w = 1$, $p_c = p_o - p_w$.

When the oil is displaced by water, the driving force mainly comes from hydraulic power, and the elastic force of liquid and rock is relatively small. In flow mechanics in porous media, the elastic effect is not considered when analyzing the two-phase flow law of oil and water by means of analytical method, that is, neither liquid nor rock has compressibility. Then the continuity equation can be simplified as follows:

$$-\left[\frac{\partial(v_{ox})}{\partial x}+\frac{\partial(v_{oy})}{\partial y}+\frac{\partial(v_{oz})}{\partial z}\right]=\phi\frac{\partial S_o}{\partial t} \tag{3.238}$$

$$-\left[\frac{\partial(v_{wx})}{\partial x}+\frac{\partial(v_{wy})}{\partial y}+\frac{\partial(v_{wz})}{\partial z}\right]=\phi\frac{\partial S_w}{\partial t} \tag{3.239}$$

In this book, the characteristics of water flooding in the case of linear flow, namely one-dimensional flow, is discussed. At this time, the continuity equations of oil and water are:

$$-\frac{\partial v_{ox}}{\partial x}=\phi\frac{\partial S_o}{\partial t} \tag{3.240}$$

$$-\frac{\partial v_{wx}}{\partial x}=\phi\frac{\partial S_w}{\partial t} \tag{3.241}$$

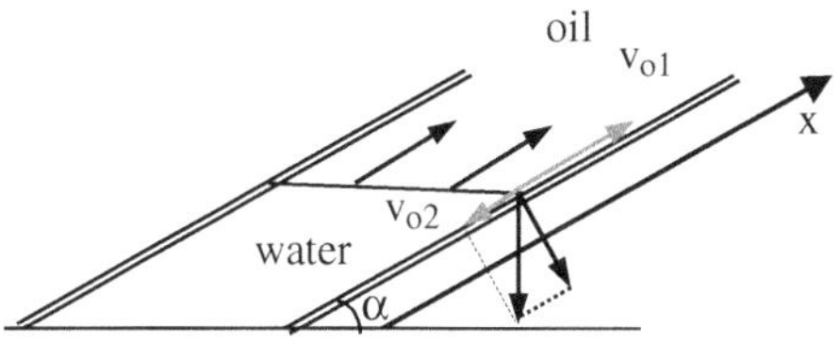

Figure 3.54. Flow diagram considering gravity and capillary forces

3.4.3.4 *Equation of iso-saturation surface advancement*

Taking linear flow as the research object, the two-phase flow equation of oil and water is established.

(1) Equation of motion

The oil and water in two-phase flow satisfy Darcy's law.

Water flooding is carried out along the x direction in an inclined reservoir, and the liquid flow velocity is composed of two parts, as shown in Figure 3.54. Taking the oil phase as an example, v_{o1} under the action of water pressure and v_{o2} under the action of gravity are expressed as follows:

$$v_{o1} = -\frac{K_o}{\mu_o}\frac{\partial p_o}{\partial x} \tag{3.242}$$

$$v_{o2} = \frac{K_o}{\mu_o}\rho_o g \sin\alpha \tag{3.243}$$

If the two velocity directions are opposite, the oil phase velocity is:

$$v_o = v_{o1} - v_{o2} \tag{3.244}$$

It can be written as:

$$v_o = -\frac{K_o}{\mu_o}\left(\frac{\partial p_o}{\partial x} + \rho_o g \sin\alpha\right) \tag{3.245}$$

Similarly, for the water phase:

$$v_w = -\frac{K_w}{\mu_w}\left(\frac{\partial p_w}{\partial x} + \rho_w g \sin\alpha\right) \tag{3.246}$$

(2) Fractional flow equation

The water cut is:

$$f_w = \frac{Q_w}{Q_w + Q_o} \tag{3.247}$$

And $Q = vA$, then:

$$f_w = \frac{v_w}{v_w + v_o} \tag{3.248}$$

From the equation of motion, it can be obtained that:

$$\frac{\mu_o}{K_o} v_o = -\frac{\partial p_o}{\partial x} - \rho_o g \sin\alpha \tag{3.249}$$

$$\frac{\mu_w}{K_w} v_w = -\frac{\partial p_w}{\partial x} - \rho_w g \sin\alpha \tag{3.250}$$

The two equations are subtracted to get:

$$\frac{\mu_w}{K_w} v_w - \frac{\mu_o}{K_o} v_o = -\left(\frac{\partial p_w}{\partial x} - \frac{\partial p_o}{\partial x}\right) - (\rho_w - \rho_o) g \sin\alpha \tag{3.251}$$

From the oil-water two-phase capillary force, it is known that:

$$p_c = p_o - p_w \tag{3.252}$$

Then:

$$\frac{\partial p_c}{\partial x} = \frac{\partial p_o}{\partial x} - \frac{\partial p_w}{\partial x} \tag{3.253}$$

Let

$$\Delta\rho = (\rho_w - \rho_o) \tag{3.254}$$

$$v_t = v_w + v_o \tag{3.255}$$

Then:

$$\frac{\mu_w}{K_w} v_w - \frac{\mu_o}{K_o}(v_t - v_w) = \frac{\partial p_c}{\partial x} - \Delta\rho g \sin\alpha \tag{3.256}$$

$$v_w\left(\frac{\mu_w}{K_w} + \frac{\mu_o}{K_o}\right) - \frac{\mu_o}{K_o} v_t = \frac{\partial p_c}{\partial x} - \Delta\rho g \sin\alpha \tag{3.257}$$

Divide them by v_t and get:

$$\frac{v_w}{v_t}\left(\frac{\mu_w}{K_w}+\frac{\mu_o}{K_o}\right)-\frac{\mu_o}{K_o}=\left(\frac{\partial p_c}{\partial x}-\Delta\rho g\sin\alpha\right)\frac{1}{v_t} \tag{3.258}$$

Divide both ends by $\frac{\mu_o}{K_o}$ and use $f_w=\frac{v_w}{v_t}$ to get:

$$f_w=\frac{1+\frac{K_o}{\mu_o}\cdot\frac{1}{v_t}\left(\frac{\partial p_c}{\partial x}-\Delta\rho g\sin\alpha\right)}{1+\frac{\mu_w}{\mu_0}\cdot\frac{K_o}{K_w}} \tag{3.259}$$

This equation is the fractional flow equation considering the influence of capillary force and gravity. The equation is analyzed as follows:

① If the direction of oil displacement by water is from bottom-up to top-down, there is:

$$f_w=\frac{1+\frac{K_o}{\mu_o}\cdot\frac{1}{v_t}\left(\frac{\partial p_c}{\partial x}+\Delta\rho g\sin\alpha\right)}{1+\frac{\mu_w}{\mu_0}\cdot\frac{K_o}{K_w}} \tag{3.260}$$

In this case, gravity makes the water cut on the same flow section rise, which has a negative effect on the water flooding. Therefore, water injection is generally applied to the bottom of the inclined reservoir to to reduce water cut.

② If the reservoir is horizontal, the gravity term can be removed, then:

$$f_w=\frac{1+\frac{K_o}{\mu_o}\cdot\frac{1}{v_t}\frac{\partial p_c}{\partial x}}{1+\frac{\mu_w}{\mu_0}\cdot\frac{K_o}{K_w}} \tag{3.261}$$

It indicates that water injected at the bottom of the inclined reservoir can inhibit the rise of water cut.

③ If the formation is horizontal and the capillary force can be ignored (when the reservoir wettability is neutral), then:

$$f_w=\frac{1}{1+\frac{\mu_w}{\mu_0}\cdot\frac{K_o}{K_w}}=\frac{1}{1+\frac{\mu_w}{\mu_0}\cdot\frac{K_{ro}}{K_{rw}}} \tag{3.262}$$

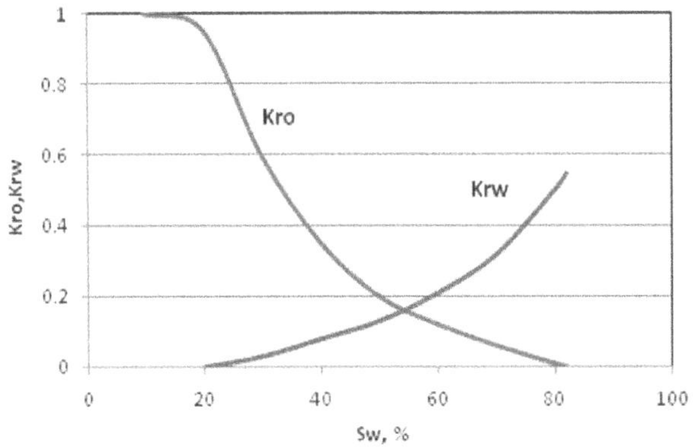

Figure 3.55. Oil-water relative permeability curve

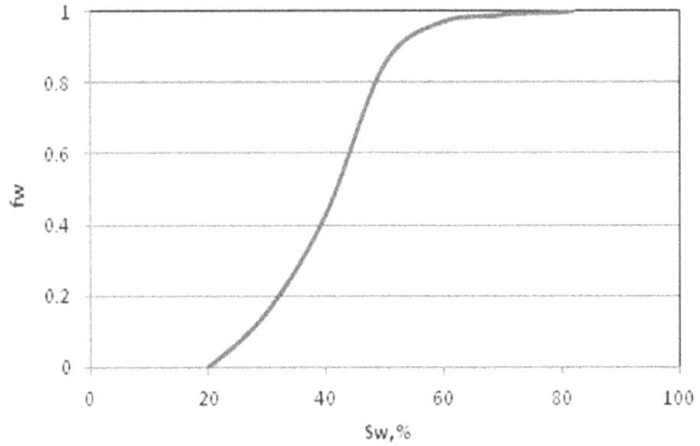

Figure 3.56. Fractional flow curve of water

In this case, the influencing factors on water cut are the viscosity ratio and relative permeability ratio. For the same reservoir, the larger the oil-water viscosity ratio $\left(\mu_r = \frac{\mu_o}{\mu_w}\right)$, the larger the water cut on the same flow section, and vice versa. The oil-water relative permeability is related to water saturation, so the fractional flow of water f_w is a function of water saturation, $f_w = f_w(S_w)$. If the oil-water phase permeability curve of the rock is known as in Figure 3.55, the fractional flow curve of water can be obtained as shown in Figure 3.56.

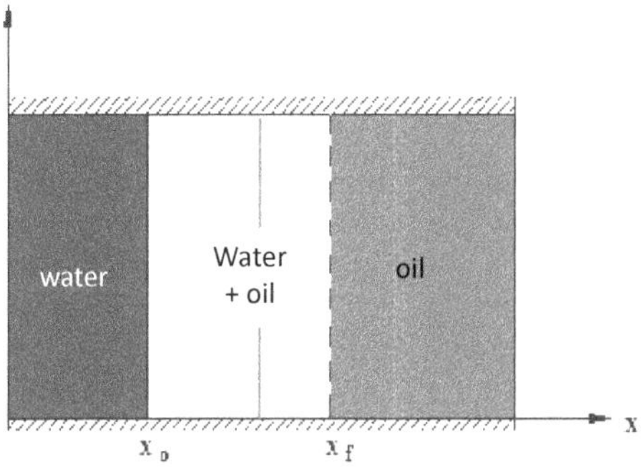

Figure 3.57. One-dimensional model for oil displacement by water

(3) Buckley-Leverett equation

As shown in Figure 3.57, in the horizontal formation, the oil is displaced by water along the x direction, starting at position x_0.

The water phase continuity equation of one-dimensional displacement is:

$$-\frac{\partial v_{wx}}{\partial x} = \phi \frac{\partial S_w}{\partial t} \tag{3.263}$$

On the left side of the equation, the fractional flow equation of water can be written as:

$$v_w = (v_w + v_o) f_w = v_t f_w \tag{3.264}$$

When the elastic energy of rock and liquid is ignored, the total flow velocity v_t is a constant. There are:

$$-\frac{\partial v_{wx}}{\partial x} = -v_t \frac{df_w}{dS_w} \frac{\partial S_w}{\partial x} \tag{3.265}$$

Then:

$$-v_t \frac{df_w}{dS_w} \frac{\partial S_w}{\partial x} = \phi \frac{\partial S_w}{\partial t} \tag{3.266}$$

It can be obtained that:

$$-\frac{v_t}{\phi} \frac{df_w}{dS_w} = \frac{\frac{\partial S_w}{\partial t}}{\frac{\partial S_w}{\partial x}} \tag{3.267}$$

For the iso-saturation surface, that is, $S_w = C$ (constant), it has:

$$dS_w = \frac{\partial S_w}{\partial x}dx + \frac{\partial S_w}{\partial t}dt = 0 \tag{3.268}$$

Then:

$$-\frac{dx}{dt} = \frac{\frac{\partial S_w}{\partial t}}{\frac{\partial S_w}{\partial x}} \tag{3.269}$$

And:

$$\frac{dx}{dt} = \frac{v_t}{\phi}\frac{df_w}{dS_w} \tag{3.270}$$

Here,

$$v_t = \frac{Q(t)}{A} \tag{3.271}$$

$Q(t)$ is the injection volume, so:

$$\frac{dx}{dt} = \frac{Q(t)}{\phi A}\frac{df_w}{dS_w} \tag{3.272}$$

This equation is the basic differential equation for the advancement of iso-saturation surface of one-dimensional oil displacement by water, and it is also the famous B-L equation (Buckley-Leverett) in the theory of oil displacement by water.

The B-L equation essentially describes the velocity of the iso-saturation surface advancing along the x direction.

By separating variables and integrating the B-L equation, the following equation can be obtained:

$$\int_{x_0}^{x} dx = \int_0^t \frac{Q(t)}{\phi A}\frac{df_w}{dS_w}dt \tag{3.273}$$

$$x - x_0 = \frac{f'_w}{\phi A}\int_0^t Q(t)dt \tag{3.274}$$

$$x - x_0 = \frac{f'_w}{\phi A}W(t) \tag{3.275}$$

where $f'_w = \frac{df_w}{dS_w}$; $W(t) = \int_0^t Q(t)dt$ is the total injection volume from the beginning to time t.

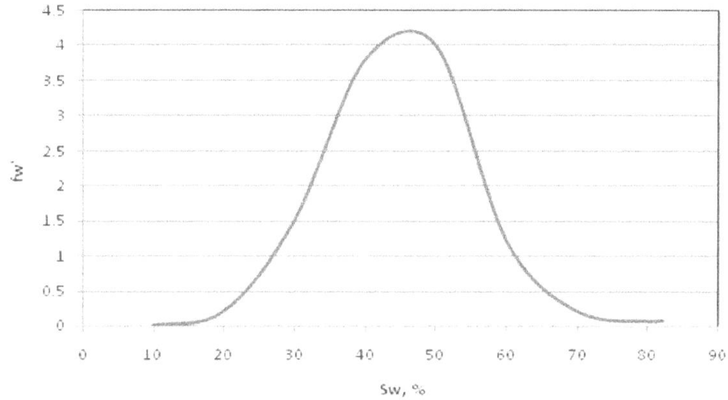

Figure 3.58. Derivative of fractional flow of water and water saturation curve

This is another expression of the B-L equation, which is often used in practical applications. f'_w can be calculated by interpolation of fractional flow curve data, as shown in Figure 3.58.

3.4.3.5 *Application of Buckley-Leverett equation*

(1) Flood front and average water saturation before water breakthrough

The flood front of oil-water mixing zone has not reached the production end, and this stage is defined as the stage before water breakthrough. At this point, the two parameters are determined as follows.

Flood front water saturation S_{wf} has a jump relative to irreducible water saturation S_{wc}, and the relationship between them can be deduced via the analysis as follows:

$$f'_w(S_{wf}) = \frac{f_w(S_{wf})}{S_{wf} - S_{wc}} \tag{3.276}$$

This equation is an implicit function with S_{wf}, which is difficult to solve directly, and can be solved by graphic method through the geometric meaning of the equation. Make a tangent line to the f_w curve through S_{wc}, and the water saturation corresponding to the tangent point is the flood front water saturation S_{wf}. See Figure 3.59 for more information.

By using this equation:

$$\overline{S_w} - S_{wc} = \frac{W(t)}{\phi A(x_f - x_0)} \tag{3.277}$$

It can be obtained:

$$\overline{S_w} - S_{wc} = \frac{1}{f'_w(S_{wf})} \tag{3.278}$$

The geometric meaning of this formula is to make a tangent to the f_w curve through S_{wc} point, and the water saturation corresponding to the intersection point of the tangent line and the $f_w = 1$ line is the average water saturation $\overline{S_w}$ in the mixing zone before water breakthrough. See Figure 3.59 for more information.

(2) Flood front and average water saturation after water breakthrough

When the flood front reaches the production end, it is called water breakthrough. If water flooding continues, the water cut at the production end will rise. At this time, S_{wf} and $\overline{S_w}$ methods are different from those before water breakthrough.

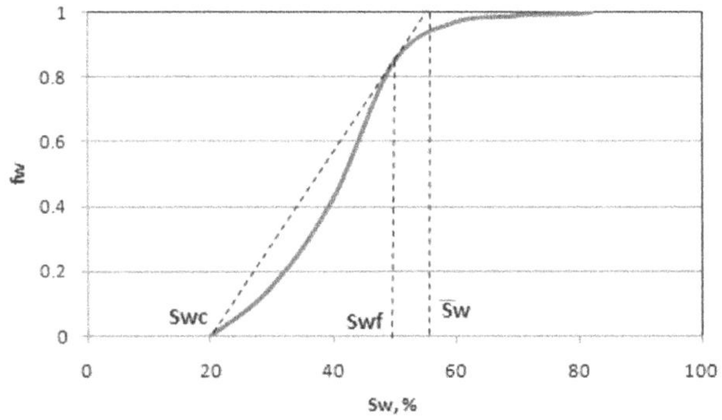

Figure 3.59. Determination of flood front water saturation and average saturation

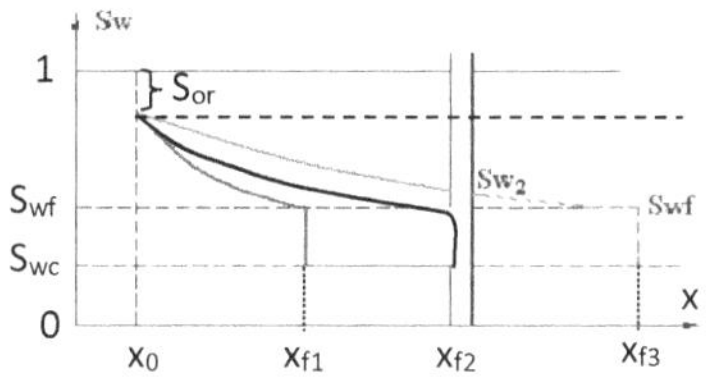

Figure 3.60. S_{wf} after water breakthrough and S_{w2} schematic diagram

When flood front x_{f1} is in the reservoir, no water is produced. S_{wf} and $\overline{S_w}$ are calculated as those before water breakthrough.

When flood front x_{f2} just reaches the production end, S_{wf} and $\overline{S_w}$ can still be calculated according to those before water breakthrough, and at this time, the water saturation at the production end is $S_{w2} = S_{wf}$.

When the flood front position reaches the production end S_{w2}, the water saturation will increase. In this case, assuming that the reservoir extends to infinity along the x direction, that is, there is no production end, the flood front position will advance in this virtual reservoir, and reach x_{f3}.

Therefore, the water saturation value of the flood front does not change after water breakthrough, and the method is the same as that before water breakthrough, but the flood front position is in the virtual reservoir.

Due to the water produced at the production end after water breakthrough, the $\overline{S_w}$ in the mixed zone of the actual reservoir changes and the water cut gradually increases.

The equation to compute saturation can be:

$$\overline{S_w} = \frac{\int_{x_0}^{x_{f2}} S_w dx}{x_{f2} - x_0} \tag{3.279}$$

After derivation and analysis, it can be obtained that:

$$\overline{S_w} = S_{w2} + Q_i f_o(S_{w2}) \tag{3.280}$$

In this equation,

$$Q_i = \frac{W(t)}{(x_{f2} - x_0)A\varphi} \tag{3.281}$$

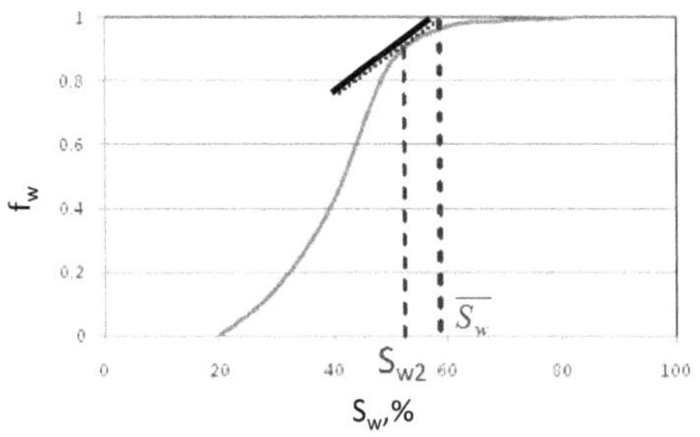

Figure 3.61. Average water saturation determination method after water breakthrough

Its physical meaning is the ratio of the cumulative injected water volume to the pore volume, also known as PV.

To obtain the physical meaning of the $\overline{S_w}$ equation: make a tangent line on the f_w curve through water saturation S_{w2} at the production end, and the water saturation corresponding to the intersection point of the tangent line and $f_w = 1$ is the average water saturation $\overline{S_w}$ in the actual reservoir, as shown in Figure 3.61.

(3) Determination of flood front position and water breakthrough time

Using B-L equation, the flood front position equation is obtained as follows:

$$x_f - x_0 = \frac{f'_w(S_{wf})}{\phi A} W(t) \tag{3.282}$$

At a certain time t, if the cumulative injected water volume $W(t)$ is known, S_{wf} can be obtained by using the graphic method. Then f'_w can be obtained by using the $f'_w(S_{wf})$ curve (see Figure 3.58), and flood front position x_f can be determined.

If $x_f < x_{f2}$, the flood front is in the reservoir and no water breakthrough occurs.

If $x_f = x_{f2}$, water breakthrough happens, and t is water breakthrough time T_{wb}.

If $x_f > x_{f2}$, the flood front is located in the virtual reservoir and water is produced.

When the water breakthrough time T_{wb} is calculated, the formula is:

$$x_{f2} - x_0 = \frac{f'_w(S_{wf})}{\phi A} W(T) \tag{3.283}$$

The unknown quantity in this case is $W(T)$, and the water breakthrough time T_{wb} can be obtained through the relationship between cumulative water injection volume and time.

(4) Determination of the distribution of water saturation

Given B-L equation, it can be obtained that:

$$x - x_0 = \frac{f'_w(S_w)}{\phi A} W(t) \tag{3.284}$$

Fractional flow curve and reservoir physical property parameters are known, and the three unknowns x, S_w and t can be solved mutually.

① If time t is known, the cumulative water injection volume $W(t)$ can be determined. Further, $f'_w(S_w)$at different positions x and S_w of water saturation corresponding to $f'_w(S_w)$ can be calculated. It may have double values, which is unreasonable.

② If S_w is known, we can obtain the time for the different positions x to reach S_w, or the position x of the S_w surface at a certain time;

③ If x is determined, we can obtain the change rule of water saturation S_w at this point with time t.

3.4.3.6 *Pressure distribution in the mixing zone of linear flow*

Ignoring gravity and capillary force, the total liquid production in any section of the mixing zone is:

$$Q_l = Q_o + Q_w \tag{3.285}$$

Production rate is:

$$Q_o = -KA\frac{K_{ro}}{\mu_o}\frac{dp}{dx} \tag{3.286}$$

The flow rate of produced water is:

$$Q_w = -KA\frac{K_{rw}}{\mu_w}\frac{dp}{dx} \tag{3.287}$$

Therefore, the total flow rate of oil and water is:

$$Q_l = -KA\left(\frac{K_{rw}}{\mu_w} + \frac{K_{ro}}{\mu_o}\right)\frac{dp}{dx} \tag{3.288}$$

Multiply the numerator and denominator of the above equation by μ_o, and the viscosity ratio is $\mu_r = \frac{\mu_o}{\mu_w}$, then:

$$Q_l = -\frac{KA}{\mu_o}(\mu_r K_{rw} + K_{ro})\frac{dp}{dx} \tag{3.289}$$

Then, the flow resistance on the length of dx in the mixed zone is:

$$-\frac{\mu_o}{KA}\cdot\frac{dx}{\mu_r K_{rw} + K_{ro}} \tag{3.290}$$

By integrating the separation variables in the total liquid production Q_1, the following equation can be obtained:

$$\int_{p_0}^{p} dp = -\int_{x_0}^{x}\frac{Q_l\mu_o}{KA(\mu_r K_{rw} + K_{ro})}dx \tag{3.291}$$

B-L equation is:

$$x - x_0 = \frac{f'_w}{\phi A}W(t) \tag{3.292}$$

Differentiate S_w on both sides and get:

$$dx = \frac{W(t)}{\phi A}f''_w(S_w)dS_w \tag{3.293}$$

Substitute this equation into the above equation and take the integration to get:

$$\int_{p_0}^{p} dp = -\int_{x_0}^{x}\frac{Q_l\mu_o W(t) f''_w}{K\phi A^2(\mu_r K_{rw} + K_{ro})}dS_w \tag{3.294}$$

$$p_0 - p = \frac{Q_l\mu_o W(t)}{K\phi A^2}\int_{x_0}^{x}\frac{f''_w}{(\mu_r K_{rw} + K_{ro})}dS_w \tag{3.295}$$

The pressure distribution in the mixing zone can be obtained by approximating the integral in the equation.

3.4.4 *Example*

1. In the one-dimensional reservoir with a porosity $\phi = 0.2$, width $B = 50\,\text{m}$, thickness $h = 8\,\text{m}$, and length $L = 100\,\text{m}$, water is injected from the supply end with a water injection rate of $50\,\text{m}^3/\text{d}$. Given the fractional flow of water and its derivative curve as shown in the figure, please find the advancement velocity of the flood front position and the water breakthrough time of the well.

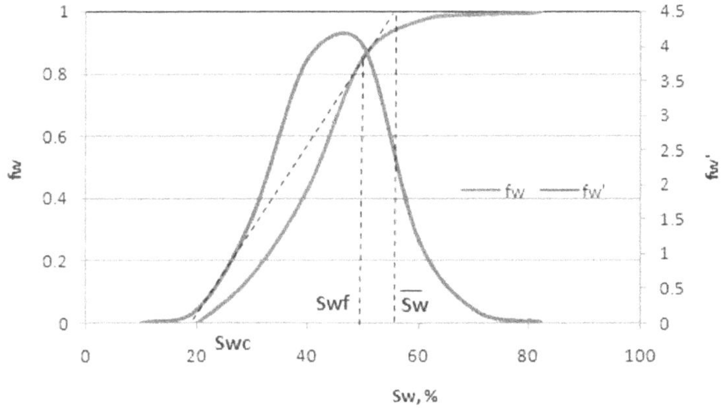

Solution:

According to the fractional flow of water curve, flood front water saturation $S_{wf} = 0.5$ and the corresponding $f'_w(S_{wf}) = 3.8$ can be obtained. According to the B-L equation, the advancement velocity of flood front position is:

$$v_{S_{wf}} = \frac{dx}{dt}\Big|_{S_{wf}} = \frac{Q(t)}{\phi A} f'_w(S_{wf}) = \frac{50}{0.2 \times 50 \times 8} \times 3.8 = 2.375m/d$$

The water breakthrough time of oil well is obtained as follows:

$$T_{wb} = \frac{L}{v_{\mathrm{S_{wf}}}} = \frac{100}{2.375} = 42.1d$$

2. With the data of oil displacement by water the same as that in question 1, please find the time when the water cut of the oil well reaches 0.95.

Solution:

According to the fractional flow curve, when the water cut is 0.95, the water saturation corresponding to it is about 0.65, and the derivative of the water cut corresponding to it is about 0.7. Then, the forward advancing speed of the water saturation is:

$$v_{S_w=0.65} = \left.\frac{dx}{dt}\right|_{S_w=0.65} = \frac{Q(t)}{\phi A} f'_w(S_w = 0.65)$$
$$= \frac{50}{0.2 \times 50 \times 8} \times 0.7 = 0.4375\, m/d$$

The time when the water cut of the oil well reaches 0.95 is:

$$T_{f_w=0.95} = \frac{L}{\mathrm{v_{S_w=0.65}}} = \frac{100}{0.4375} = 228.57d$$

3.5 Natural gas flow theory

3.5.1 *Introduction*

In 2000, China's natural gas production was only $260 \times 10^8\,\mathrm{m}^3$, but in 2010, it increased to $1000 \times 10^8\,\mathrm{m}^3$, and in 2016, it reached $1370 \times 10^8\,\mathrm{m}^3$, which is equivalent to 1.23×10^8t oil. Compared with the crude oil production of 1.97×10^8t in that year, the production of natural gas has accounted for 38.4% of China's total oil and gas production. With the development of unconventional resources such as shale gas, coal-bed gas, tight gas and natural gas hydrate, the proportion of natural gas will also increase.

The main characteristic of natural gas is its high compressibility, which is very different from that of liquid. For example, its viscosity and compressibility are both functions of pressure, the compression factor is a special parameter and also a function of pressure, the pressure in the liquid flow equation is replaced by a pseudo-pressure function, and the pressure square difference in the steady state flow is linearly related to the gas flow, etc.

The parameters marked "sc" in this section refer to those at standard temperature and pressure condition (20°C, 1 atm).

3.5.2 *Concepts*

Ideal gas: *the ideal gas refers to gas molecules that have mass, no volume, independent motion, and no intermolecular force.*

Natural gas pseudo-pressure function: *it can replace the pressure p in the liquid flow equation, and the formula is*

$$m^* = 2\int_{p_0}^{p} \frac{p}{\mu_g(p)Z(p)}dp \tag{3.296}$$

where m^* is the pseudo-pressure function; p_0 is the initial pressure; $\mu_g(p)$ is a function of gas viscosity as a function of pressure; and $Z(p)$ is a function of gas compression factor as a function of pressure.

Absolute open flow potential of gas well: *natural gas flow rate at the bottom of a gas well that is fully open ($p_w = 0.1\,MPa$) is an important parameter used to evaluate the production capacity of a gas well, commonly known as q_{AOF}.*

Binomial productivity formula of gas well: $p_e^2 - p_w^2 = aQ_{sc} + bQ_{sc}^2$.

3.5.3 *Analysis*

3.5.3.1 *Governing equation of natural gas flow*

(1) Continuity equation

Similar to the derivation of the continuity equation of single-phase liquid flow, the continuity equation of natural gas flow can be obtained as follows:

$$-\left[\frac{\partial(\rho_g v_x)}{\partial x} + \frac{\partial(\rho_g v_y)}{\partial y} + \frac{\partial(\rho_g v_z)}{\partial z}\right] = \frac{\partial(\rho_g \phi)}{\partial t} \tag{3.297}$$

(2) Equation of motion

The viscosity of natural gas is very small. The flow velocity of natural gas in the pores is much faster than that of liquid, and many of them reach non-linear flow. However, in the pores of low permeability, tight and shale reservoirs, the flow velocity of natural gas is still within the range of Darcy flow.

The motion equation in line with Darcy's law is:

$$v_x = -\frac{K}{\mu_g}\frac{\partial p}{\partial x}, v_y = -\frac{K}{\mu_g}\frac{\partial p}{\partial y}, v_z = -\frac{K}{\mu_g}\frac{\partial p}{\partial z} \tag{3.298}$$

(3) Equation of state

The equation of state of an ideal gas is:

$$pV = nRT \tag{3.299}$$

Then:

$$\rho_g = \frac{pM}{RT} \tag{3.300}$$

If there is a deviation between the actual gas and the ideal gas, the equation of state of the real gas is:

$$pV = nZRT \tag{3.301}$$

Then:

$$\rho_g = \frac{pM}{ZRT} \tag{3.302}$$

or

$$\rho_g = \frac{T_{sc}Z_{sc}\rho_{gsc}}{p_{sc}}\frac{p}{ZT} \tag{3.303}$$

Assuming that the temperature of the gas layer remains unchanged, the isothermal compressibility of the gas can be obtained as follows:

$$C_g(p) = \frac{1}{p} - \frac{1}{Z}\frac{dZ}{dp} \tag{3.304}$$

For an ideal gas, the isothermal compressibility is:

$$C_g(p) = \frac{1}{p} \tag{3.305}$$

Since the elastic energy of gas is very large, the elasticity of rock is negligible.

(4) Basic differential equations

① Ideal gas

By substituting the equation of motion and the equation of state of ideal gas into the continuity equation, the basic differential equation of ideal gas can be obtained:

$$\frac{\partial^2 p^2}{\partial x^2}+\frac{\partial^2 p^2}{\partial y^2}+\frac{\partial^2 p^2}{\partial z^2}=\frac{\varphi\mu_g(\overline{p})}{Kp}\frac{\partial p^2}{\partial t} \tag{3.306}$$

For the coefficient term, take $C_g(p)=\frac{1}{p}$ and $p=\overline{p}$, and define the gas diffusivity as:

$$\eta=\frac{K}{\varphi\mu_g(\overline{p})C_g(\overline{p})} \tag{3.307}$$

Its physical meaning is similar to the liquid diffusivity coefficient, that is, the area of pressure wave propagation in unit time.

Then, the basic differential equation of ideal gas is:

$$\frac{\partial^2 p^2}{\partial x^2}+\frac{\partial^2 p^2}{\partial y^2}+\frac{\partial^2 p^2}{\partial z^2}=\frac{1}{\eta}\frac{\partial p^2}{\partial t} \tag{3.308}$$

It can be re-written as:

$$\nabla^2 p^2=\frac{1}{\eta}\frac{\partial p^2}{\partial t} \tag{3.309}$$

② Real gas

The equation of motion and the state equation of real gas are substituted into the continuity equation, and the incompressibility of rock is also considered, then the following equation can be obtained:

$$\frac{\partial}{\partial x}\left[\frac{p}{\mu_g(p)Z(p)}\frac{\partial p}{\partial x}\right]+\frac{\partial}{\partial y}\left[\frac{p}{\mu_g(p)Z(p)}\frac{\partial p}{\partial y}\right]+\frac{\partial}{\partial z}\left[\frac{p}{\mu_g(p)Z(p)}\frac{\partial p}{\partial z}\right]=\frac{\phi}{K}\frac{\partial}{\partial t}\left[\frac{p}{Z(p)}\right]$$

The right side becomes:

$$\frac{\partial}{\partial t}\left[\frac{p}{Z(p)}\right]=\left[\frac{1}{p}-\frac{1}{Z(p)}\frac{\partial p}{\partial Z(p)}\right]\frac{p}{Z(p)}\frac{\partial p}{\partial t}=C_g(p)\frac{p}{Z(p)}\frac{\partial p}{\partial t}$$

Then:

$$\nabla\left[\frac{p}{\mu_g(p)Z(p)}\nabla p\right]=\frac{\varphi\mu_g(p)C_g(p)}{K}\left[\frac{p}{\mu_g(p)Z(p)}\frac{\partial p}{\partial t}\right]$$

Similarly, in terms of coefficients, if you take $p=\overline{p}$, you get the gas diffusivity. Then, it is:

$$\nabla\left[\frac{p}{\mu_g(p)Z(p)}\nabla p\right]=\frac{1}{\eta}\left[\frac{p}{\mu_g(p)Z(p)}\frac{\partial p}{\partial t}\right] \tag{3.310}$$

In order to be similar to the mathematical model of liquid flow, the pseudo-pressure function is introduced and defined as:

$$m^*=2\int_{p_0}^{p}\frac{p}{\mu_g(p)Z(p)}dp \tag{3.311}$$

Then:

$$\frac{\partial^2 m^*}{\partial x^2}+\frac{\partial^2 m^*}{\partial y^2}+\frac{\partial^2 m^*}{\partial z^2}=\frac{1}{\eta}\frac{\partial m^*}{\partial t} \tag{3.312}$$

This is the basic differential equation for a real gas.

This equation is the same as the basic differential equation of single-phase liquid. The difference is that pseudo-pressure is used in gas flow.

In the pseudo-pressure function, $\mu_g(p)Z(p)$ is a function of pressure and temperature. In practical applications, its value can be approximately equal to the corresponding value under $\overline{p}$, then the pseudo-pressure difference is:

$$m_1^*-m_2^*=\frac{1}{\mu_g(\overline{p})Z(\overline{p})}(p_1^2-p_2^2) \tag{3.313}$$

3.5.3.2 *Steady state flow of natural gas*

(1) Darcy flow

Here, we only take steady radial flow in a real gas well as an example. The object of study is real gas, which conforms to Darcy's law.

The expression in polar coordinates of radial flow is similar to that of liquid. The basic differential equation of steady state flow is:

$$\frac{d^2m^*}{dr^2}+\frac{1}{r}\frac{dm^*}{dr}=0 \tag{3.314}$$

Then, the pseudo-pressure distribution formula can be obtained as follows:

$$m^*=m_e^*-\frac{m_e^*-m_w^*}{\ln\frac{R_e}{R_w}}\ln\frac{R_e}{r} \tag{3.315}$$

If the corresponding value of average formation pressure $\overline{p}$ is substituted by $\mu_g(p)Z(p)$, then:

$$p^2=p_e^2-\frac{p_e^2-p_w^2}{\ln\frac{R_e}{R_w}}\ln\frac{R_e}{r} \tag{3.316}$$

The volumetric flow rate of natural gas varies with the pressure, and the result cannot be obtained directly by applying the formula of steady flow for single-phase liquid.

When it is steady state flow, the mass flow rate does not change. According to Darcy's law, the mass flow rate is:

$$Q_m=\frac{2\pi Kh}{\mu_g}r\frac{dp}{dr}\rho_g \tag{3.317}$$

Substitute into the real gas the equation of state is:

$$Q_m=\frac{2\pi Kh}{\mu_g}r\frac{dp}{dr}\frac{T_{sc}Z_{sc}\rho_{gsc}}{p_{sc}}\frac{p}{Z(p)T} \tag{3.318}$$

According to the definition of pseudo-pressure function, the following equation can be obtained:

$$Q_m=\frac{\pi KhT_{sc}Z_{sc}\rho_{gsc}}{p_{sc}T}r\frac{dm^*}{dr} \tag{3.319}$$

By the integration of the separated variables, it can be obtained that:

$$Q_m=\frac{\pi KhT_{sc}Z_{sc}\rho_{gsc}}{p_{sc}T}\frac{m_e^*-m_w^*}{\ln\frac{R_e}{R_w}} \tag{3.320}$$

When $\mu_g(p)Z(p)$is taken as $\overline{p}$, the formula of volumetric flow rate for steady state flow in a gas well can be obtained:

$$Q_{sc} = \frac{\pi K h T_{sc} Z_{sc}}{p_{sc} T \mu_g(\overline{p}) Z(\overline{p})} \frac{p_e^2 - p_w^2}{\ln \frac{R_e}{R_w}} \tag{3.321}$$

Differentiate the pressure distribution formula to get the pressure gradient:

$$\frac{dp}{dr} = \frac{p_e^2 - p_w^2}{\ln \frac{R_e}{R_w}} \frac{1}{2p \cdot r} \tag{3.322}$$

Therefore, under the condition of the same pressure difference, gas well bottom hole pressure gradient is larger than oil well, which indicates that most of the energy consumption occurs near the well a few meters away. The average formation pressure is close to the supply pressure, generally it uses $\overline{p} = p_e$.

(2) Non-linear flow

The basic differential equation of non-linear steady state flow of natural gas can be written as follows:

$$\begin{cases} \nabla\left[\delta(\nabla m^*)\right] = 0 \\ \delta = \dfrac{1}{1 + \frac{\alpha_2 \rho_g K v}{\mu_g}} \end{cases} \tag{3.323}$$

where α_2 is a constant related to the shape of the pores, which is generally less than 0.1; δ is the turbulence coefficient, which is a decimal closer to 1 when the velocity is not too high.

The flow of natural gas in the gas reservoir is often non-linear, and the two forms of non-linear flow equations are used:

Binomial flow formula:

$$\begin{cases} p_e^2 - p_w^2 = aQ_{sc} + bQ_{sc}^2 \\ a = \dfrac{p_{sc} T \mu_g(\overline{p}) Z(\overline{p})}{\pi K h T_{sc} Z_{sc}} \ln \dfrac{R_e}{R_w} \\ b = \dfrac{\alpha_2 \rho_{gsc}}{2\pi^2 h^2} \dfrac{p_{sc} T Z(\overline{p})}{T_{sc} Z_{sc} Z} \left(\dfrac{1}{R_w} - \dfrac{1}{R_e}\right) \end{cases} \tag{3.324}$$

Exponential flow formula:

$$Q_{sc} = C(p_e^2 - p_w^2)^n \tag{3.325}$$

where $C = 2\pi KhC_2\left(\frac{1-n}{n}\right)^n \rho_{gsc}^{n-1}\left(\frac{T_{sc}}{2Z(\overline{p})p_{sc}T}\right)^n\left(R_w^{1-\frac{1}{n}} - R_e^{1-\frac{1}{n}}\right)^{-n}$ and C_2 is a coefficient related to gas reservoir and fluid properties; n is a flow index, $0.5 \leq n < 1$.

The above coefficients are very complex and difficult to be calculated accurately with the formula. In practical applications, they are determined according to well test data.

3.5.3.3 *Steady state well test of natural gas*

The purpose of steady state gas well test is to determine the productivity, judge the effect of well stimulation and determine the reasonable working schedule.

If the linear flow production formula is applied, the relationship curve between $p_e^2 - p_w^2$ and Q_{sc} can be drawn by measuring the bottom pressure and then calculating $p_e^2 - p_w^2$. Theoretically, the formation parameters can be calculated by taking the slope of the straight line through the origin. Its application is similar to the single-phase liquid steady state well test.

If the non-linear flow formula is applied, it can be divided into binomial and exponential expressions.

① Binomial

The binomial transformation of non-linear gas flow is:

$$\frac{p_e^2 - p_w^2}{Q_{sc}} = a + bQ_{sc} \tag{3.326}$$

The binomial characteristic curve is shown in Figure 3.62.

Coefficients a and b can be obtained from the measured curve, and the binomial productivity equation can be obtained.

If $p_w = p_{sc}$ and p_{sc} is atmospheric pressure (1×10^5 Pa), then:

$$p_e^2 - p_{sc}^2 = aQ_{AOF} + bQ_{AOF}^2 \tag{3.327}$$

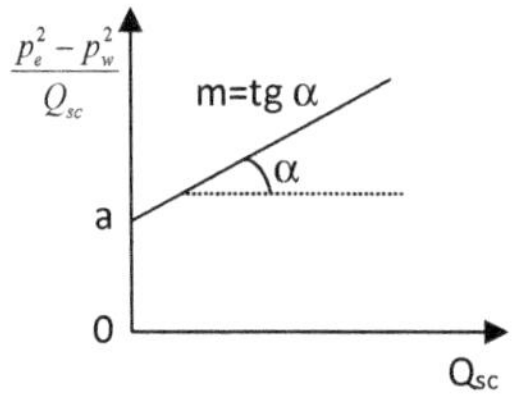

Figure 3.62. Binomial characteristic curve

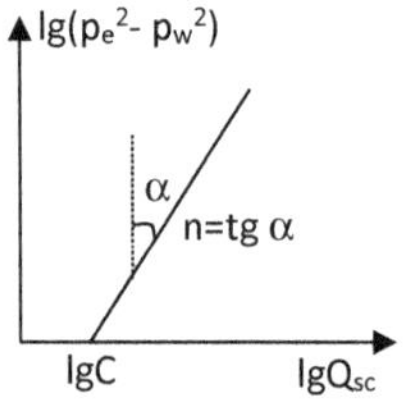

Figure 3.63. Exponential characteristic curve

Thus, the absolute open flow potential of a gas well can be calculated.

② Exponential

The exponential form of non-linear flow in a gas well is:

$$\lg Q_{sc} = \lg C + n \lg(p_e^2 - p_w^2) \tag{3.328}$$

The exponential characteristic curve is shown in Figure 3.63.

Coefficient n can be obtained on the measured curve, and then find a point on the curve and substitute it into the exponential characteristic equation to get C, so as to obtain the exponential productivity equation.

Similarly, if $p_w = p_{sc}$, then:

$$Q_{AOF} = C(p_e^2 - p_{sc}^2)^n \tag{3.329}$$

The absolute open flow potential of a gas well can be obtained.

On the exponential characteristic curve, when the straight line moves to the left, it indicates that C value decreases and formation permeability decreases, and vice versa.

3.5.3.4 *Transient flow of natural gas*

Given the basic differential equation of real gas, the expression in polar coordinates of radial flow is similar to that of liquid. The basic differential equation of transient flow is:

$$\frac{\partial^2 m^*}{\partial r^2} + \frac{1}{r}\frac{\partial m^*}{\partial r} = \frac{1}{\eta}\frac{\partial m^*}{\partial t} \tag{3.330}$$

The pressure distribution is:

$$p^2 = p_i^2 - \frac{Q_{sc}\mu_g(\overline{p})}{2\pi Kh}\frac{p_{sc}Z(\overline{p})T}{Z_{sc}T_{sc}}\left[-Ei\left(-\frac{r^2}{4\eta t}\right)\right] \tag{3.331}$$

Since the diffusivity coefficient of gas is much larger than that of liquid (up to 10 times), $\frac{r^2}{4\eta t} < 0.01$ can be easily achieved. Therefore, the pressure distribution can be written as follows:

$$p^2 = p_i^2 - \frac{Q_{sc}\mu_g(\overline{p})}{2\pi Kh}\frac{p_{sc}Z(\overline{p})T}{Z_{sc}T_{sc}}\ln\frac{2.25\eta t}{r^2} \tag{3.332}$$

The bottom hole pressure is:

$$p_w^2 = p_i^2 - \frac{Q_{sc}\mu_g(\overline{p})}{2\pi Kh}\frac{p_{sc}Z(\overline{p})T}{Z_{sc}T_{sc}}\ln\frac{2.25\eta t}{R_w^2} \tag{3.333}$$

The multi-well interference of transient flow in a gas well is the same as that of single-phase liquid, and the transient well test can be divided into pressure drawdown well test and pressure buildup well test.

3.5.4 *Example*

In the steady state radial gas flow, there is a production well in the center of the circular supply boundary. Given that the boundary radius $R_e = 150\,\text{m}$, $p_e = 20\,\text{MPa}$, $T = 323\,\text{K}$, $K = 5\,\text{mD}$, $h = 5\,\text{m}$, $\mu_g(\overline{p} = p_e) = 0.022\,\text{mPa}\cdot\text{s}$, $Z(\overline{p} = p_e) = 0.81$, $T_{sc} = 293.15\,\text{K}$, $Z_{sc} = 1$, $p_{sc} = 0.1\,\text{MPa}$, $R_w = 0.1\,\text{m}$, $p_w = 17\,\text{MPa}$, $\phi = 0.2$, $\rho_g(\overline{p} = p_e) = 0.15\,\text{g/cm}^3$, $B_g = 0.0045$, please find the flow rate $Q(\times 10^4\,\text{m}^3/\text{d})$ and the actual flow rate of gas at the well wall $v_{w\phi}(\mu\text{m/s})$.

Solution:

$$Q_{sc} = \frac{\pi K h T_{sc} Z_{sc}}{p_{sc} T \mu_g(\overline{p}) Z(\overline{p})} \frac{p_e^2 - p_w^2}{\ln \frac{R_e}{R_w}}$$

$$= \frac{3.14 \times 5 \times 10^{-15} \times 5 \times 293.15 \times 1}{0.1 \times 10^6 \times 323 \times 0.022 \times 10^{-3} \times 0.81}$$

$$\times \frac{(20^2 - 17^2) \times 10^{12}}{\ln \frac{150}{0.1}} \times 86400$$

$$= 5.24 \times 10^4 m^3/d$$

$$v_{w\varphi} = \frac{Q_{sc} B_g}{2\pi R_w h \varphi} = \frac{5.24 \times 10^4 \times 0.0045}{86400} \times \frac{1}{2 \times 3.14 \times 0.1 \times 5 \times 0.2}$$

$$= 4345.8\, \mu m/s$$

$$R_e = \frac{v_{w\varphi}\sqrt{K}\rho}{1750 \mu \varphi^{\frac{3}{2}}} = \frac{4345.8 \times 10^{-4} \times \sqrt{5 \times 10^{-3}} \times 0.15}{1750 \times 0.022 \times 0.2^{1.5}} = 0.0134$$

The viscosity of natural gas is one thousandth that of oil, $(p_e^2 - p_w^2) > (p_e - p_w)$, so natural gas can flow in the lower porosity medium at a speed hundreds of times that of oil.

The Reynolds number in this problem is still within the range of Darcy flow. When the well production is high, the gas flow no longer completely conforms to Darcy's law.

3.6 Gas-water two-phase flow theory

3.6.1 *Introduction*

Natural gas and water almost co-exist in underground gas reservoirs. In the process of water flooding, when the water-gas interface advances at a fast speed, a large amount of gas stays in the water phase, forming a closed gas. The Jamin effect and water-lock effect will make it difficult to extract natural gas. However, when the water-gas interface advances at a slow speed, the water flooding performance will be poor due to the elastic effect of gas.

When the saturation of closed gas is high, water drainage is needed to improve recovery. Therefore, dewatering gas production increases the pressure difference and promotes the flow of closed gas.

3.6.2 *Concepts*

Water-lock effect: *when the water phase co-exists with gas, it will greatly reduce relative permeability of the gas phase; this phenomenon is called water-lock effect.*

Jamin effect: *when air bubbles or oil droplets want to pass through smaller pore throat radius, they must change shape to become elongated. This deformation will consume energy, thus it slows bubbles or oil droplets movement by adding an additional resistance; this phenomenon is called Jamin effect.*

The gas-phase pseudo-pressure function of gas-water two-phase is similar to that of natural gas, and the formula is

$$H_g(p) = \int_0^p \frac{K_{rg}}{B_g(p)\mu_g(p)} dp \tag{3.334}$$

where H is the pseudo-pressure function; K_{rg} is the relative permeability of gas; and B_g is the gas isothermal compressibility.

Produced water-gas ratio: *the ratio of water and gas produced under surface conditions. Symbol is WGR, and $m^3/10^4\,m^3$ is commonly used as the unit.*

3.6.3 *Analysis*

(1) Establishment of governing equations

Still applying the mass conservation law of the REV, the fluid in and out of the REV has two phases, gas and water, and the two phases should be analyzed respectively. Similar to oil-water two-phase continuity equation, the equations for gas and water phases are:

$$-\left[\frac{\partial(\rho_g v_{gx})}{\partial x} + \frac{\partial(\rho_g v_{gy})}{\partial y} + \frac{\partial(\rho_g v_{gz})}{\partial z}\right] = \frac{\partial(\phi \rho_g S_g)}{\partial t} \tag{3.335}$$

$$-\left[\frac{\partial(\rho_w v_{wx})}{\partial x} + \frac{\partial(\rho_w v_{wy})}{\partial y} + \frac{\partial(\rho_w v_{wz})}{\partial z}\right] = \frac{\partial(\phi \rho_w S_w)}{\partial t} \tag{3.336}$$

The auxiliary equations are: $S_g + S_w = 1$, $p_{cgw} = p_g - p_w$.

Assume that Darcy's law is valid, gas and water are immiscible, and rock is incompressible. Then the equations of motion are:

$$v_{gx} = -\frac{K_g}{\mu_g}\frac{\partial p_g}{\partial x}, \quad v_{gy} = -\frac{K_g}{\mu_g}\frac{\partial p_g}{\partial y}, \quad v_{gz} = -\frac{K_g}{\mu_g}\frac{\partial p_g}{\partial z} \tag{3.337}$$

$$v_{wx} = -\frac{K_w}{\mu_w}\frac{\partial p_w}{\partial x}, \quad v_{wy} = -\frac{K_w}{\mu_w}\frac{\partial p_w}{\partial y}, \quad v_{wz} = -\frac{K_w}{\mu_w}\frac{\partial p_w}{\partial z} \tag{3.338}$$

Define:

$$\rho_g = \frac{\rho_{gsc}}{B_g(p)} \tag{3.339}$$

$$\rho_w = \rho_{wsc} \tag{3.340}$$

$$K_g = KK_{rg} \tag{3.341}$$

$$K_w = KK_{rw} \tag{3.342}$$

Substitute the equations into the continuity equation and we can get the basic differential equation.

Gas phase:

$$\nabla\left[\frac{K_{rg}}{B_g(p)\mu_g(p)}\nabla p_g\right] = \frac{\phi}{K}\frac{\partial}{\partial t}\left[\frac{S_g}{B_g(p)}\right] \tag{3.343}$$

Water phase:

$$\nabla[K_{rw}\nabla p_w] = \frac{\phi\mu_w}{K}\frac{\partial}{\partial t}[S_w] \tag{3.344}$$

(2) Gas-water two-phase steady state flow

If the capillary pressure is not considered, $p_g = p_w$. The pseudo-pressure function is:

$$H_g(p) = \int_0^p \frac{K_{rg}}{B_g(p)\mu_g(p)} dp \tag{3.345}$$

If it is steady state flow, the gas phase equation becomes:

$$\nabla^2 H = 0 \tag{3.346}$$

This is the Laplace equation for single-phase liquid flow. For gas wells, the expression in polar coordinates of radial flow is similar to that of liquid. The basic differential equation for steady state flow is:

$$\frac{d^2 H_g}{dr^2} + \frac{1}{r}\frac{dH_g}{dr} = 0 \tag{3.347}$$

Then the pseudo-pressure distribution formula can be obtained as:

$$H_g = H_{ge} - \frac{H_{ge} - H_{gw}}{\ln \frac{R_e}{R_w}} \ln \frac{R_e}{r} \tag{3.348}$$

Just like natural gas flow, the volumetric flow of natural gas varies with the pressure. When the formula of steady state flow of single-phase liquid is directly applied, the result cannot be obtained. Therefore, another method is needed.

When the flow is steady state, the mass flow rate does not change. According to Darcy's law, the mass flow rate is:

$$Q_m = \frac{2\pi K K_{rg} h}{\mu_g(p)} r \frac{dp}{dr} \rho_g \tag{3.349}$$

The gas density is:

$$\rho_g = \frac{\rho_{gsc}}{B(p)} \tag{3.350}$$

Combining the above equations, we can get:

$$Q_m = \frac{2\pi K K_{rg} h}{\mu_g(p)} r \frac{dp}{dr} \frac{\rho_{gsc}}{B(g)} \tag{3.351}$$

According to the definition of pseudo-pressure function, the following equation can be obtained:

$$Q_m = 2\pi K h \rho_{gsc} r \frac{dH_g}{dr} \tag{3.352}$$

Separate variables and take the integration on both sides to obtain:

$$Q_m = 2\pi K h \rho_{gsc} \frac{H_{ge} - H_{gw}}{\ln \frac{R_e}{R_w}} \tag{3.353}$$

Then, when the flow is steady state, the volumetric flow of gas well is:

$$Q_g = 2\pi Kh \frac{H_{ge} - H_{gw}}{\ln \frac{R_e}{R_w}} \tag{3.354}$$

As can be seen from the above formula, the physical meaning of H_g is as follows: when gas and water flow from the reservoir to the bottom of the well at the same time, the actual total energy consumed is $\Delta p = p_e - p_w$, and the energy consumed for gas flow is $\Delta H_g = H_{ge} - H_{gw}$.

Both the pressure distribution formula and the flow formula involve the difference of pseudo-pressure. Its calculation can be given in integral form, namely:

$$H_{ge} - H_{gw} = \int_{p_w}^{p_e} \frac{K_{rg}}{\mu_g(p)B_g(p)} dp \tag{3.355}$$

In the formula, $\mu_g(p)$ and $B_g(p)$ are functions of pressure, which can be obtained through experiments or empirical equation curves. However, K_{rg} is a function of S_w in the gas-water phase permeability curve, and S_w is a function of pressure. If the relationship between K_{rg} and pressure is obtained, it needs to be obtained with the aid of the produced water-gas ratio WGR.

It is assumed that water flows out of the wellhead during the production of the gas well, and the content of condensed water is not considered (or the amount of condensed water shall be deducted if considered).

The produced water-gas ratio is defined as:

$$WGR = \frac{Q_{wsc}}{Q_{gsc}} \tag{3.356}$$

It is known that:

$$Q_{wsc} = Q_w = \frac{KK_{rw}}{\mu_w} A \frac{dp}{dr} \tag{3.357}$$

$$Q_{gsc} = Q_g \frac{p}{p_{sc}} = \frac{KK_{rg}}{\mu_g(p)} A \frac{dp}{dr} \frac{p}{p_{sc}} \tag{3.358}$$

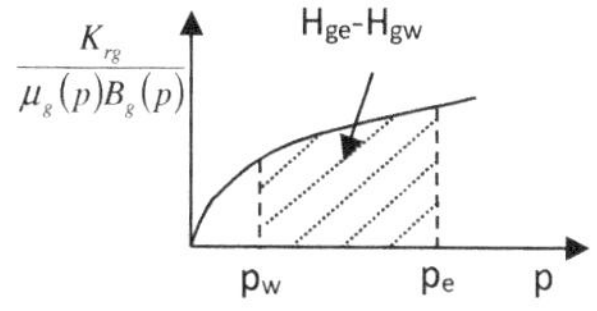

Figure 3.64. Pseudo-pressure difference of gas-water flow

Then:

$$WGR = \frac{\mu_g(p)K_{rw}}{\mu_w K_{rg}} \frac{p_{sc}}{p} \tag{3.359}$$

In the case of steady state flow, relative to an average formation pressure, the water-gas ratio of production is a constant, and the relationship curve of $WGR \sim p$ can be obtained. Then, combining with $\mu_g(p) \sim p$, the relationship of $\frac{K_{rw}}{K_{rg}} \sim p$ can be obtained. Then, the relationship of $K_{rg} \sim p$ can be obtained by means of gas-water phase permeability curve, and finally the relationship of $\frac{K_{rg}}{\mu_g(p)B_g(p)} \sim p$ can be obtained.

As shown in Figure 3.64, if the shaded area is taken from the relation curve of $\frac{K_{rg}}{\mu_g(p)B_g(p)} \sim p$, the pseudo-pressure difference can be obtained. Or if the numerical integration method is applied, $\frac{K_{rg}}{\mu_g(p)B_g(p)}$ can be obtained by interpolation method, and the pseudo-pressure difference can still be obtained.

3.7 Oil-gas two-phase flow theory

3.7.1 *Introduction*

In the single-phase oil elastic drive process, if the pressure is lower than the saturation pressure, the natural gas dissolved in oil, i.e., the solution gas, will escape from the oil and become free gas. The free gas has different physical properties from the solution gas, and the single-phase flow changes to oil and gas two-phase flow. The free gas has much larger elastic energy than the liquid, and drive mechanism changes from the elastic drive to the solution gas drive.

Two-phase flow of oil and gas is different from single-phase flow, and it is also quite different from oil-water and gas-water two-phase flow. For oil and gas phases, the gas phase can be dissolved in the oil phase, while the oil phase is insoluble in the gas phase. Therefore, the basic differential equations established will be very different.

In field practice, oil and gas are usually associated, that is, oil in the reservoir condition is always a mixture of oil and gas after being extracted and separated on the ground surface. The oil after separation is the reservoir oil phase, and the gas after separation comprises two parts: solution gas in the reservoir oil phase and free gas in the reservoir gas phase. Solution gas and oil have the same flow characteristics in the reservoir condition, while free gas is different from them.

3.7.2 *Concepts*

Produced gas-oil ratio: *the ratio of the total amount of gas produced by an oil well to the amount of pure oil under surface conditions. Symbol is GOR.*

The oil-phase pseudo-pressure function in two-phase flow of oil and gas: $H_o(p) = \int_0^p \frac{K_{ro}}{B_o(p)\mu_o(p)} dp$

3.7.3 *Analysis*

3.7.3.1 *Physical process of oil and gas flow*

Figure 3.65 shows the recovery curve of a closed unsaturated reservoir when the reservoir pressure is lower than the saturation pressure p_b. *GOR* is divided into three phases.

Stage I: When the formation pressure is just below the saturation pressure, free gas is in the form of individual bubbles dispersed in the oil phase, and has not yet become a continuous phase. The expansion of the bubbles from the release of elastic energy is the main force to drive the oil flow. Therefore, with the *GOR* changing slowly, the pressure basically remains the same. This stage is very short.

Stage II: With a decrease in pressure, the free gas escaping from the oil increases. Because the gas viscosity is very small, the dispersed free gas gradually forms a continuous phase, and soon it flows from the oil into the bottom hole, and *GOR* increases rapidly, which

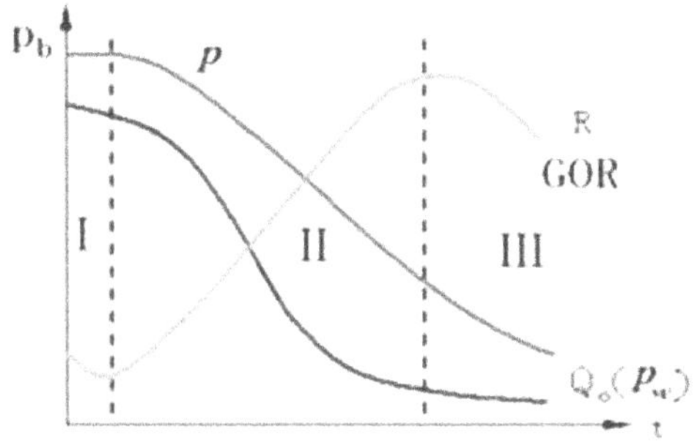

Figure 3.65. Solution gas drive recovery curve

causes rapid pressure drop, increased oil viscosity, and reduced oil production. The performance of the elastic drive of the solution gas is poor.

Stage III: *GOR* drops rapidly, energy is nearly exhausted since it is the last stage of the production.

From the above three processes, it can be seen that the performance of the solution gas drive is the best in the first stage, then gradually becomes poor. Therefore, generally the formation pressure will be maintained through water injection or gas injection to provide additional energy.

3.7.3.2 *Governing equation for two-phase flow of oil and gas*

(1) Continuity equation

In the same way as before, the continuity equation is deduced by using REV and the law of material balance.

The continuity equation of oil phase is derived as shown in Figure 3.66.

The oil phase at this point is pure oil after degassing and is called oil component in multiphase multicomponent flow. If ρ_D is the density of underground mixed gas and oil and G_1 is the density of solution gas, in the x direction the mass flow velocity of the oil phase at the center point M is:

$$(\rho_D - G_1)v_{0x} \tag{3.360}$$

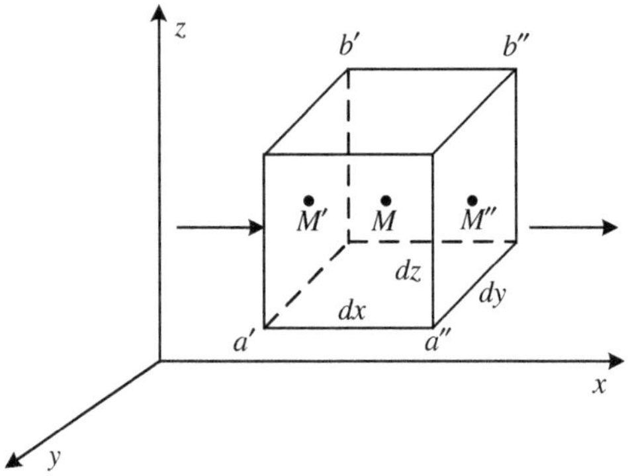

Figure 3.66. Schematic diagram of REV for oil-gas two-phase flow

Then, the mass difference of oil phase flowing in and out with dt time in the x direction is:

$$-\frac{\partial[(\rho_D - G_1)v_{0x}]}{\partial x}dxdydzdt \tag{3.361}$$

Similarly, the mass difference in y and z directions can be obtained, and the mass difference in the inflow and outflow oil phases of the whole REV can be obtained as follows:

$$-\left\{\frac{\partial[(\rho_D - G_1)v_{0x}]}{\partial x} + \frac{\partial[(\rho_D - G_1)v_{0y}]}{\partial y} + \frac{\partial[(\rho_D - G_1)v_{0z}]}{\partial z}\right\}dxdydzdt \tag{3.362}$$

The change of oil phase mass in dt time of the REV is:

$$\frac{\partial}{\partial t}[(\rho_D - G_1)\varphi S_o]dxdydzdt \tag{3.363}$$

The oil phase continuity equation is obtained according to the mass conservation law:

$$-\left\{\frac{\partial[(\rho_D - G_1)v_{0x}]}{\partial x} + \frac{\partial[(\rho_D - G_1)v_{0y}]}{\partial y} + \frac{\partial[(\rho_D - G_1)v_{0z}]}{\partial z}\right\} = \frac{\partial}{\partial t}[(\rho_D - G_1)\varphi S_o] \tag{3.364}$$

or

$$-\nabla \cdot [(\rho_D - G_1)\overrightarrow{v_o}] = \frac{\partial[(\rho_D - G_1)\varphi S_o]}{\partial t} \tag{3.365}$$

The gas phase continuity equation is derived as follows.

The gas phase consists of solution gas in the oil phase and free gas in the gas phase.

In the x direction, the velocity component of gas phase flow at the center point M is:

$$\rho_g v_{gx} + G_1 v_{0x} \tag{3.366}$$

Then, the mass difference of the gas phase flowing in and out with dt time in the x direction is:

$$-\frac{\partial(\rho_g v_{gx} + G_1 v_{0x})}{\partial x} dxdydzdt \tag{3.367}$$

Similarly, the mass difference in y and z directions can be obtained, and the mass difference in the inflow and outflow gas phases of the whole REV can be obtained as:

$$-\left\{\frac{\partial(\rho_g v_{gx} + G_1 v_{0x})}{\partial x} + \frac{\partial(\rho_g v_{gy} + G_1 v_{0y})}{\partial y} + \frac{\partial(\rho_g v_{gz} + G_1 v_{0z})}{\partial z}\right\} dxdydzdt \tag{3.368}$$

The change of gas phase mass of the REV in dt time is:

$$\frac{\partial}{\partial t}[\varphi(\rho_g S_g + G_1 S_o)]dxdydzdt \tag{3.369}$$

The gas phase continuity equation is obtained according to the mass conservation law:

$$-\left[\frac{\partial(\rho_g v_{gx} + G_1 v_{0x})}{\partial x} + \frac{\partial(\rho_g v_{gy} + G_1 v_{0y})}{\partial y} + \frac{\partial(\rho_g v_{gz} + G_1 v_{0z})}{\partial z}\right] = \frac{\partial[\varphi(\rho_g S_g + G_1 S_o)]}{\partial t} \tag{3.370}$$

or

$$-\nabla \cdot [(\rho_g \overrightarrow{v_g} + G_1 \overrightarrow{v_0})] = \frac{\partial\,[\varphi(\rho_g S_g + G_1 S_o)]}{\partial t} \tag{3.371}$$

(2) Equation of motion

Assuming that both of them conform to Darcy's law and that there is no capillary force action in the two phases of oil and gas, then the equation of motion can be obtained as follows.

Oil phase:

$$v_{ox} = -\frac{K_o}{\mu_o}\frac{\partial p}{\partial x}, \quad v_{oy} = -\frac{K_o}{\mu_o}\frac{\partial p}{\partial y}, \quad v_{oz} = -\frac{K_o}{\mu_o}\frac{\partial p}{\partial z} \tag{3.372}$$

Gas phase:

$$v_{gx} = -\frac{K_g}{\mu_g}\frac{\partial p}{\partial x}, \quad v_{gy} = -\frac{K_g}{\mu_g}\frac{\partial p}{\partial y}, \quad v_{gz} = -\frac{K_g}{\mu_g}\frac{\partial p}{\partial z} \tag{3.373}$$

(3) Equation of state

① Free gas

$$\rho_g = \frac{\rho_{gsc}}{B_g(p)} \tag{3.374}$$

② Solution gas

$$G_1 = \frac{R_s(p)\rho_{gsc}}{B_o(p)} \tag{3.375}$$

③ Oil

$$\rho_D = \frac{\rho_{osc} + R_s(p)\rho_{gsc}}{B_o(p)} \tag{3.376}$$

④ Rock

Elastic compressibility of rock is not considered.

(4) Basic differential equations

The equation of motion and the equation of state are substituted into the continuity equation to obtain the basic differential equation:

Oil phase:

$$\nabla\left[\frac{K_{ro}}{B_o(p)\mu_o(p)}\nabla p\right]=\frac{\phi}{K}\frac{\partial}{\partial t}\left[\frac{S_o}{B_o(p)}\right] \tag{3.377}$$

Gas phase:

$$\nabla\left[\frac{K_{rg}}{B_g(p)\mu_g(p)}\nabla p\right]+\nabla\left[\frac{R_s(p)K_{ro}}{B_o(p)\mu_o(p)}\nabla p\right]$$
$$=\frac{\phi}{K}\frac{\partial}{\partial t}\left[\frac{S_g}{B_g(p)}+\frac{R_s(p)}{B_o(p)}S_o\right] \tag{3.378}$$

3.7.3.3 *Steady state two-phase flow of oil and gas*

Similar to the gas-water two-phase flow, a pseudo-pressure function is introduced as:

$$H_o(p)=\int_0^p \frac{K_{ro}}{B_o(p)\mu_o(p)}dp \tag{3.379}$$

Let the oil well be produced by radial flow. If the flow is steady state, the oil phase equation becomes:

$$\nabla^2 H_o=0 \tag{3.380}$$

Write it in polar coordinate form:

$$\frac{d^2H_o}{dr^2}+\frac{1}{r}\frac{dH_o}{dr}=0 \tag{3.381}$$

Then, the pseudo-pressure distribution formula and flow rate formula can be obtained as:

$$H_o=H_{oe}-\frac{H_{oe}-H_{ow}}{\ln\frac{R_e}{R_w}}\ln\frac{R_e}{r} \tag{3.382}$$

$$Q_o=2\pi Kh\frac{H_{oe}-H_{ow}}{\ln\frac{R_e}{R_w}} \tag{3.383}$$

As can be seen from the above formula, the physical meaning of the pseudo-pressure function H_o is: when oil and gas flow from the reservoir to the bottom of the well at the same time, the actual total energy consumed is $\Delta p = p_e - p_w$, and the energy consumed to make oil flow is $\Delta H_o = H_{oe} - H_{ow}$.

The pseudo-pressure difference is:

$$H_{oe} - H_{ow} = \int_{p_w}^{p_e} \frac{K_{ro}}{\mu_o(p)B_o(p)} dp \tag{3.384}$$

In the formula, $\mu_g(p)$and $B_g(p)$ are both functions of pressure, which can be obtained through experiment or empirical equation curve. However, K_{ro} is a function of S_o in the oil-gas phase permeability curve, and S_o is a function of pressure. The relationship between K_{ro} and pressure needs to be obtained by means of GOR.

The gas-oil ratio of production is:

$$GOR = \frac{Q_{gsc}}{Q_{osc}} \tag{3.385}$$

Gas is composed of two parts, namely solution gas and free gas, and the formulae are:

$$Q_{gsc} = Q_g \frac{p}{p_{sc}} + Q_o C \tag{3.386}$$

$$Q_{osc} = \frac{Q_o}{B_o(p)} \tag{3.387}$$

By using Darcy's law:

$$Q_g = \frac{KK_{rg}}{\mu_g(p)} A \frac{dp}{dr} \tag{3.388}$$

$$Q_o = \frac{KK_{ro}}{\mu_o(p)} A \frac{dp}{dr} \tag{3.389}$$

Then:

$$GOR = \frac{K_{rg}\mu_o(p)}{K_{ro}\mu_g(p)} B_o(p) \frac{p}{p_{sc}} + R_s(p) \tag{3.390}$$

In the case of steady state flow, relative to an average formation pressure, the oil-gas ratio of production is a constant, and the relation curve of $GOR \sim p$ can be obtained. Then, the relation curve of $\frac{K_{rg}}{K_{ro}} \sim p$ can be obtained by combining the four curves of $\mu_g(p) \sim p$,

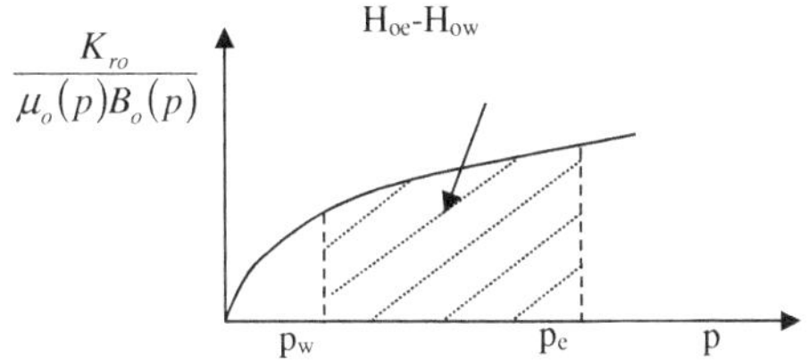

Figure 3.67. Pseudo-pressure difference of gas-water flow

$\mu_o(p) \sim p$, $B_o(p) \sim p$ and $R_s(p) \sim p$, and the relation of $K_{ro} \sim p$ can be obtained by means of the oil-gas phase permeability curve. Finally, the relation of $\frac{K_{ro}}{\mu_o(p)B_o(p)} \sim p$ can be obtained as a result.

As shown in Figure 3.67, if the shaded area is taken from the $\frac{K_{ro}}{\mu_o(p)B_o(p)} \sim p$ relation curve, the pseudo-pressure difference can be obtained. Or if the numerical integration method is applied and $\frac{K_{ro}}{\mu_o(p)B_o(p)}$ is obtained by interpolation method, the pseudo-pressure difference can still be obtained.

Sometimes, when the pressure is not too low, the relationship curve can be approximated to a straight line, that is:

$$\frac{K_{ro}}{\mu_o(p)B_o(p)} = a + bp \tag{3.391}$$

Then:

$$\begin{aligned} H_{oe} - H_{ow} &= \int_{p_w}^{p_e} \frac{K_{ro}}{\mu_o(p)B_o(p)} dp \\ &= \int_{p_w}^{p_e} (a + bp)dp \\ &= a(p_e - p_w) + \frac{b}{2}(p_e^2 - p_w^2) \end{aligned} \tag{3.392}$$

The pseudo-pressure difference can be calculated from the intercept and slope of the straight line.

Chapter 4

Flow in complex porous media

The flow characteristics in complex media are different from those in single porosity media as has been discussed in the classical flow mechanics of porous media. Complex media mainly include: the fracture-pore-matrix medium, the anisotropic medium, the deformable medium, the heterogeneous medium, the fractal medium and the horizontal well flow medium. Each part focuses on the flow characteristics, key parameters and factors considered in the establishment of the mathematical model.

4.1 Flow in dual porosity media

4.1.1 *Introduction*

The media in classical flow theory all refer to pure inter-granular pore media or pure fractured media.

The dual porosity medium can be often encountered in many reservoirs, such as natural fractured reservoirs and hydraulic fractured medium. In addition, with the development of unconventional resources, such as tight, shale and geothermal hot dry rocks, hydraulic fracturing has become a necessary means of developing the fractured formation, and the flow medium is dominated by fractures and pores.

There are two flow systems, namely, fracture system f and matrix system m. The permeability of fracture system (K_f) is much higher than that of rock matrix (K_m). Fracture system is the main flow channel. Due to the space volume of fractures in the system being far less than the matrix pore volume, the porosity of the rock matrix (ϕ_m) is obviously higher than that of fracture system porosity (ϕ_f), which are the main storage space for the reservoir fluid. The separation of flow storage capacity and fluid flow capacity is the basic characteristic of fracture-pore media.

For the convenience of research, the fractured porous dual porosity media reservoirs are abstractly simplified into different geological models. There are four typical models: Warren-Root model, Kazemi model, De Swaan model, and Fractal model.

(1) Warren-Root model

A dual porosity medium reservoir is simplified as a geological model with the matrix blocks cut into hexahedrons, as shown in Figure 4.1. It is assumed that the direction of fractures is consistent with the direction of main permeability and all the fractures have uniform aperture.

To further relax the assumption, the fracture networks can be treated as non-uniform distributed apertures. In this case, the anisotropy of fracture networks can be studied by using non-uniform fracture grids.

This is a commonly used dual porosity model and is discussed in this chapter as an example.

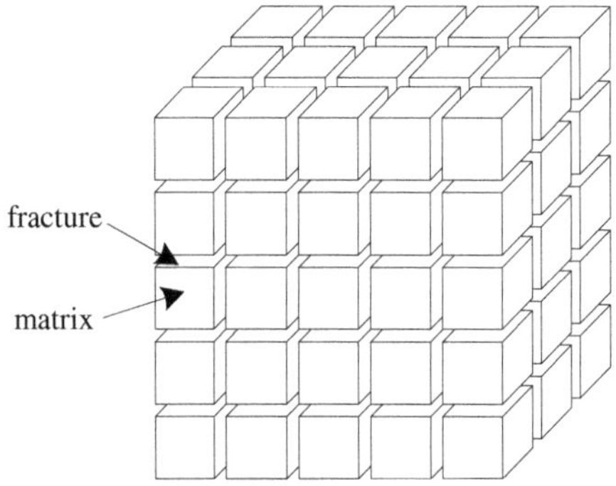

Figure 4.1. Warren-Root model

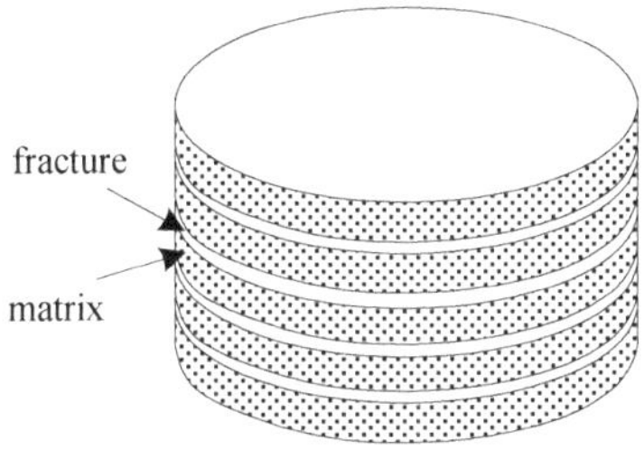

Figure 4.2. Kazemi model

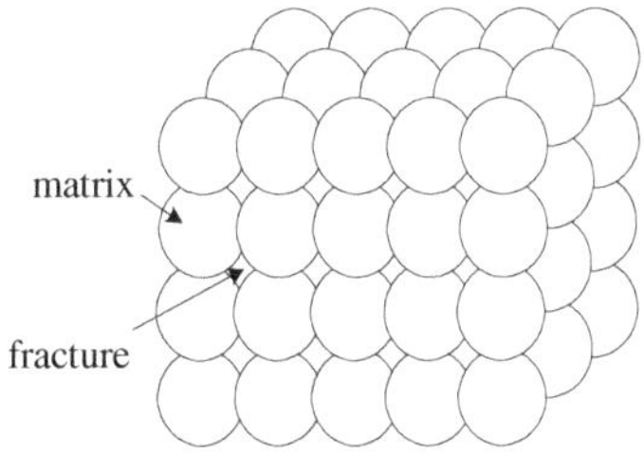

Figure 4.3. De Swaan model

(2) Kazemi model

This model simplifies an actual dual porosity reservoir to a set of parallel stratified fractures that divide the matrix blocks into layers, as shown in Figure 4.2. In other words, the model is composed of horizontal fractures and horizontal matrix layers.

(3) De Swaan model

The matrix block is changed from a parallel hexahedron into a sphere as shown in Figure 4.3. The matrix blocks distribution and arrangement are still consistent with those in the Warren-Root model. Fractures are represented by pores between spheres and matrix blocks are represented by spheres.

(4) Fractal model

The fractal model shown in Figure 4.4 is a model with a certain similarity between the whole object and the individual parts. The fractal

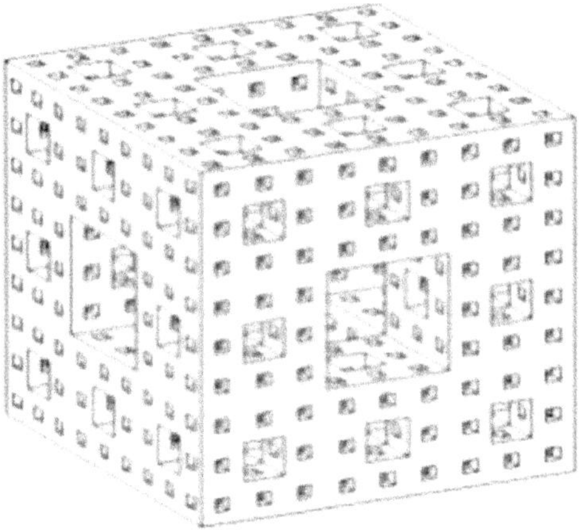

Figure 4.4. Fractal model

model of fractured reservoir considers that fracture distribution and pore structure of matrix belong to fractal system. Fractal dimension varies with reservoir heterogeneity.

This model is often used to simulate natural fracture distribution.

4.1.2 *Concepts*

Interporosity flow: *if there are two flow fields, there will be fluid exchange between them. This flow is called the interporosity flow.*

Elastic storativity ratio: *it refers to the ratio of elastic storage capacity of fracture system to total elastic storage capacity of reservoir. The symbol is ω, and the equation is:*

$$\omega = \frac{\phi_f C_f}{\phi_f C_f + \phi_m C_m} \tag{4.1}$$

where ω is the elastic storage capacity ratio; ϕ_f is the fracture porosity; C_f is the elastic compressibility of fracture; ϕ_m is the matrix porosity; and C_m is the elastic compressibility of matrix.

Interporosity flow coefficient: *the interporosity flow coefficient is used to describe the physical quantity of fluid exchange between different media. It reflects the ability of fluid in the matrix to flow to* *fractures. Interflow coefficient is defined as:*

$$\lambda = \alpha \frac{k_m}{k_f} R_w^2 \tag{4.2}$$

where λ *is the interporosity flow coefficient;* α *is the shape factor;* k_m *is the matrix permeability;* k_f *is the fracture permeability; and* R_w *is the wellbore radius.*

Shape factor: *it is related to the size of the matrix block and the number of orthogonal fracture groups. The smaller the block, the greater the fracture density and shape factor. The symbol is* α*, and its formula is:*

$$\alpha = \frac{4n(n+2)}{L^2} \tag{4.3}$$

where α *is shape factor;* n *is the number of orthogonal fracture groups; and* L *is the characteristic length of a rock block.*

Characteristic length: *it is a representative length that defines the scale of the object. For example, the characteristic length of a sphere is the radius or diameter of the sphere.*

4.1.3 *Analysis*

4.1.3.1 *Governing equation for dual porosity model*

In the Warren-Root model, the flow conforms to Darcy's law. Fractures and matrix are evenly distributed and orthogonal to each other. Fluid flows from matrix pores to fractures and then into the bottom of the well.

Two flow systems are formed by the dual porosity. The subscript f represents the fracture system and the subscript m represents the matrix system.

(1) Equation of motion

The two systems satisfy their equations of motion respectively.

Fracture:

$$\vec{v}_f = -\frac{k_f}{\mu} gradp_f \tag{4.4}$$

Matrix:

$$\vec{v}_m = -\frac{k_f}{\mu} gradp_m \tag{4.5}$$

(2) Interporosity flow equation

There is a pressure distribution between the matrix and the fracture, and fluid exchange occurs, but this fluid exchange occurs slowly and can be regarded as a steady state process. Then, the equation of interporosity flow is:

$$q = \frac{\alpha \rho_0 k_m}{\mu}(p_m - p_f) \tag{4.6}$$

where q is the mass of fluid flowing out of rock per unit volume per unit time; p_m is the matrix pressure; and p_f is the fracture pressure.

(3) Equation of state

The porous media, fracture media and formation fluid are considered as being slightly compressible.

The equation of state for fracture media is:

$$\phi_f = \phi_{f0} + C_{\phi_f}(p_f - p_0) \tag{4.7}$$

The equation of state for porous media is:

$$\phi_m = \phi_{m0} + C_{\phi_m}(p_m - p_0) \tag{4.8}$$

The equation of state for formation fluid is:

$$\rho = \rho_0 e^{C_L(p-p_0)} \tag{4.9}$$

(4) Continuity equation

From the mass conservation law, the continuity equation is obtained as:

Fracture:

$$\frac{\partial}{\partial t}(\rho\phi_f) + \operatorname{div}(\rho\vec{v}_f) - q = 0 \tag{4.10}$$

Matrix:

$$\frac{\partial}{\partial t}(\rho\phi_m) + \operatorname{div}(\rho\vec{v}_m) - q = 0 \tag{4.11}$$

(5) Establishment of the governing equations

The equation of motion, the equation of state and the equation of interporosity flow are substituted into the continuity equation.

Fracture system:

$$\phi_f C_f \frac{\partial p_f}{\partial t} - \frac{K_f}{\mu}\operatorname{div}(\operatorname{grad} p_f) - \frac{\alpha K_m}{\mu}(p_m - p_f) = 0 \tag{4.12}$$

Matrix system:

$$\phi_m C_m \frac{\partial p_m}{\partial t} - \frac{K_m}{\mu}\operatorname{div}(\operatorname{grad} p_m) + \frac{\alpha K_m}{\mu}(p_m - p_f) = 0 \tag{4.13}$$

In this model, $K_m \neq 0$ and $\varphi_f \neq 0$, that is, the media have both the storage capacity flowing capacity of fluid, so it is called the dual permeability and dual porosity model.

4.1.3.2 *Governing equation for single porosity and single permeability model*

Since $K_m \ll K_f$ and $\varphi_f \ll \varphi_m$, they can be set as $K_m = 0$, $\varphi_f = 0$, and the governing equation can be simplified as:

Fracture system:

$$\frac{K_f}{\mu}\operatorname{div}(\operatorname{grad} p_f) + \frac{\alpha K_m}{\mu}(p_m - p_f) = 0 \tag{4.14}$$

Matrix system:

$$\phi_m C_m \frac{\partial p_m}{\partial t} + \frac{\alpha K_m}{\mu}(p_m - p_f) = 0 \tag{4.15}$$

The above equation only considers the matrix storability and fracture flow capability.

When it is the flow at constant production rate in the infinite reservoir, the bottom hole pressure can be obtained as:

$$p_w(t) = p_i - \frac{Q\mu}{4\pi Kh}\left[\ln\frac{\eta_f t}{R_w^2} + Ei(-bt) - Ei(-b\omega t) + 0.809\right] \quad (4.16)$$

where $\eta_f = \frac{K_f}{\mu(C_t)_{f+m}}$, $\omega = \frac{(\varphi C_t)_f}{(\varphi C_t)_{f+m}}$, $b = \frac{\lambda\eta_f}{R_w^2\omega(1-\omega)}$ and $\lambda = \frac{\alpha R_w^2}{K_f}$. This formula can be used for well test of the transient flow in fractured and porous media.

4.1.3.3 *Description of fracture system*

In the dual porosity model of fractures and pores, it is often difficult to get accurate results on the description of fractures. Generally, the distribution of fractures is estimated and predicted, and then the distribution characteristics of fractures are corrected through late productivity data.

(1) Natural fracture prediction method

There are many methods to predict natural fractures, including outcrop, core, earthquake, well logging and well testing, etc. The factors affecting natural fractures mainly include lithology, structure, thickness and stress, etc.

For example, the empirical formula for calculating the spacing of natural fractures in an oil field is:

$$D_0 = 0.02h + \frac{H}{1000B_{RIT-T}} \quad (4.17)$$

where D_0 is fracture spacing, m; h is sandstone thickness, m; B_{RIT-T} is brittleness coefficient; and H is the formation depth, m.

The calculation method of brittleness coefficient is:

$$B_{RIT-T} = \frac{V_{qa} + V_{ca} + V_{do}}{V_{qa} + V_{ca} + V_{do} + V_{cl}} \times 100\% \quad (4.18)$$

where V_{qa} is the percentage of quartz volume; V_{ca} is the percentage of calcite volume; V_{do} is the percentage of dolomite volume; and V_{cl} is the percentage of clay.

The fracture development of each oil field is different, and the empirical formula may be different.

If fracture spacing is known, fracture aperture and fracture density can be calculated:

$$b = 2 \ln D_0 + 5 \tag{4.19}$$

$$D_{lf} = \frac{1}{D_0} \tag{4.20}$$

where b is fracture aperture, μm; and D_{lf} is the fracture line density, 1/m.

Thus, fracture permeability and fracture conductivity can be calculated as:

$$K_f = 0.0837 b^2 \tag{4.21}$$

$$E_f = 10^{-4} K_f b \tag{4.22}$$

where K_f is fracture permeability, D; and E_f is fracture conductivity, D · cm.

If the fracture length L_f is known, the porosity of the fracture can be calculated approximately, and the formula is:

$$\varphi_f = \frac{b D_{lf}}{L_f^2}$$

(2) Fractal description of fractures

According to the fracture branching characteristics, there are several laws.

Law of quantity:

$$N_u = R_B^{(S-u)} \tag{4.23}$$

$$R_B = \frac{N_u}{N_{u-1}} \tag{4.24}$$

where N_u is the number of the current u^{th} level fracture branching; R_B is the bi-branching ratio of u and $u-1$ levels; S is the highest branching level of fracture; and u is the current branching level of the fracture.

Aperture law:

$$W_u = R_u^{(u-1)} W_1 \tag{4.25}$$

$$R_u = \frac{W_u}{W_{u-1}} \tag{4.26}$$

where W_u is the fracture aperture of level u; R_u is the aperture ratio of u and $u-1$ levels; and W_1 is the first level fracture aperture.

Area law:

$$A_u = R_a^{(u-1)} A_1 \tag{4.27}$$

$$R_a = \frac{A_u}{A_{u-1}} \tag{4.28}$$

where A_u is the control area of level u fractures; R_a is the area ratio; and A_1 is the control area of the first level fractures.

Fractal dimension of fracture network:

$$D_s = 2\frac{\ln R_L}{\ln R_B} \tag{4.29}$$

where R_L is the length ratio, $R_L = \frac{L_u}{L_{u-1}}$.

The relationship between fracture density, fracture length and fractal dimension is:

$$D_{lf} = L_f^{-D_s} \tag{4.30}$$

This formula can also be used to calculate the fractal dimension.

4.1.4 *Example*

1. Given that the fracture spacing is 0.3 m and fractal dimension is 1.68, please find the relevant fracture parameters.

Solution:

The fracture aperture is $b = 2\ln D_0 + 5 = 2 \times \ln 0.3 + 5 = 2.59\,\mu\text{m}$.

The fracture density is $D_{lf} = \frac{1}{D_0} = \frac{1}{0.3} = 3.33(l/m)$.

The fracture permeability is $K_f = 0.0837b^2 = 0.0837 \times 2.59^2 = 0.56D$.

The fracture conductivity is

$$E_f = 10^{-4} K_f b = 10^{-4} \times 0.56 \times 2.59 = 0.000146 D \cdot \text{cm}.$$

The fracture length is

$$L_f = D_{lf}^{-\frac{1}{D_s}} = 3.33^{-\frac{1}{1.68}} = 0.5\,\text{m}.$$

The length of the fracture is

$$L_f = D_{lf}^{-\frac{1}{D_s}} = 3.33^{-\frac{1}{1.68}} = 0.5\,\text{m}.$$

The fracture porosity is

$$\varphi_f = \frac{bD_{lf}}{L_f^2} = \frac{2.59 \times 10^{-6} \times 3.33}{0.5^2} = 3.45 \times 10^{-5}.$$

Natural fractures have small conductivity and porosity.

2. Please find the permeability when the pore diameter and fracture aperture are both 0.1 μm, 1 μm and 10 μm.

Solution:

When the pore radius of the matrix is 0.1 μm, 1μm and 10 μm, the corresponding porosity is 0.05, 0.1 and 0.2, respectively.

According to $K = 30.625\varphi r^2$ (K unit mD, r unit μm), we can get:

$$K(r = 0.1\,\mu\text{m}) = 0.015\,\text{mD}$$

$$K(r = 1\,\mu\text{m}) = 3.06\,\text{mD}$$

$$K(r = 10\,\mu\text{m}) = 612\,\text{mD}$$

In the fracture system, according to $K_f = 83.7b^2$ (K unit mD, b unit μm), the following equation can be obtained:

$$K_f(r = 0.1\mu\text{m}) = 0.837\,\text{mD}$$

$$K_f(r = 1\mu\text{m}) = 83.7\,\text{mD}$$

$$K_f(r = 10\mu\text{m}) = 8370\,\text{mD}$$

It can be seen that for the same pore size, the permeability of the fracture is much larger than that of the pore.

4.2 Flow in anisotropic media

4.2.1 *Introduction*

The reservoir parameters are anisotropic due to different sedimentary conditions and internal and external stress. There are two types of anisotropic reservoirs. One type is formed by sedimentation, which is called sedimentary anisotropic reservoir. The other type is caused by fracture action, called fracture anisotropic reservoir.

Anisotropic reservoir is a reservoir with directional permeability. The direction of such permeability is as follows: at the same point in the formation, the permeability of fluid flowing in one direction is larger than that in other directions, while the permeability is minimized in the direction perpendicular to this direction, as shown in Figure 4.5. Directional permeability is called anisotropic permeability. Therefore, anisotropic reservoir is also called anisotropic permeability reservoir. This section mainly considers reservoir permeability anisotropy.

4.2.2 *Concepts*

Anisotropy: *the properties of a medium in different directions at the same point are different.*

Homogeneous isotropic reservoir: *it means that the reservoir permeability is constant, independent of spatial coordinates and direction.*

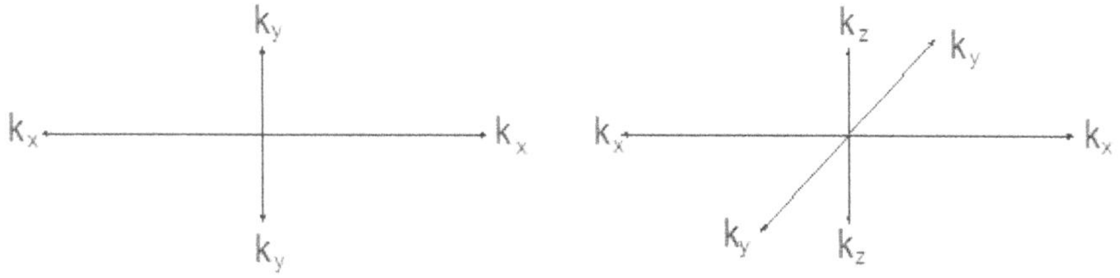

Figure 4.5. Schematic diagram of anisotropic permeability

Homogeneous anisotropic reservoir: *permeability is not a function of spatial coordinates, but a function of direction.*

Heterogeneous isotropic reservoir: *it means that the reservoir permeability is a function of spatial coordinates, but independent of the direction of each point.*

Heterogeneous anisotropic reservoir: *it means that the permeability is a function of both spatial coordinates and the direction of each point.*

Primary permeability direction: *it usually refers to the direction with the largest permeability.*

Permeability tensor K: *generally, it is a diagonal matrix which contains the complete properties of reservoir permeability and determines the characteristics of reservoir flow. Its expression is:*

$$K = \begin{bmatrix} K_x & 0 & 0 \\ 0 & K_y & 0 \\ 0 & 0 & K_z \end{bmatrix} \tag{4.31}$$

4.2.3 *Analysis*

4.2.3.1 *Basic flow characteristics in anisotropic reservoir*

In isotropic reservoirs, the size and direction of the flow velocity are determined by the size and direction of the pressure gradient, the direction of the flow velocity is consistent with the direction of the pressure gradient, and the size of the flow velocity is independent of the direction of the pressure gradient.

In anisotropic reservoirs, the direction of flow velocity is determined by the direction of pressure gradient, but the two are generally inconsistent. When the direction of pressure gradient is on the main permeability axis, the direction of flow velocity is consistent with that of pressure gradient. The magnitude of flow velocity is determined by the magnitude and direction of the pressure gradient, that is, whether the magnitude or the direction of the pressure gradient changes, the magnitude of flow velocity will change.

4.2.3.2 *Governing equation*

Let the permeability tensor take its diagonal form in the rectangular coordinate system, and write it as:

$$K = \begin{bmatrix} K_x & 0 & 0 \\ 0 & K_y & 0 \\ 0 & 0 & K_z \end{bmatrix} \tag{4.32}$$

(1) Steady state flow of single-phase incompressible liquid in homogeneous anisotropic media

Equation of motion is:

$$v = -\frac{1}{\mu} K grad p \tag{4.33}$$

Continuity equation is:

$$\text{div}\,(v) = 0 \tag{4.34}$$

Since it's an incompressible liquid, you don't have to worry about state changes. By substituting the equation of motion into the equation of continuity, the following equation can be obtained:

$$K_x \frac{\partial^2 p}{\partial x^2} + K_y \frac{\partial^2 p}{\partial y^2} + K_z \frac{\partial^2 p}{\partial z^2} = 0 \tag{4.35}$$

The above equation is the governing equation of steady state flow of single-phase incompressible liquid in homogeneous anisotropic reservoirs.

The conditions to derive this governing equation are: the flow medium is anisotropic and homogeneous, the liquid is single phase, Darcy's law is satisfied, the compressibility of porous media and liquids is not considered, and the flow process is under isothermal steady state flow regime.

(2) Elastic transient flow of single-phase slightly compressible liquid in anisotropic media

The equation of motion is the same as shown for steady state flow.

The equations of state are:

$$\varphi = \varphi_0 + C_f(p - p_0) \tag{4.36}$$

$$\rho = \rho_0[1 + C_L(p - p_0)] \tag{4.37}$$

Continuity equation is:

$$\frac{\partial(\varphi\rho)}{\partial t} + \mathrm{div}(\rho v) = 0 \tag{4.38}$$

The governing equation can be derived as:

$$K_x\frac{\partial^2 p}{\partial x^2} + K_y\frac{\partial^2 p}{\partial y^2} + K_z\frac{\partial^2 p}{\partial z^2} = \frac{1}{\eta}\frac{\partial p}{\partial t} \tag{4.39}$$

where η is the pseudo conductivity coefficient, $\eta = \frac{1}{\mu C_t}$.

The above equation is the governing equation of elastic transient flow of single-phase slightly compressible liquid in anisotropic media.

4.2.3.3 *Anisotropic and isotropic conversion*

In the governing equation established above, it is difficult to directly solve it because the permeability principal values are not equal. The anisotropic space can be changed into equivalent isotropic space by coordinate transformation.

In addition to the cartesian coordinate system (x, y, z), a new cartesian coordinate system (x_1, y_1, z_1) is introduced, and the permeability in the new coordinate system is K, and the transformation relationship between the two is:

$$x_1 = x\sqrt{\frac{K}{K_x}}, \quad y_1 = y\sqrt{\frac{K}{K_y}}, \quad z_1 = z\sqrt{\frac{K}{K_z}} \tag{4.40}$$

$$K = (K_xK_yK_z)^{1/3}, \quad \beta_1 = \sqrt{\frac{K_y}{K_z}}, \quad \beta_2 = \sqrt{\frac{K_z}{K_x}},$$

$$\beta_3 = \sqrt{\frac{K_x}{K_y}} \tag{4.41}$$

Then, the governing equations of steady state flow and transient flow of single-phase liquid in the new coordinate system are:

$$\frac{\partial^2 p}{\partial x_1^2} + \frac{\partial^2 p}{\partial y_1^2} + \frac{\partial^2 p}{\partial z_1^2} = 0 \tag{4.42}$$

$$\frac{\partial^2 p}{\partial x_1^2} + \frac{\partial^2 p}{\partial y_1^2} + \frac{\partial^2 p}{\partial z_1^2} = \frac{K}{\mu C_t}\frac{\partial p}{\partial t} \tag{4.43}$$

In the old coordinate system, the angles between the length of a straight line and the axes x, y and z are α, θ and γ, respectively. In the new coordinate system, there is:

$$l_1 = l\sqrt{K}\sqrt{\frac{\cos^2\alpha}{K_x} + \frac{\cos^2\theta}{K_y} + \frac{\cos^2\gamma}{K_z}} \tag{4.44}$$

$$\alpha_1 = arctg\left(\frac{\sqrt{\beta_3^2\cos^2\theta + \frac{\cos^2\gamma}{\beta_2^2}}}{\cos\alpha}\right) \tag{4.45}$$

$$\theta_1 = arctg\left(\frac{\sqrt{\beta_1^2\cos^2\gamma + \frac{\cos^2\alpha}{\beta_3^2}}}{\cos\theta}\right) \tag{4.46}$$

$$\gamma_1 = arctg\left(\frac{\sqrt{\beta_2^2\cos^2\alpha + \frac{\cos^2\theta}{\beta_1^2}}}{\cos\gamma}\right) \tag{4.47}$$

If it is a two-dimensional case, then:

$$K = \sqrt{K_x K_y}, \quad \beta = \sqrt{\frac{K_x}{K_y}} \tag{4.48}$$

$$x_1 = \frac{x}{\sqrt{\beta}}, \quad y_1 = y\sqrt{\beta} \tag{4.49}$$

$$l_1 = l\sqrt{\frac{\cos^2\alpha}{\beta} + \beta\sin^2\alpha} \tag{4.50}$$

$$\alpha_1 = arctg(\beta tg\alpha) \tag{4.51}$$

$$\theta_1 = 90 - \alpha_1 \tag{4.52}$$

After coordinate transformation, the flow equations can be changed into equations discussed before in the isotropic cases.

4.2.3.4 *Equations of production rate*

For the single well in isotropic reservoirs, the production rate is:

$$Q = \frac{2\pi Kh(P_e - P_w)}{\mu \ln \frac{R_e}{R_w}} \tag{4.53}$$

For the single well in anisotropic reservoirs, the production rate is:

$$Q = \frac{2\pi Kh(P_e - P_w)}{\mu \ln \frac{R'_e}{R'_w}} \tag{4.54}$$

In the formula, K, R'_e and R'_w are calculated according to the coordinate transformation formula.

For the two point sinks with equal production rate in the isotropic reservoir, the production rate is:

$$Q = \frac{2\pi Kh(P_e - P_w)}{\mu \ln \frac{R_e^2}{2aR_w}} \tag{4.55}$$

Anisotropic production rate equation is:

$$Q = \frac{2\pi Kh(P_e - P_w)}{\mu \ln \frac{R_e'^2}{2a'R'_w}} \tag{4.56}$$

where K, R'_e, R'_w and a' are calculated according to the coordinate transformation equation.

For the one point source and one point sink with equal production rate in the isotropic reservoir, the production rate is:

$$Q = \frac{2\pi Kh(P_e - P_w)}{\mu \ln \frac{2a}{R_w}} \tag{4.57}$$

For the one point source and one point sink with equal production rate in the anisotropic reservoir, the production rate is:

$$Q = \frac{2\pi Kh(P_e - P_w)}{\mu \ln \frac{2a'}{R'_w}} \tag{4.58}$$

Other pressure equations and transient flow equations can be converted from anisotropic permeability problem to isotropic permeability problem by changing coordinate form, and then they can be solved by applying the basic equations we have learned.

4.3 Flow in deformed media

4.3.1 *Introduction*

Reservoir rock has a small compressibility; in the classical theory of flow mechanics in porous media, it is assumed to be slightly compressible. In actual oil fields, especially low permeability or high pressure reservoir, the decrease in reservoir pressure not only reduces the pore volume, but also significantly reduces the permeability. Sometimes, they can be reduced by 10 times. In the late stage of reservoir development, the effect of permeability change cannot be ignored. In this case, the deformation of the reservoir rock cannot be ignored.

Permeability decreases with decrease in pressure, and this phenomenon is often called stress sensitive effect. A medium with obvious stress sensitivity is the deformed medium. A reservoir with this property is also called variable permeability formation.

4.3.2 *Concepts*

Deformable medium: *the rock deforms with decrease in formation pressure, and the pore space and permeability decrease. The porous medium with deformation in the pore space is called deformable medium.*

Stress sensitivity: *the pore volume (porosity) and permeability in porous media change with the effective stress.*

Permeability deformation coefficient: *the ratio of the decreased permeability to the initial permeability caused by unit pressure difference, and the equation is:*

$$\alpha_K = \frac{1}{K_0}\frac{dK}{dp} \tag{4.59}$$

where α_K *is permeability deformation coefficient,* MPa^{-1}*, whose value is on the order of* 10^{-1} *and* 10^{-2}*;* K_0 *is rock permeability at initial state, mD.*

4.3.3 *Analysis*

4.3.3.1 *Flow characteristics in deformed media*

Assuming that the fluid flow in the deformed medium conforms to Darcy's law, the equation is:

$$v = -\frac{K(p)}{\mu}\frac{dp}{dx} \tag{4.60}$$

When the pressure change in the reservoir is relatively small, the equation of state can be written as:

$$K(p) = K_0[1 - \alpha_K(p_0 - p)] \tag{4.61}$$

Substitute Equation (4.61) into Equation (4.60), and the linear motion equation of the deformed medium can be written as:

$$v = -\frac{K_0}{\mu}[1 - \alpha_K(p_0 - p)]\frac{dp}{dx} \tag{4.62}$$

If a new function U is introduced and we set $U = [1-\alpha_K(p_0-p)]$, then the equation can be written as:

$$v = -\frac{K_0}{2\alpha_k\mu}\frac{dU^2}{dx} \tag{4.63}$$

Let $u_1 = \frac{U^2}{2\alpha_K}$, and it can be written in exactly the same form as Darcy's law:

$$v = -\frac{K_o}{\mu}\frac{du_1}{dx} \tag{4.64}$$

Suppose the viscosity μ and density ρ of fluid are also functions of pressure, and they are linearly proportional to the pressure change:

$$\mu = \mu_0[1 - \alpha_\mu(p_0 - p)] \tag{4.65}$$

$$\rho = \rho_0[1 - \alpha_\rho(p_0 - p)] \tag{4.66}$$

Combining with (4.64), (4.65) and (4.66), the mass flow velocity can be expressed as:

$$v_\rho = \rho v = \frac{-K_0[1 - \alpha_K(p_0 - p)]\rho_0[1 - \alpha_\rho(p_0 - p)]}{\mu_0[1 - \alpha_\mu(p_0 - p)]}\frac{dp}{dx} \tag{4.67}$$

where α_p is the density change coefficient, $\alpha_\rho = \frac{1}{\rho_0}\frac{d\rho}{dp}$; and α_μ is the viscosity change coefficient, $\alpha_\mu = \frac{1}{\mu_0}\frac{d\mu}{dp}$.

When the pressure drops greatly in the reservoir, the change of $K(p)$ with the pressure conforms to the negative exponential attenuation equation:

$$K(p) = K_0 \exp[-\alpha_K(p_0 - p)] \tag{4.68}$$

If the change of the density and viscosity with pressure both conform to the negative exponential attenuation equation, the motion equation can be expressed as:

$$v_\rho = -\frac{K_0\rho_0}{\mu_0}\exp[-\alpha(p_0 - p)]\frac{dp}{dx} \tag{4.69}$$

The coefficient α is defined as:

$$\alpha = \alpha_K + \alpha_\rho - \alpha_\mu \tag{4.70}$$

If a new variable u_2 is introduced and we let $u_2 = \frac{1}{2}\exp[-\alpha(p_0 - p)]$, then the flow motion equation of the deformed medium is written as:

$$v_\rho = -\frac{K_0\rho_0}{\mu_0}\frac{d\mu_2}{dx} \tag{4.71}$$

Thus, the flow equation of motion in the deformed media is completely similar to the form of Darcy's law. The only difference is replacing the volume velocity v with the mass velocity v_ρ, and replacing pressure p with u_1 and u_2.

4.3.3.2 *Steady state flow in deformed media*

Suppose there is a well in the deformed medium, liquid is single-phase and incompressible, and the flow regime is steady state radial flow, then its continuity equation can be expressed as:

$$\frac{\partial v_{\rho x}}{\partial x}+\frac{\partial v_{\rho y}}{\partial y}=0 \tag{4.72}$$

where $v_{\rho x}$ is the mass flow velocity in the x direction; and $v_{\rho y}$ is the mass flow velocity in the y direction.

Since it is an incompressible liquid, the density and viscosity of the liquid are both constants, so the equation of motion is substituted into the above equation to obtain:

$$\frac{\partial}{\partial x}\left\{[1-\alpha_K(p_0-p)]\frac{\partial p}{\partial x}\right\}+\frac{\partial}{\partial y}\left\{[1-\alpha_K(p_0-p)]\frac{\partial p}{\partial y}\right\}=0 \tag{4.73}$$

Introduce a new variable u_1:

$$u_1=\frac{1}{2\alpha}[1-\alpha_K(p_0-p)]^2 \tag{4.74}$$

It can be written as:

$$\frac{\partial^2 u_1}{\partial x^2}+\frac{\partial^2 u_1}{\partial y^2}=0 \tag{4.75}$$

For radial flow, polar coordinates are introduced. According to $r=\sqrt{x^2+y^2}$, it can be written as:

$$\frac{\partial^2 u_1}{\partial r^2}+\frac{1}{r}\frac{\partial u_1}{\partial r}=0 \tag{4.76}$$

This is the Laplace equation, and according to Darcy's law, we can get:

$$Q=\frac{2\pi r h K}{\mu}\frac{du_1}{dr} \tag{4.77}$$

Then:

$$\int_{u_{1w}}^{u_{1e}} K du_1=\int_{R_w}^{R_e}\frac{Q\mu}{2\pi h}\frac{1}{r}dr \tag{4.78}$$

By substituting the formula (4.60) of permeability changing with pressure into the above equation, it can be obtained that:

$$\int_{u_{1w}}^{u_{1e}} K_0[1-\alpha_K(p_o-p)]du_1 = \int_{R_w}^{R_e} \frac{Q\mu}{2\pi h}\frac{1}{r}dr \tag{4.79}$$

After calculation, the production rate can be expressed as:

$$Q = \frac{2\pi K_0 h}{\mu \ln \frac{R_e}{R_w}}\left[(p_e-p_w) - \frac{\alpha_K}{2}(p_e-p_w)^2\right] \tag{4.80}$$

When considering the negative exponential decay of permeability with the change of pressure, Equation (4.73) is converted into:

$$\frac{\partial}{\partial x}\left\{\exp[-\alpha(p_0-p)]\frac{\partial p}{\partial x}\right\} + \frac{\partial}{\partial y}\left\{\exp[-\alpha(p_0-p)]\frac{\partial p}{\partial x}\right\} = 0 \tag{4.81}$$

Introduce a new variable:

$$u_2 = \frac{1}{\alpha}\exp[-\alpha(p_0-p)] \tag{4.82}$$

Then:

$$\frac{\partial^2 u_2}{\partial x^2} + \frac{\partial^2 u_2}{\partial y^2} = 0 \tag{4.83}$$

It can be rewritten in polar coordinates:

$$\frac{\partial^2 u_2}{\partial r^2} + \frac{1}{r}\frac{\partial u_2}{\partial r} = 0 \tag{4.84}$$

After solving Equation (4.84), the radial flow production rate formula in the deformed medium can be obtained as:

$$Q = \frac{2\pi K_0 h}{\mu}\frac{1-\exp[-\alpha(p_e-p_w)]}{\alpha \ln \frac{R_e}{R_w}} \tag{4.85}$$

The continuity equation of transient flow can usually be expressed as:

$$\frac{\partial(\phi\rho)}{\partial t} = \frac{\partial}{\partial x_i}\left[\frac{K\rho}{\mu}\frac{\partial P}{\partial x_i}\right] \tag{4.86}$$

where $i = 1, 2, 3$ represent the x, y and z coordinate directions respectively.

For the deformed medium, the equations of state of all the parameters in the above equation are:

$$K(p) = K_0 \exp[\alpha_k(p - p_0)], \quad \alpha_k = \frac{1}{K_0}\frac{dK}{dp} \tag{4.87}$$

$$\phi(p) = \phi_0 \exp[\alpha_\phi(p - p_0)], \quad \alpha_\phi = \frac{1}{\phi_0}\frac{d\phi}{dp} \tag{4.88}$$

$$\rho(p) = \rho_0 \exp[\alpha_\rho(p - p_0)], \quad \alpha_\rho = \frac{1}{\rho_0}\frac{d\rho}{dp} \tag{4.89}$$

$$\mu(p) = \mu_0 \exp[\alpha_\mu(p - p_0)], \quad \alpha_\mu = \frac{1}{\mu_0}\frac{d\mu}{dp} \tag{4.90}$$

And:

$$\frac{\partial}{\partial t}\exp[\beta(p - p_0)] = \frac{K_0}{\mu_0\phi_0}\frac{\partial}{\partial x_i}\left\{\exp[\alpha(p - p_0)]\frac{\partial p}{\partial x_i}\right\} \tag{4.91}$$

$$\alpha = \alpha_K + \alpha_\rho - \alpha_\mu, \quad \beta = \alpha_\phi + \alpha_\rho \tag{4.92}$$

It can be rewritten as:

$$\frac{1}{\eta}\frac{\partial\varphi}{\partial t} = \nabla^2(\varphi) \tag{4.93}$$

where $\varphi = \exp[\beta(p - p_0)]$; $\eta = \gamma D^2 = \frac{K_0}{\mu_0\phi_0\beta}$; $D^2 = \frac{K_0}{\pi_0\phi_0\alpha}$; and $\gamma = \frac{\alpha}{\beta}$.

The equation has the same form as the governing equation for elastic transient flow of single-phase liquid.

If the well is set to produce at a constant mass flow rate G, the pressure distribution and bottom hole pressure can be obtained as:

$$p(r, t) = p_i + \frac{1}{\alpha}\ln\left[1 + \frac{\alpha\lambda}{2}Ei\left(-\frac{r^2}{4\eta t}\right)\right] \tag{4.94}$$

$$e^{-\alpha[p_i - p_w(t)]} - 1 = \frac{\alpha}{2}\lambda\ln\frac{2.25\eta t}{R_w^2} \tag{4.95}$$

where $\lambda = \frac{G\mu_0}{2\pi K_0\rho_0 h}$.

4.4 Flow in heterogeneous reservoirs

4.4.1 *Introduction*

The mathematical model of flow mechanics in porous media is established and solved by analytical methods. In order to solve the equation analytically, reservoirs are regarded as homogeneous, that is, the distribution of permeability, porosity, thickness and so on is the same at any point in a reservoir. This is the ideal model.

In actual oil fields, reservoir physical properties cannot be equal everywhere, and the problems described by the ideal flow model are different from the actual situation. The effect of reservoir heterogeneity cannot be ignored.

The most effective method to resolve the heterogeneity issue is the numerical solution of flow model, namely reservoir numerical simulation. The probability and statistical method can also be used to solve the flow problem analytically.

In the heterogeneous reservoir, the major factor considered to be heterogeneous is permeability. It is a function of coordinates, that is, $K = K\ (x,\ y)$. According to the probability distribution of K, the best method that can represent the flow characteristics of the real reservoir is to consider the issue of heterogeneity.

4.4.2 *Concepts*

Reservoir heterogeneity: *the spatial distribution of the reservoir physical characteristics is not uniform. Generally, it has two types of heterogeneity: heterogeneity within the same layer and heterogeneity in different layers. The key heterogeneous parameter is permeability.*

Permeability ratio: *It is the ratio of maximum to minimum permeability, indicating the range and degree of permeability difference.* $J_k > 2$ *is heterogeneous. Its equation is:*

$$J_k = \frac{K_{\max}}{K_{\min}} \tag{4.96}$$

where J_k *is the permeability level difference;* K_{max} *is the maximum permeability; and* K_{min} *is the minimum permeability.*

Permeability variation coefficient: *it is the ratio of standard deviation to mean value of the permeability sample.* $V_k > 0.5$ *can be considered as being heterogeneous, and the equation is:*

$$V_k = \frac{\sqrt{\frac{1}{n}\sum_{i=1}^{n}(K_i - \overline{K})^2}}{\overline{K}} \tag{4.97}$$

where V_k *is the permeability variation coefficient;* n *is the number of cores or samples;* K_i *is the permeability of the* i^{th} *sample; and* $\overline{K}$ *is the mean value of permeability.*

Permeability homogeneity coefficient: *it is the ratio of average to maximum permeability,* $T_p = \frac{\overline{K}}{K_{\max}}$. *When* $T_p < 0.5$, *it is regarded as being heterogeneous.*

Expected value: *in probability theory and statistics, expected value is the probability of each possible outcome in an experiment with a discrete random variable, multiplied by the sum of its outcomes. The symbol is E.*

Variance: *a measure of the degree of dispersion of a random variable in probability theory. The variance in the statistics is the mean square of the difference between each sample value and the mean of all sample values. The formula is:*

$$\sigma^2 = \frac{\sum (x - E)^2}{n} \tag{4.98}$$

where σ^2 *is variance;* x *is the sample value;* E *is the mean of all sample values; and* n *is the number of samples.*

Standard deviation: *it is the arithmetic square root of variance. The formula is:*

$$\sigma = \sqrt{\frac{1}{n}\sum (x - E)^2} \tag{4.99}$$

Normal distribution: *also known as Gaussian distribution, its curve is bell-shaped. The density distribution function is:*

$$f(x) = \frac{1}{\sqrt{2\pi}\sigma} e^{-\frac{(x-\mu)^2}{2\sigma^2}} \tag{4.100}$$

where f(x) is the density distribution function; σ is the standard deviation; and μ is the expected value of a normal distribution.

$N(\mu, \sigma^2)$ is the notation of normal distribution with the expected value μ, which determines its position and the standard deviation σ, which determines the spreading level of the distribution. When $\mu = 0$, $\sigma = 1$, it is a normal distribution.

4.4.3 *Analysis*

4.4.3.1 *Average permeability method*

Heterogeneous reservoirs are converted to homogeneous reservoirs by means of average permeability.

The steps are as follows:

① Take cores extensively in the whole oil reservoir, test their permeabilities, and record samples at different positions;

② Calculate the percentage of the number of cores in the total number of blocks for each permeability range, and obtain the density distribution. The equation formula is:

$$f(K) = \frac{N_i}{N} \tag{4.101}$$

③ Draw the density distribution curve; the normal density distribution curve is shown in Figure 4.6.

④ Judge the type of density distribution, mainly including normal distribution, logarithmic normal distribution, and so on.

⑤ According to the determined curve type, the expected value E can be obtained, that is, the average permeability.

⑥ Set the permeability of the homogeneous reservoir with $K = E$, and calculate production rate and pressure distribution.

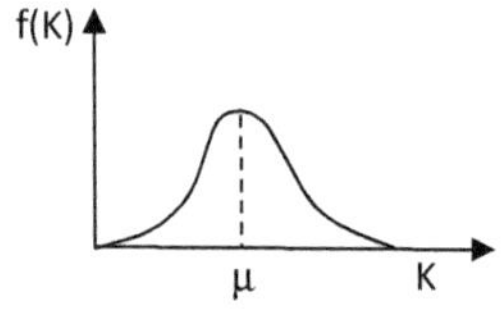

Figure 4.6. Normal distribution curve

4.4.3.2 *Formation parameter estimation method*

The maximum and minimum values of known formation parameter j are $x_{j\max}$ and $x_{j\min}$, respectively, and it follows the normal distribution. The generated random value on formation parameter j is:

$$x_j^r = x_{j\min} + R^r(x_{j\max} - x_{j\min}) \tag{4.102}$$

where x_j^r is the rth value randomly generated for formation parameter j; $x_{j\min}$ is the minimum value of formation parameter j;R^r is the random number generated from the standard normal distribution; $x_{j\max}$ is the maximum value of formation parameter j.

The random parameter values are used to calculate the formation pressure and production, and the expected values of the results are calculated to obtain the parameters.

In general, it requires hundreds or thousands of Monte Carlo simulations to obtain a stable result.

However, there is an error between the parameters generated randomly and the actual well parameters. If the parameters generated randomly each time are corrected with known parameters, the random parameters will be more accurate. Therefore, it is necessary to fit the random formation parameters with the actual measurement data.

For each formation parameter j, N times simulation can be run. The parameter at the well location with hard-data is determined, denoted as X_{jM}, and the number of wells is denoted as M. Therefore, the error of each simulation can be calculated.

$$\sum_{i=1}^{M} \varepsilon_{ji}^r = |x_{jM}^r - X_{jM}| \tag{4.103}$$

For a single well, the sum of errors of all formation parameters is linearly correlated with the value of each formation parameter X_{jM}, and L parameters are set as follows:

$$\varepsilon_M = \sum_{j=0}^{L} a_{jM} X_{jM} \tag{4.104}$$

where X_{jM} is the actual value, ε_M can be obtained, and α_{jM} is unknown.

There are M equations for each well, but there are $M \times L$ unknowns, which needs to be solved by the least squares method. If the coefficient α_{jM} is determined, the parameter X_{jM} can be solved in turn. The linear programming method can be used to solve the equation.

4.4.3.3 *Empirical equation method*

Let the random function of permeability be expressed as:

$$K(x) = \overline{K}(x) + \varepsilon(x) \tag{4.105}$$

where $K(x)$ is a random function of permeability; $\overline{K}(x)$ is the expected value; and $\varepsilon(x)$ is the random fluctuation value of permeability.

If the formation is one-dimensional, the flow resistance of the formation can be written as:

$$R_k(l) = \int_0^l \frac{dx}{\overline{K}(x) + \varepsilon(x)} \tag{4.106}$$

Expand the above equation to get:

$$R_k(l) = \int_0^l \frac{dx}{\overline{K}(x)} \left[1 - \int_0^l \frac{\varepsilon(x)dx}{\overline{K}^2(x)} \int_0^l \frac{dx}{\overline{K}(x)}\right] \tag{4.107}$$

Then the production rate is:

$$Q = \frac{h(p_e - p_w)}{\mu R_k(l)} \tag{4.108}$$

It can be rewritten as:

$$Q = Q_0 \left[1 + \frac{1}{I_0} \int_0^l \frac{\varepsilon(x)}{\overline{K}^2(x)} dx\right] \tag{4.109}$$

where $Q_0 = \frac{h(p_e - p_w)}{\mu I_0}$ and $I_0 = \int_0^l \frac{dx}{\overline{K}(x)}$.

$\overline{K}(x)$ can be expressed by using empirical equation, which is:

$$\overline{K}(x) = Ce^{-\frac{x}{\alpha}} \cos \beta x \tag{4.110}$$

where α and β are empirical parameters which are related to rock properties and how much information obtained in the reservoir.

The expected value of production rate can be obtained as:

$$\overline{Q} = Q_0 = \frac{h(p_e - p_w)}{\mu I_0} \tag{4.111}$$

4.4.3.4 *Perturbation method*

Taking the radial steady state flow of a single-phase liquid as an example, its governing equation is:

$$\frac{\partial}{\partial x}\left[K(x,y)\frac{\partial p}{\partial x}\right] + \frac{\partial}{\partial y}\left[K(x,y)\frac{\partial p}{\partial y}\right] = 0 \tag{4.112}$$

Let

$$K(x,y) = \overline{K}(x,y) + \varepsilon_k(x,y) \tag{4.113}$$

$$p(x,y) = \overline{p}(x,y) + \varepsilon_p(x,y) \tag{4.114}$$

First, the expected value of the pressure is calculated by applying the expected value of the permeability, which is denoted as p_0, then:

$$\frac{\partial}{\partial x}\left[\overline{K}(x,y)\frac{\partial p_0}{\partial x}\right] + \frac{\partial}{\partial y}\left[\overline{K}(x,y)\frac{\partial p_0}{\partial y}\right] = 0 \tag{4.115}$$

This equation can solve for p_0.

The perturbation term is solved by:

$$\frac{\partial}{\partial x}\left[\overline{K}(x,y)\frac{\partial \varepsilon_p}{\partial x}\right] + \frac{\partial}{\partial y}\left[\overline{K}(x,y)\frac{\partial \varepsilon_p}{\partial y}\right] = \varphi_0(x,y) \tag{4.116}$$

The term on the left side of the equation is known, and the term on the right is:

$$-\varphi_0(x,y) = \frac{\partial}{\partial x}\left[\varepsilon_k(x,y)\frac{\partial \varepsilon_{p_0}}{\partial x}\right] + \frac{\partial}{\partial y}\left[\varepsilon_k(x,y)\frac{\partial \varepsilon_{p_0}}{\partial y}\right] \tag{4.117}$$

This equation is a random function whose characteristic is determined by the term $\varepsilon_k(x, y)$. If p_0 is the deterministic solution, then the equation can be solved.

The pressure obtained is:

$$p(x, y) = p_0(x, y) + \varepsilon_p(x, y) \tag{4.118}$$

4.5 Fractal flow

4.5.1 *Introduction*

Reservoir space is actually composed of a large number of pores of different sizes and shapes, and the pore structure distribution is extremely complex. In the theory of flow mechanics in porous media, this complex distribution is basically simplified and some parameters are introduced to quantitatively describe the properties of the reservoir, such as permeability, porosity, specific surface, etc.

Fractal theory can be used to describe extremely complex distribution, such as the distribution of cosmic objects and nebulae, as well as crystal growth and molecular movement. In oil and gas development, fractal theory is often used to describe pore structure characteristics and natural fracture distribution.

Lines, planes and bodies correspond to one dimension, two dimensions and three dimensions respectively. Fractal geometry, characterized by fractal dimension, plays an important role in describing very complex objects.

Fractal has two basic properties: self-similarity and scale invariance.

The pore model based on Euclidean geometry is characterized by the study of continuous, asymptotic and smooth objects with "characteristic length". The model based on fractal geometry is usually characterized by the study of discontinuity, roughness, mutation and other objects, and there is no "characteristic length".

From the perspective of actual reservoir characteristics, fractal pores are more reasonable and closer to reality, especially for reservoirs containing fractures, because natural fractures are intermittent, rough and abrupt, and their distribution is not planned.

This chapter takes fracture-pore media as an example, but only establishes the fractal model for fracture systems.

4.5.2 *Concepts*

Fractal: *a form characteristic of filling space in non-integral dimensions. That is, a rough or fragmentary geometric shape can be divided into several parts, and each part is the approximate shape of the reduced whole shape, which is also the self-similarity of fractal. Figure 4.7 shows the classic fractal curve (Koch curve).*

Fractal dimension: *it is the parameter to describe the fractal geometry. Normally in Euclidean geometry, lines are one dimensional, planes are two dimensional, and bodies are three dimensional. As shown in Figure 4.7, the Koch curve has dimensions between 1 and 2. The symbol is D_f. In general, the fractal dimension of reservoir fractures is* $1.5 \sim 2.0$.

Self-similarity: *it refers to the fact that the characteristics of a structure or process are similar in different spatial or temporal scales, or that the local characteristics are similar to the whole.*

Scale invariance: *it refers to the fact that any selected enlarged local area shows the characteristics of the original graphics.*

Characteristic length: *it refers to the representative length of the object. The characteristic length of the sphere is its radius or*

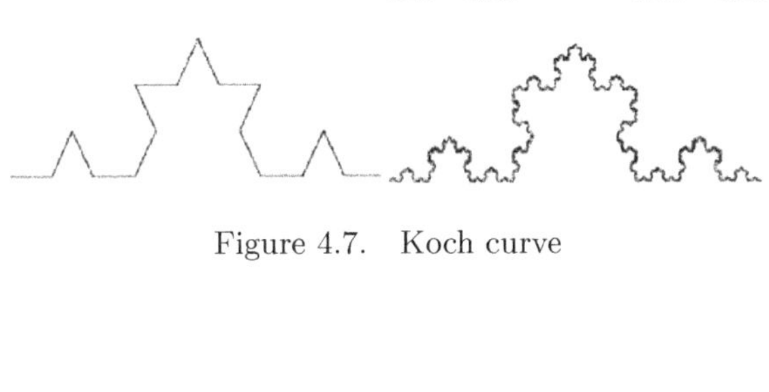

Figure 4.7. Koch curve

diameter, the characteristic length of the pore is the diameter of the pore throat, and the characteristic length of the fracture is the length or aperture of the fracture.

4.5.3 *Analysis*

4.5.3.1 *Fractal dimension*

There are many kinds of fractal dimensions, but only two common ones are introduced.

(1) Similarity dimension D_r

$$D_r = \frac{\ln b}{\ln c} \tag{4.119}$$

where D_r is the similarity dimension; b is the number of self-similarity pieces; and c is the magnification factor.

Therefore, the similarity dimension of Koch curve is $\frac{\ln 4}{\ln 3} = 1.2618$.

(2) Hausdorff dimension D_H

$$N = r^{-D_H} \tag{4.120}$$

$$D_H = -\frac{\ln N}{\ln r} \tag{4.121}$$

where D_H is the Haussdorff dimension; r is a unit of measure (think of it as a ruler); and N is the number of measurements with the ruler r.

If fracture density N and fracture length L are known, this relation can be applied:

$$N = L^{-D} \tag{4.122}$$

Calculate the fractal dimension D as:

$$D = -\frac{\ln N}{\ln L} \tag{4.123}$$

If the average fracture length is 0.5 m and the fracture density is 3/m, then the fractal dimension is 1.585.

4.5.3.2 *Parameters of fractal fractures*

$$\varphi_f(r) = \varphi_{R_w}\left(\frac{r}{R_w}\right)^{D-d} \tag{4.124}$$

$$K_f(r) = K_{R_w}\left(\frac{r}{R_w}\right)^{D-d-\theta} \tag{4.125}$$

where φ_{R_w} is the porosity at the borehole wall; K_{R_w} is the permeability at borehole wall; D is fractal dimension; d is Euclidean spatial dimension embedded in the fracture fractal; θ is the abnormal index, which indicates the connectivity of the network. The closer the value is to 0, the better the connectivity; the larger the value, the worse the connectivity.

4.5.3.3 *Governing equation of fractal flow*

Let the fractal fracture system be embedded in the Euclidean geometric rock block with the planar radial flow $d = 2$, and take the cylindrical REV of the fractal geological body, namely the small ring of the planar radial flow, from r to $r + dr$.

Set V_s as the volume of fluid in a fractal REV (equivalent to the scale), Q_r as the flow rate of the REV, and n_r as the density, i.e., the number of V_s in the cylindrical REV.

(1) Continuity equation

The mass difference between the inflow and outflow per unit time is:

$$\left[\rho Q_r + \frac{1}{2}\frac{\partial(\rho Q_r)}{\partial r}\right] - \left[\rho Q_r - \frac{1}{2}\frac{\partial(\rho Q_r)}{\partial r}\right] = \frac{\partial(\rho Q_r)}{\partial r} \tag{4.126}$$

The fluid mass inside the REV at unit time changes is:

$$\frac{\partial(V_s n_r \rho)}{\partial t} \tag{4.127}$$

According to the law of mass conservation, there is:

$$\frac{\partial(\rho Q_r)}{\partial r} = \frac{\partial(V_s n_r \rho)}{\partial t} \tag{4.128}$$

V_s and n_r have nothing to do with time, so:

$$\frac{\partial(\rho Q_r)}{\partial r} = V_s n_r \frac{\partial \rho}{\partial t} \tag{4.129}$$

For slightly compressible liquid, its elastic compressibility is:

$$C_L = \frac{1}{\rho}\frac{\partial \rho}{\partial p} \tag{4.130}$$

C_L is generally small, so:

$$\frac{1}{\rho}\frac{\partial(\rho Q_r)}{\partial r} = V_s n_r \frac{1}{\rho}\frac{\partial \rho}{\partial p}\frac{\partial p}{\partial t} \tag{4.131}$$

Then

$$\frac{1}{\rho}\left(\frac{\partial \rho}{\partial r}Q_r + \rho\frac{\partial Q_r}{\partial r}\right) = V_s n_r C_L \frac{\partial p}{\partial t} \tag{4.132}$$

The fluid density changes very little along the radial direction, $\frac{\partial \rho}{\partial r} = 0$ so that:

$$\frac{\partial Q_r}{\partial r} = V_s n_r C_L \frac{\partial p}{\partial t} \tag{4.133}$$

This is the continuity equation under fractal conditions.

(2) Equation of motion

Darcy's law cannot be directly applied to fractal oil reservoirs, but it is applicable to REV. The equation of motion is:

$$v = \frac{K_f}{\mu}\frac{\partial p}{\partial r} \tag{4.134}$$

where K_f is fractal permeability.

Substitute $d = 2$ into the fractal permeability formula, and obtain:

$$v = \frac{K_{R_w}}{\mu}\left(\frac{r}{R_w}\right)^{D-2-\theta}\frac{\partial p}{\partial r} \tag{4.135}$$

(3) Equation of state

$$\rho = \rho_0[1 + C_L(p - p_0)] \tag{4.136}$$

(4) Governing equations

Substitute the equation of motion and the equation of state into the continuity equation, and get:

$$\frac{\partial}{\partial r}\left\{2\pi rh\left[\frac{K_{R_w}}{\mu}\left(\frac{r}{R_w}\right)^{D-2-\theta}\frac{\partial p}{\partial r}\right]\right\} = V_s n_r C_L \frac{\partial p}{\partial t} \tag{4.137}$$

And it is known:

$$V_s n_r = 2\pi rh\varphi_f \tag{4.138}$$

$$\varphi_f = \varphi_{R_w}\left(\frac{r}{R_w}\right)^{D-2} \tag{4.139}$$

The fractal flow model is obtained as:

$$r^{\theta}\frac{\partial}{\partial r}\left[r^{\beta}\frac{\partial p}{\partial r}\right] = \frac{\varphi_{R_w}\mu C_L}{K_{R_w}R_w^{\theta}}\frac{\partial p}{\partial t} \tag{4.140}$$

where $\beta = D - \theta - 1$ if it is considered that the fracture is normal, that is, the communication is good, $\theta = 0$, and the flow equation becomes:

$$\frac{\partial}{\partial r}\left[r^{D-1}\frac{\partial p}{\partial r}\right] = \frac{\varphi_{R_w}\mu C_L}{K_{R_w}}\frac{\partial p}{\partial t} \tag{4.141}$$

If $D = 2$, it is an integer dimension, and the equation becomes a diffusivity equation.

4.5.3.4 *Steady state flow in the fractal reservoir*

If $\theta = 0$ when the flow is steady state, the governing equation is:

$$\frac{\partial}{\partial r}\left[r^{\beta}\frac{\partial p}{\partial r}\right] = 0 \tag{4.142}$$

where:

$$p = p_e + \frac{(p_e - p_w)(r^{1-\beta} - R_e^{1-\beta})}{R_e^{1-\beta} - R_w^{1-\beta}} \tag{4.143}$$

$$Q = \frac{2\pi hK_{R_w}(1-\beta)}{\varphi_{R_w}\mu}\frac{p_e - p_w}{(R_e^{1-\beta} - R_w^{1-\beta})} \tag{4.144}$$

4.6 Horizontal well flow

4.6.1 *Introduction*

The main advantages of horizontal wells are:

① The contact area between the oil layer and the wellbore is greatly increased, the flow resistance near the well is reduced, the production pressure difference is reduced, and the production rate of single well is improved;

② Drainage area for the single well is greatly increased, and the drilling and surface facility costs are reduced;

③ The horizontal well can connect to fractures more effectively and it can improve production rate;

④ The small production pressure difference used is conducive to inhibit the rise of water cut and improve recovery;

⑤ In the bottom water-driven or gas cap-driven reservoir, the coning speed of water or gas can be reduced and the water-free production period can be prolonged;

⑥ Be it water injection, gas injection, miscible flooding or polymer flooding, horizontal well as an injection well can improve the sweep volume and oil displacement efficiency;

⑦ In unconventional reservoirs, the fractured horizontal or multi-lateral wells open up a network of connections with the reservoir and ensures sufficient production.

Therefore, horizontal wells are mainly used in: heavy oil reservoirs, low permeability reservoirs, tight oil and gas reservoirs, shale oil and gas reservoirs, fractured reservoirs, bottom water reservoirs, thin reservoirs, offshore reservoirs and so on.

Horizontal well technology was proposed in 1863. In the 1940s, horizontal drilling was successfully used. It was industrialized in the 1980s in the United States, Canada and other countries. In the 1990s, the horizontal well was applied in China rapidly. Nowadays, starting from 2010, the fractured horizontal well is the only effective means to develop unconventional oil and gas resources.

Many types of horizontal wells have been developed for different formation characteristics, as shown in Figure 4.8.

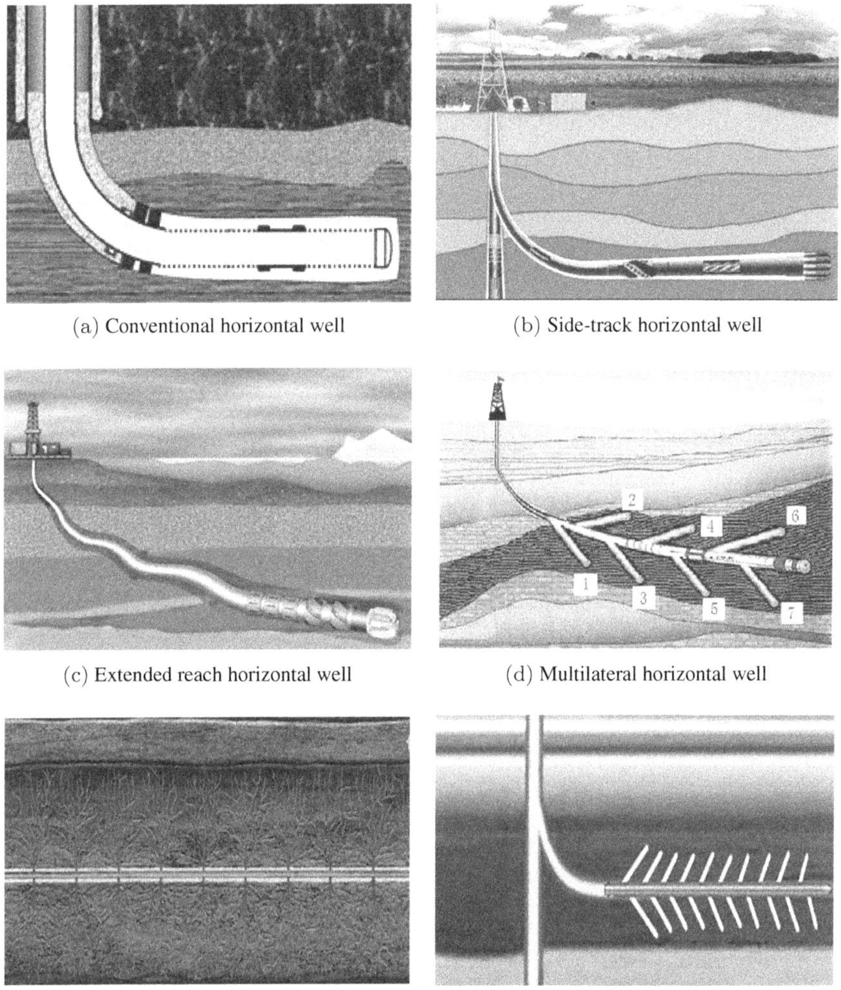

(a) Conventional horizontal well
(b) Side-track horizontal well
(c) Extended reach horizontal well
(d) Multilateral horizontal well
(e) Fractured horizontal well
(f) Fish-bone wells

Figure 4.8. Different types of horizontal wells

At present, there are two main methods to calculate the productivity of horizontal wells. The first one is the theoretical calculation method, which involves directly establishing a mathematical model and solving it by the equivalent flow resistance method, mirror image principle and the superposition method of potential function, mainly for single-phase flow in the formation. The other method is numerical simulation, which studies the dynamic relationship between productivity and inflow of horizontal wells and it can calculate the two-phase flow of oil and gas in the formation. In addition, there are different empirical methods. This chapter mainly introduces the flow characteristics of horizontal wells, the pressure drawdown and the distribution of potential function.

4.6.2 *Concepts*

Horizontal well: *it is a special well that has a maximum inclination of 90° and maintains a certain length of horizontal section in the target formation. The length of horizontal section is mostly about 1000 m, although ultra-long horizontal wells can reach more than 6000 m.*

Extended reach well: *wells with a ratio of horizontal distance to depth not less than 2 and a true vertical depth more than 3000 m, or wells with a horizontal displacement more than 3000 m.*

Penetration ratio: *ratio between length of the horizontal section and effective length of the drainage area.*

Vertical permeability: *it is the permeability perpendicular to the reservoir, which describes the capacity of fluid flowing perpendicular to the reservoir, symbol K_z or K_v, whose value is generally $0.1 \sim 0.5$ times less than the horizontal permeability.*

Horizontal permeability: *it is the permeability in all directions along the horizontal directions of the reservoir, which describes the fluid's ability to flow along the reservoir, symbols K_x and K_y, or K_h.*

Infinite conductivity flow: *there is no pressure loss when the fluid is flowing in a horizontal wellbore, which is an ideal case.*

Finite conductivity flow: *the pressure loss is considered when flowing in a horizontal wellbore, which is the real situation.*

Frictional resistance coefficient: *it is used to represent the friction between the tube wall and the fluid in the process of fluid flow, which can be determined by the Reynolds number N_{Re} of the* *liquid phase and the relative roughness of the tube wall $\frac{\varepsilon}{D}$, and its value is between 0.01 and 0.1.*

Relative roughness of pipe wall: *it is the ratio of absolute roughness of pipe wall ε to pipe diameter D, and usually it is $\varepsilon = 4.57 \times 10^{-5}\, m$.*

Liquid phase Reynolds number: *it is defined by*

$$N_{Re} = \frac{D v_l \rho_l}{\mu_l} \tag{4.145}$$

where N_{Re} is the liquid phase Reynolds number; D is the tubing diameter, m; ν_l is the fluid velocity in the pipe, m/s; ρ_l is the fluid density in the pipe, kg/m^3; and μ_l is the fluid viscosity in the tube, $mPa \cdot s$.

4.6.3 *Analysis*

4.6.3.1 *Flow characteristics of horizontal well*

Depending on reservoir and well properties, the flow may pass through 4 stages before reaching pseudo-steady state.

(1) Early radial flow

As shown in Figure 4.9(a), this stage is the initial production period of the horizontal well. Vertically, the pressure wave does not reach the upper and lower bottom of the reservoir.

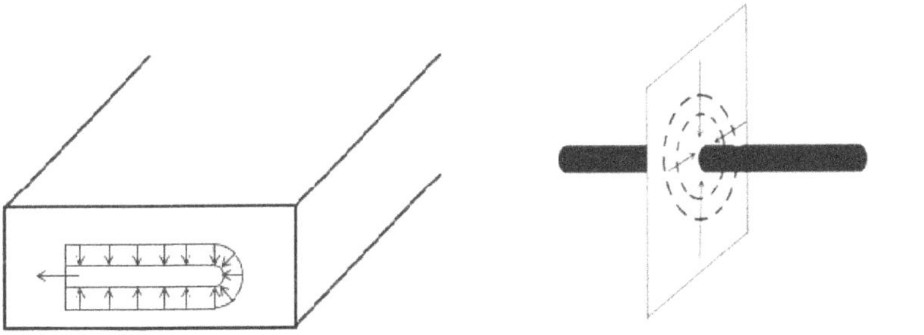

(a) along the wellbore direction (b) perpendicular to the wellbore direction

Figure 4.9. Early radial flow stage for horizontal well flow

Seen from the vertical wellbore plane, as shown in Figure 4.9(b), it is similar to the single-well radial flow, so it is called early radial flow.

If the vertical and horizontal permeabilities are very different, the equipotential lines appear elliptic rather than circular. If the formation thickness is small or the ratio of vertical to horizontal permeability is small, this flow stage may not occur.

(2) Early linear flow stage

If the horizontal well penetration ratio is relatively large, the flow of fluid to the horizontal well is similar to the linear flow on the horizontal plane after the early radial flow of the horizontal well, so it is called early linear flow, as shown in Figure 4.10.

At this stage, the pressure wave has reached the upper and lower boundaries, and the fluid basically flows laterally along the horizontal well. If the ratio of vertical to horizontal permeability is small or the formation thickness is very thick, this stage of flow may not occur.

(3) Late pseudo-radial flow stage

As shown in Figure 4.11, after the early linear flow, the horizontal well can be regarded as a point source, and the flow on the horizontal plane is similar to the radial flow on the plane, so it is called late radial flow.

It is largely dependent on the horizontal well penetration ratio. This stage is also affected by the upper and lower boundaries of the formation and the drainage radius should be many times larger

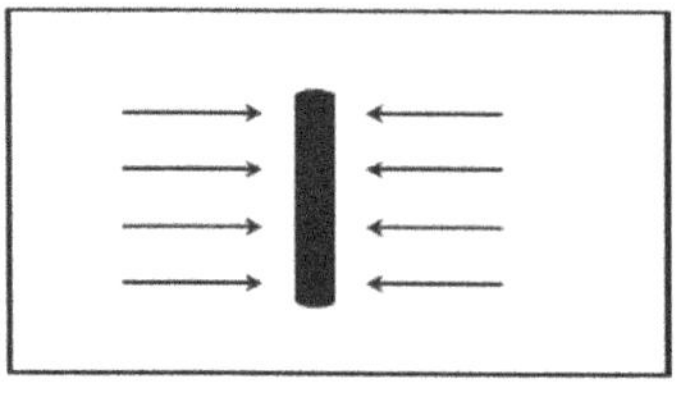

(a) along the wellbore direction

(b) perpendicular to the wellbore direction

Figure 4.10. Early linear flow stage for horizontal well flow

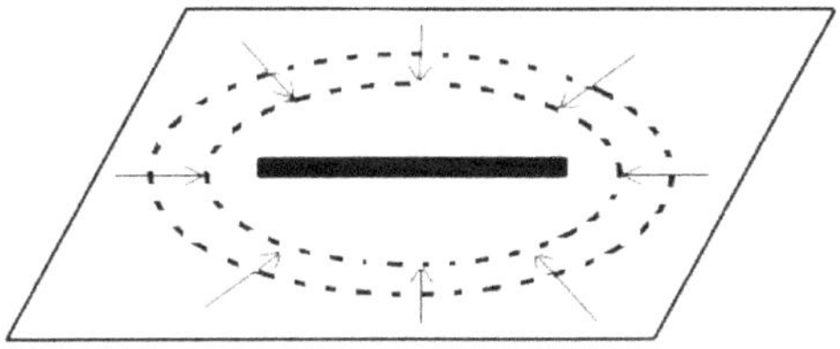

Figure 4.11. Late pseudo-radial flow for horizontal well flow

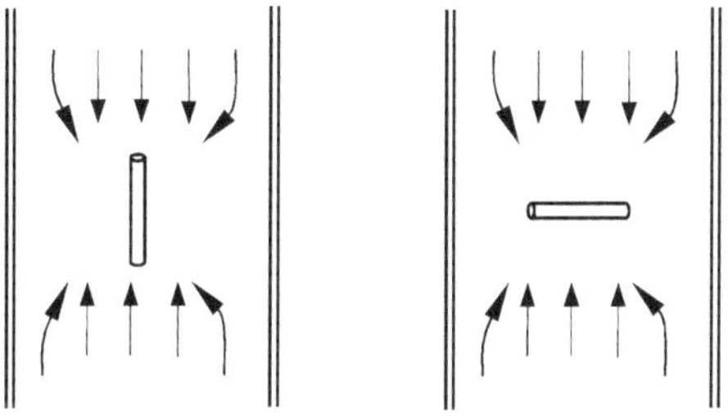

(a) along the horizontal well direction

(b) perpendicular to horizontal well direction

Figure 4.12. Late linear flow for horizontal well flow

than the horizontal section. If the pressure wave spreads to the outer boundary, this phase will disappear.

(4) Late linear flow stage

When the radial flow pressure wave propagates to the boundary, if the reservoir has lateral boundaries, the linear flow can be formed in a large area along or perpendicular to the horizontal well direction, and it is called late linear flow, as shown in Figure 4.12.

(5) Flow net diagram of horizontal well

As shown in Figure 4.13, the flow net diagram of planar linear flow is two groups of evenly distributed parallel lines, and these two groups are perpendicular to each other. On the other hand, radial flow is a group of evenly distributed radial lines (streamlines) and a group of concentric circles (equipotential lines) that become increasingly close to the well. The flow net of horizontal wells is a set of hyperbolic

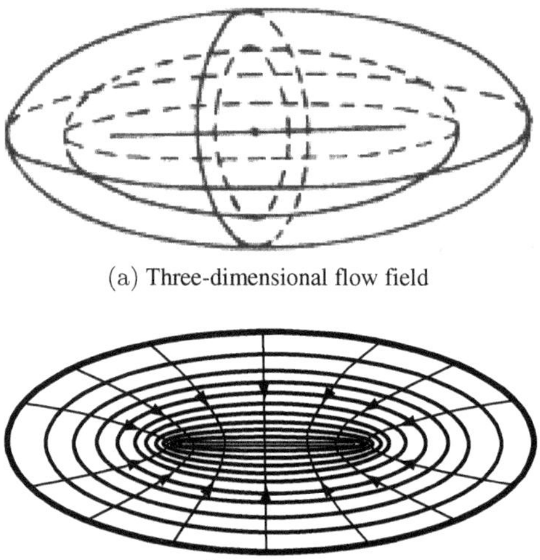

(a) Three-dimensional flow field

(b) Planar view

Figure 4.13. Three-dimensional and two-dimensional flow net of horizontal wells

curves (streamlines) and a set of ellipses (equipotential) that are getting closer and closer to the end of horizontal wells.

4.6.3.2 *Infinite conductivity flow of horizontal well*

If there is no pressure loss of fluid flow in the wellbore of the horizontal well, the infinite conductivity flow can be assumed.

Many scholars proposed different solutions of production rate, which are listed here.

(1) Joshi equation

$$Q_h = \frac{0.54287 K_h h \Delta p / (\mu_0 B_0)}{\ln\left[\frac{a+\sqrt{a^2-(L/2)^2}}{L/2}\right] + \left(\frac{\beta h}{L}\right)\ln\left(\frac{\beta h}{2\pi R_w}\right)} \tag{4.146}$$

$$a = (L/2)[0.5 + \sqrt{(2R_e/L)^4 + 0.25}]^{0.5} \tag{4.147}$$

where Q_h is the horizontal well production rate, m^3/d; K_h is the vertical permeability, mD; $\beta = \sqrt{K_h/K_v}$; L is the horizontal well length, m; and a is the half length of the major axis, m.

However, the Joshi equation considers horizontal well flows to be decomposed into horizontal and vertical flows, and simply adds them together. In fact, horizontal and vertical flows need to be added up by considering them as vectors.

(2) Giger equation

Giger derived the equation by using the electric analog method, and it is:

$$Q_h = \frac{0.54287 K_h h \Delta p/(\mu_0 B_0)}{\ln\left[1+\sqrt{1-\left(\frac{L}{2R_e}\right)^2}/\left(\frac{L}{2R_e}\right)\right] + \left(\frac{h}{L}\right)\ln\left(\frac{h}{2\pi R_w}\right)} \tag{4.148}$$

(3) Borisov equation

Borisov equation assumes that the oil reservoir is isotropic, the horizontal well is located in the middle of the oil reservoir, and the liquid in the oil reservoir is incompressible, then it is:

$$Q_h = \frac{0.54287 K_h h \Delta p/(\mu_0 B_0)}{\ln\left[\frac{4R_e}{L}\right] + \left(\frac{h}{L}\right)\ln\left(\frac{h}{2\pi R_w}\right)} \tag{4.149}$$

For conventional horizontal wells without fracturing, the Joshi and Borisov formulae have similar calculation results. These two formulae can be used for the initial productivity prediction of horizontal wells.

(4) Renard and Depuy equation

$$Q_h = \frac{0.54287 K_h h \Delta p}{\mu_o B_o}\left[\frac{1}{\cosh^{-1}\left(\frac{2a}{L}\right) + (h/L)\ln[h/(2\pi R_w)]}\right] \tag{4.150}$$

(5) Albertus Retnanto and M.J. Economides equation

A semi-analytical model for productivity prediction of vertical wells, horizontal wells and inclined wells in the late stage of pseudo-steady flow is proposed:

$$J = \frac{q}{\overline{p} - p_w} = \frac{\overline{K} x_e}{887.22 B \mu_o \left(p_D + \frac{x_e}{2\pi L} \sum S\right)} \tag{4.151}$$

where p_D is the dimensionless mean pressure; x_e is the length of the drainage area in the x-axis, m; $\overline{K}$ is the average reservoir permeability ($\overline{K} = \sqrt[3]{K_x K_y K_z}$), mD; S is the skin factor; and L is the wellbore length, m.

(6) Leif Larsen equation

$$J = \frac{q}{\overline{p} - p_{wf}} = \frac{Kh}{1.842 B \mu_o \left(\frac{1}{2} \ln \frac{4A}{e^{\gamma} C_A R_w^2} + S\right)} \tag{4.152}$$

where γ is the Euler constant, whose value is equal to 0.57722156; C_A is the Dietz shape factor, reflecting the shape of the drainage area and the influence of well location; and A is the drainage area, m^2.

(7) Fan and Lin equation

The corresponding productivity formula is proposed for edge water driven reservoir:

$$Q_h = \frac{0.54287 K L \Delta p}{B_o \mu_o \left(\ln \frac{\beta h}{2\pi R_w} + \frac{1.338\pi b}{\beta h} - \ln\left(\sin \frac{\pi z_w}{h}\right)\right)} \tag{4.153}$$

where z_w is the distance between horizontal segment and bottom of reservoir, m; and b is the distance between horizontal section borehole and edge water, m.

(8) Zhang and Han equation

Based on the Poisson equation for steady state flow of horizontal wells in homogeneous, anisotropic and single-phase fluid reservoirs, the three-dimensional steady state solution of horizontal wells is directly solved, and the accurate productivity formula of horizontal infinite reservoir and bottom water driven reservoir is established.

The production rate of horizontal infinite reservoir is:

$$Q_h = \frac{0.54287 K_h h \Delta p}{\mu B \left(\ln R + \frac{1}{2L_D} \ln \frac{\beta h}{2\pi R_w \sin(\pi z_{wD})} - \frac{1}{2L_D^2} \left(z_{wD}(1 - z_{wD}) + \frac{1}{3} \right) + R_2 \right)} \tag{4.154}$$

where z_{wD} is the dimensionless distance from horizontal to bottom of reservoir; L_D is the dimensionless length of horizontal segment; R_2 is the dimensionless coefficient; $R = \sqrt{\frac{a+b}{a-b}}$.

The production rate of bottom water driven reservoir is:

$$Q = \frac{0.54287 L \sqrt{K_h K_v} \Delta p}{B_o \mu_o \left\{ \ln \left[\frac{4\beta h}{\pi R_w} tg \left(\frac{\pi z_{wD}}{2} \right) \right] \frac{z_{wD}}{2} + R_1 \right\}} \tag{4.155}$$

where R_1 is the dimensionless coefficient.

4.6.3.3 *Finite conductivity flow of horizontal well*

When a horizontal well is being produced, reservoir fluid flows into the wellbore along the length of the horizontal well, as shown in Figure 4.14.

From the toe to the heel of the horizontal wellbore, the fluid mass flow rate increases gradually. In this case, the flow velocity gradually increases along the main streamline direction, and the pressure loss is no longer equal to 0.

The fluid flowing into the reservoir along the horizontal wellbore interferes with the boundary layer of the main wellbore wall and affects its velocity profile, thus changing the wall friction resistance determined by the velocity distribution.

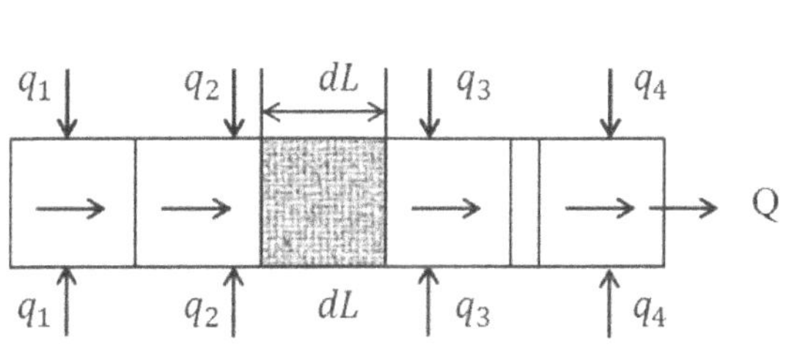

Figure 4.14. Schematic diagram of flow in horizontal wells with finite conductivity

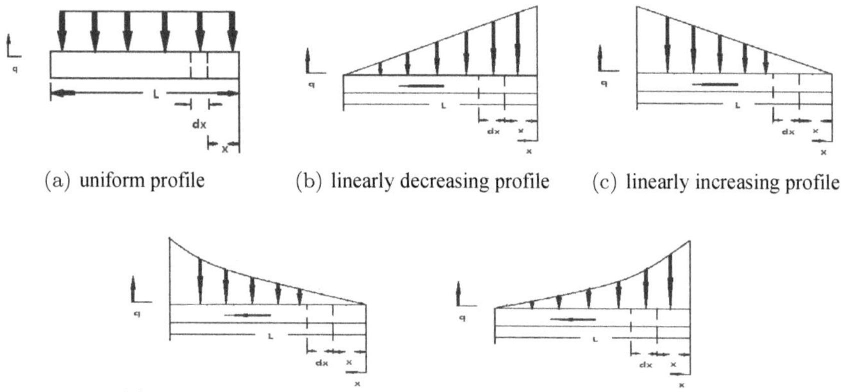

(a) uniform profile (b) linearly decreasing profile (c) linearly increasing profile

(d) parabolic increasing profile (e) parabolic decreasing profile

Figure 4.15. Schematic diagram of five inflow profiles in horizontal wellbores

On the other hand, the radial flow affects the pressure distribution and pressure drop in the horizontal wellbore. In turn, the pressure distribution in the wellbore also affects the radial flow from the reservoir into the wellbore, so there is a coupling relationship between the flow in the reservoir and the flow in the horizontal wellbore.

Assuming that fluid enters the wellbore throughout the horizontal segment (i.e., the flow in the horizontal wellbore is variable mass flow), the five possible inflow profiles for fluid entering the horizontal wellbore are shown in Figure 4.15.

(1) Pressure drop distribution of uniform inflow profile is:

$$\Delta p_a(x) = 0.81\lambda \frac{\rho q_t^2}{D^5 L^2} \cdot \frac{x^3}{3} \tag{4.156}$$

The total pressure drop along the path is:

$$\Delta p_a = 0.81\lambda \frac{\rho q_t^2}{D^5 L^2} \cdot \frac{L^3}{3} \tag{4.157}$$

where λ is the coefficient of friction resistance; D is the wellbore diameter, m; q_t is the total wellbore flow rate, m^3/d; L is the length of horizontal segment, m.

(2) Pressure drop distribution of the linear decline inflow profile is:

$$\Delta p_b(x) = 0.81\lambda\frac{\rho q_t^2}{D^5L^2}\left[\frac{3}{4}L^2x^3 - Lx^4 + \frac{x^5}{5}\right] \tag{4.158}$$

The total pressure drop along the path is:

$$\Delta p_b = 0.81\lambda\frac{\rho q_t^2}{D^5L^2}\cdot\frac{8L^3}{15} = \frac{8}{15}\Delta p \tag{4.159}$$

(3) Pressure drop distribution of the linear increase inflow profile is:

$$\Delta p_c(x) = 0.81\lambda\frac{\rho q_t^2}{D^5L^4}\frac{x^5}{5} \tag{4.160}$$

The total pressure drop along the path is:

$$\Delta p_c = 0.81\lambda\frac{\rho q_t^2}{D^5L^2}\cdot\frac{L^3}{5} = \frac{1}{5}\Delta p \tag{4.161}$$

(4) Pressure drop distribution of parabolic increasing inflow profile is:

$$\Delta p_d(x) = 0.81\lambda\frac{\rho q_t^2}{D^5L^6}\frac{x^7}{7} \tag{4.162}$$

The total pressure drop along the path is:

$$\Delta p_d = 0.81\lambda\frac{\rho q_t^2}{D^5}\cdot\frac{L}{7} = \frac{1}{7}\Delta p \tag{4.163}$$

(5) Pressure drop distribution of parabolic decreasing inflow profile is:

$$\Delta p_e(x) = 0.81\lambda\frac{\rho q_t^2}{D^5L^6}\left[3L^4x^3 - \frac{9}{2}L^3x^4 + 3L^2x^6 - Lx^6 + \frac{x^7}{7}\right] \tag{4.164}$$

The total pressure drop along the path is:

$$\Delta p_e = 0.81\lambda\frac{\rho q_t^2}{D^5}\cdot\frac{9L}{14} = \frac{9}{14}\Delta p \tag{4.165}$$

4.6.3.4 *Mathematical flow model of horizontal well*

The horizontal well length L is divided into N segments and the radial flow (reservoir flow into wellbore) of segment I is assumed to be q_r (i) and the flow pressure is p_w (i) $(1 \leq i \leq N)$, and the following assumptions are made:

① The reservoir is homogeneous, uniformly thick and infinite formation with isotropic horizontal permeability K_h and vertical permeability $K_v \neq K_h$;

② The fluid is the single-phase incompressible fluid, and the reservoir flow conforms to Darcy's law;

③ The influence of variable mass flow in the wellbore of the production section on reservoir flow is considered, that is, the pressure loss along the production section.

Set the coordinates of the heel end of the horizontal well as M (x_0, y_0, z_0), and the coordinates of any point on the REV are:

$$x_p(i,t) = x_0 + \frac{L}{N}\left(\sum_{s=0}^{i-1} \sin\theta_s \cos\alpha_s + t\sin\theta_i \cos\alpha_i\right) \tag{4.166}$$

$$y_p(i,t) = y_0 + \frac{L}{N}\left(\sum_{s=0}^{i-1} \sin\theta_s \sin\alpha_s + t\sin\theta_i \sin\alpha_i\right) \tag{4.167}$$

$$z_p(i,t) = z_0 + \frac{L}{N}\left(\sum_{s=0}^{i-1} \cos\theta_s + t\cos\theta_i\right) \tag{4.168}$$

where i is the ith digit segment; N is the number of segments; L is the length of the horizontal segment; θ_i is the inclination angle; α_i is the azimuth angle; t is the length ratio of section $i (0 \leq t \leq 1)$ to the total length, and if $t = 0.5$, it is the middle point on section i.

The steady state flow of the horizontal well production in an infinite reservoir is in accordance with the Laplace equation:

$$\frac{\partial^2 p}{\partial x^2} + \frac{\partial^2 p}{\partial y^2} + \frac{\partial^2 p}{\partial z^2} = 0 \tag{4.169}$$

Outer boundary condition is:

$$p(x, y, z)|_{x\to\infty, y\to\infty, z\to\infty} = p_e \tag{4.170}$$

Inner boundary condition is:

$$p(x, y, z)|_{[x-x_p(i,t)]^2+[y-y_p(i,t)]^2+[z-z_p(i,t)]^2=R_w^2} = p_w(i)\ (1 \le i \le N)$$

Set $\beta = \sqrt{\frac{K_h}{K_v}}$, and define the potential function as:

$$\Phi = \frac{\sqrt{K_v K_h}}{\mu_0} p + C \tag{4.171}$$

Perform linear transformation $Z' = \beta Z$, then:

$$\frac{\partial^2 \Phi}{\partial x^2} + \frac{\partial^2 \Phi}{\partial y^2} + \frac{\partial^2 \Phi}{\partial z'^2} = 0 \tag{4.172}$$

The principle of superposition of potential is applied to solve the equation, and the potential distribution formula is obtained as:

$$\Phi(x, y, z) = \frac{Q}{4\pi} \ln \frac{R_1 + R_2 + L}{R_1 + R_2 - L} + C \tag{4.173}$$

where $R_1 = \sqrt{(x_0 - x)^2 + (y_0 - y)^2 + (z_0 - z)^2}$; and $R_2 = \sqrt{(x_0 + L - x)^2 + (y_0 - y)^2 + (z_0 - z)^2}$.

By using the definition of potential and its distribution in space and the principle of potential superposition, the production rate of horizontal wells can be calculated.

Chapter 5

Physicochemical flow in porous media

Different from the simple Newtonian fluid in the classical flow theory of porous media, there will be some physical property changes and chemical reactions in the physicochemical flow process. These physical properties can be viscosity, surface tension, phase state, concentration, temperature, and so on. The flow problems mainly include non-Newtonian fluid flow, mass transfer by diffusion, and the multiphase and multicomponent fluid flow. In this chapter, we will introduce the flow characteristics and key parameters considered in the establishment of the mathematical model, so as to provide the basis to quantitatively describe the physicochemical flow problem in porous media.

5.1 Non-Newtonian liquid flow

5.1.1 *Introduction*

In classical flow mechanics in porous media, the viscosity of liquid does not change with the flow velocity, pressure and time. The change of viscosity of crude oil is caused by the separation or dissolution of natural gas in crude oil, rather than properties of the liquid itself. However, the viscosity of some liquids, such as polymer solution, micro-porous fluid and fracturing fluid, will change greatly with the flow velocity and time.

Some liquids become thinner or thicker when the shear rate or stirring time changes. Some liquids are only viscous, but others are both viscous and elastic. All these changes are related to the rheological characteristics of the liquid.

5.1.2 *Concepts*

Non-Newtonian fluid: *It refers to fluids that do not conform to Newton's law. For a Newtonian fluid, the shear stress and shear rate follow a linear relationship and the viscosity caused by the internal friction is constant. The non-Newtonian fluid is the opposite of the Newtonian fluid.*

Law of internal friction: *as shown in Figure 5.1, the distance between a pair of parallel plates is dr. The space between the plates is filled with the liquid. If the upper plate is moving with a velocity v, it causes a velocity difference du to the two fluid layers with a distance dy. du/dy is the shear rate, and it has a linear relationship with the shear stress,* $\tau = \mu \frac{du}{dy} = \mu\dot{\gamma}$, *where the viscosity coefficient* μ *caused by the internal friction of different fluid layers is constant. The relationship is called the law of internal friction, also known as Newton's laws of internal friction.*

Rheology of liquid: *it is the study of the flow and deformation when an external force is acted upon the fluid.*

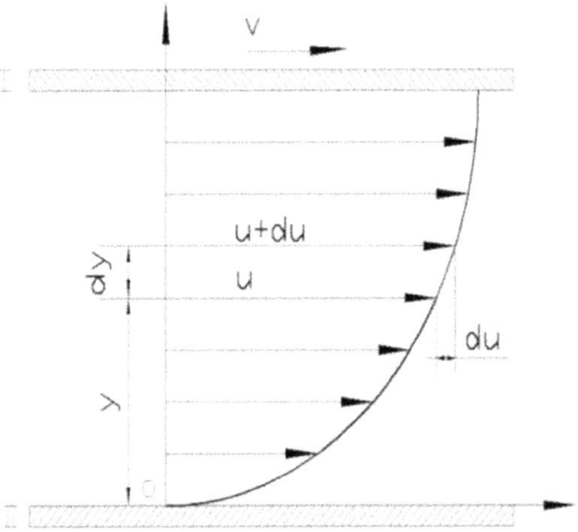

Figure 5.1. Schematic diagram of the law of internal friction

Constitutive equation: *it is also known as rheological equation and it is the state equation of fluid flow describing the relationship between shear stress and shear rate.*

Shear stress: *The direction of internal friction between the moving layers in the laminar flow is generally the tangent line along the liquid layer, and the deformation of the liquid during the flow process is caused by this force, and the shear force on the unit area is the shear stress. It also refers to viscous forces in flow mechanics in porous media, symbol* τ*, unit Pa.*

Shear rate: *it is the rate of flow velocity change in the fluid moving perpendicular to the plane, i.e.,* du/dy*, symbol* $\dot{\gamma}$*, unit* s^{-1}.

Rheological curve: *it is a curve composed of shear stress and shear rate.*

Power-law fluid: *it is a liquid whose constitutive equation can be expressed by a power function.*

$$\tau = H\dot{\gamma}^n \tag{5.1}$$

where H *is the consistency coefficient, whose value is the viscosity when the shear rate is 1* s^{-1}*, which is about* $10 \sim 50$ *for gel fracturing fluids and* $10 \sim 150$ *for polymer solutions;* n *is the power law index or viscosity index, and for polymer solutions it is* $0.1 < n < 1$.

Generalized Darcy's law: *the flow law of non-Newtonian liquids can be expressed in the form of Darcy formula, and the formula is:*

$$u_r^n = \frac{K}{\mu_{eff}}\frac{\partial p}{\partial r} \tag{5.2}$$

where u_r^n *is radial flow velocity; and* μ_{eff} *is the effective viscosity of a non-Newtonian liquid.*

Apparent viscosity: *it is the slope corresponding to the tangent line at each point on the rheological curve, that is, the ratio of shear stress to shear velocity; symbol is* μ_α*. When its value is constant,* *it is a Newtonian liquid, Otherwise, it is a non-Newtonian liquid. The equation is:*

$$\mu_a = \frac{\tau}{\dot{\gamma}} \tag{5.3}$$

Effective viscosity: *it is viscosity expressed by generalized Darcy's law, whose value is:*

$$\mu_{eff} = \frac{H}{12}\left(9 + \frac{3}{n}\right)^n (150Kf)^{\frac{1-n}{2}} \tag{5.4}$$

$$\mu_s = \frac{KA\Delta p}{QL} \tag{5.5}$$

Viscoelastic liquid: *it is the fluid having both the viscous properties of a fluid and the elastic properties of a solid.*

5.1.3 *Analysis*

5.1.3.1 *Rheology of non-Newtonian liquids*

Newtonian fluids conform to the law of internal friction:

$$\tau = \mu\frac{d\mu}{dy} = \mu\dot{\gamma} \tag{5.6}$$

or

$$\mu = \frac{\tau}{\dot{\gamma}} \tag{5.7}$$

The characteristic of a "Newtonian liquid" is that the shear stress acting on the liquid is proportional to the shear rate, and the proportionality constant is called "kinetic viscosity". It is also referred to as "viscosity", which is independent of time.

The relation between shear rate and shear stress for the non-Newtonian fluid is not a straight line through the origin, that is, it does not conform to Newton's law of internal friction. It has three main categories: pure viscous non-Newtonian liquid, unsteady non-Newtonian liquid, and viscoelastic non-Newtonian liquid.

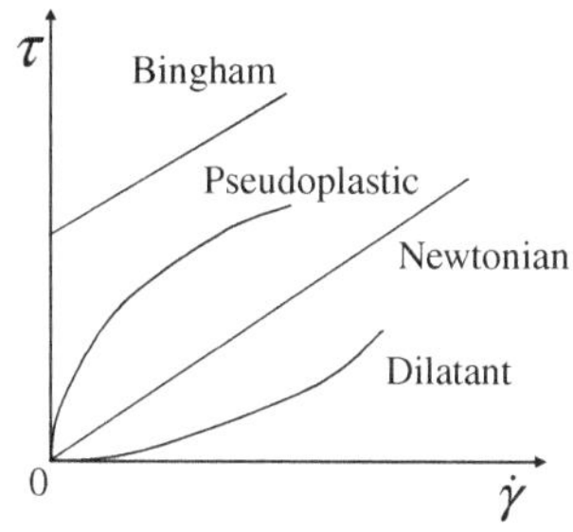

Figure 5.2. Rheological curves of non-Newtonian fluids

(1) Pure viscous non-Newtonian liquid

As shown in Figure 5.2, the viscosity of such a liquid is only related to shear stress and shear rate, independent of time, and can be further divided into power-law fluid and Bingham fluid.

① Power-law fluid

This kind of non-Newtonian fluid introduces the concept of "apparent viscosity", namely:

$$\mu_a = \frac{\tau}{\dot{\gamma}} \tag{5.8}$$

"Power-law fluid" is a common type of "non-Newtonian liquid", and its expression is:

$$\tau = H\dot{\gamma}^n \tag{5.9}$$

When $n = 1$, it is a Newtonian liquid, such as water and oil;

When $0 < n < 1$, it is a pseudo-plastic liquid, characterized by shear thinning, such as blood and polymer solution;

When $n > 1$, it is a dilatant fluid, which is characterized by shear thickening, such as concentrated sugar solution.

The power-law fluid is expressed by the constitutive equation:

$$\mu_a = H\dot{\gamma}^{n-1} \tag{5.10}$$

② Bingham fluid

The plastic fluid in Figure 5.2 is characterized by that the fluid flows only when the shear stress exceeds a certain stress value, and then the rheological properties are the same as those of Newtonian liquid. Its constitutive equation is:

$$\tau = \tau_0 + \mu\dot{\gamma} \tag{5.11}$$

Mud and toothpaste are generally Bingham fluids.

(2) Unsteady non-Newtonian liquid

As shown in Figure 5.3, the viscosity of such liquids is related to action time and can be divided into two categories:

① Thixotropic fluid

Apparent viscosity decreases with shear time, also known as shear thinning, such as paint.

② Rheopectic fluid

Apparent viscosity increases with shear time, also known as shear thickening, such as cement slurry.

(3) Viscoelastic non-Newtonian fluids

This kind of liquid has both the viscosity of liquid and the elasticity of solid, such as flour ball, asphalt and high molecular weight polymer.

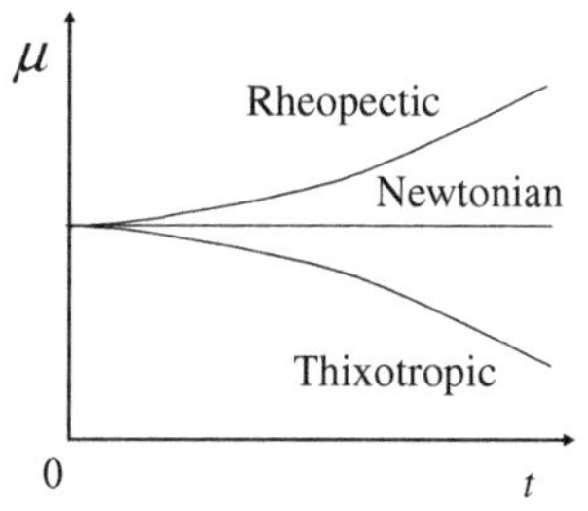

Figure 5.3. Viscosity-time curve of unsteady non-Newtonian liquid

5.1.3.2 *Generalized Darcy's law*

For one-dimensional power-law fluid flowing through porous media, generalized Darcy's law can be expressed as:

$$u_r^n = \frac{K}{\mu_{eff}} \frac{\partial p}{\partial r} \tag{5.12}$$

where u_r is radial surface velocity; and μ_{eff} is the effective viscosity which can be expressed as:

$$\mu_{eff} = \frac{H}{12}\left(9 + \frac{3}{n}\right)^n (150Kf)^{\frac{1-n}{2}} \tag{5.13}$$

When $n = 1$, $\mu_{eff} = H$ is Newtonian viscosity.

Sometimes in the oilfields, engineers also study the rheology of liquid by using Darcy formula to inversely calculate the viscosity, which is:

$$\mu_s = \frac{KA\Delta p}{QL} \tag{5.14}$$

5.1.3.3 *Steady-state flow of power-law fluids*

Applying generalized Darcy's law, the formula for radial steady state flow of power-law fluid is:

$$\frac{dp}{dr} = \frac{\mu_{eff}}{K} v^n \tag{5.15}$$

$$\mu_{eff} = \frac{H}{12}\left(9 + \frac{3}{n}\right)^n (150Kf)^{\frac{1-n}{2}} \tag{5.16}$$

$$v = \frac{Q}{2\pi r h} \tag{5.17}$$

Substitute Equations (5.16) and (5.17) into Equation (5.15), and integrate both sides after separating variables:

$$\int_{p_w}^{p_e} dp = \frac{\mu_{eff}}{K}\left(\frac{Q}{2\pi h}\right)^n \int_{R_w}^{R_e} \frac{dr}{r^n} \tag{5.18}$$

The equation of power-law fluid under steady state flow is obtained as:

$$Q = 2\pi h \left[\frac{(1-n)(p_e - p_w)}{\frac{\mu_{\mathit{eff}}}{K}(R_e^{1-n} - R_w^{1-n})} \right]^{\frac{1}{n}} \tag{5.19}$$

5.1.3.4 *Transient flow of power-law fluids*

The mathematical model of radial transient flow can be expressed as:

$$\frac{1}{r}\frac{\partial}{\partial r}\left(\frac{r}{\mu}\frac{\partial p}{\partial r}\right) = \frac{\phi C_t}{K}\frac{\partial p}{\partial t} \tag{5.20}$$

Liquid viscosity is a variable.

Let

$$\frac{r}{R_w} = e^u \tag{5.21}$$

Then:

$$r = R_w e^u \tag{5.22}$$

$$dr = R_{\mathrm{w}} e^u du = r du \tag{5.23}$$

$$\frac{\partial p}{\partial r} = \frac{\partial p}{\partial u}\frac{\partial u}{\partial r} = \frac{\partial p}{\partial u}\frac{1}{r} \tag{5.24}$$

$$\frac{1}{r}\frac{\partial}{\partial r}\left(\frac{r}{\mu}\frac{\partial p}{\partial r}\right) = \frac{1}{r}\frac{\partial}{\partial u}\left(\frac{1}{u}\frac{\partial p}{\partial u}\right)\frac{du}{dr} = \frac{1}{r^2}\frac{\partial}{\partial u}\left(\frac{1}{u}\frac{\partial p}{\partial u}\right) \tag{5.25}$$

And:

$$\frac{\partial}{\partial u}\left(\frac{1}{\mu}\frac{\partial p}{\partial u}\right) = R_w^2 e^{2u}\frac{\phi C_t}{K}\frac{\partial p}{\partial t} \tag{5.26}$$

Write the above equation in the finite differential form:

$$\frac{1}{\Delta u}\left[\frac{1}{\mu_{i-\frac{1}{2},n+1}}\left(\frac{p_{i-1,n+1} - p_{i,n+1}}{\Delta u}\right) - \frac{1}{\mu_{i+\frac{1}{2},n+1}}\left(\frac{p_{i,n+1} - p_{i+1,n+1}}{\Delta u}\right)\right] = R_w^2 e^{2\mu_{i,n}}\frac{\phi C_t}{K}\left(\frac{p_{i,n+1} - p_{i,n}}{\Delta t}\right) \tag{5.27}$$

Initial and boundary conditions are as follows:

$$p(r, 0) = p_0 \tag{5.28}$$

$$q = \frac{2\pi K h}{\sqrt{\mu}} \left(\frac{p_{1,n+1} - p_{0,n+1}}{\Delta u} \right) \tag{5.29}$$

$$\left. \frac{\partial p}{\partial r} \right|_{r=R_e} = 0 \tag{5.30}$$

Then the numerical simulation method is used to solve the problem.

5.2 Non-isothermal flow

5.2.1 *Introduction*

In classical flow mechanics in porous media, one of the assumptions is that the formation is isothermal. In the actual process of reservoir development, thermal conduction and exchange are often involved, such as thermal recovery of heavy oil reservoirs, gas non-isothermal flow, underground heating recovery of shale oil, and natural gas hydrate recovery.

In the case of non-isothermal flow, the physical parameters of the fluid and the formation are often functions of pressure and temperature, accompanied by changes in phase state and phase permeability. The flow field and temperature field exist simultaneously and affect each other, making the flow law more complicated. The establishment of the flow mathematical model requires the flow equation and energy equation.

The flow equation is based on Darcy's law and mass conservation law, while the energy equation is based on Fourier's law (fundamental law of thermal conduction) and energy conservation equation.

5.2.2 *Concepts*

Fundamental law of thermal conduction: *it is also known as Fourier law, and it shows that the thermal velocity of heat in the* *medium is linearly related to the temperature gradient, namely*

$$J_h = -\lambda_h gradT \tag{5.31}$$

where J_h *the heat velocity per unit area,* $J/(m^2 \cdot s)$; λ_h *is the thermal conductivity,* $J/m \cdot s \cdot K$.

Law of conservation of energy: *it is also the first law of thermodynamics, which can be expressed as the change in the total energy of a system is equal to the amount of energy coming in or out of the system. The total energy is the sum of the system's mechanical energy, thermal energy and any form of internal energy other than thermal energy.*

Specific heat capacity: *it is referred to as specific heat, and it is the absorption or release of heat for the unit mass of the body when the unit temperature changes. The symbol is* C_h, *and the unit is* $J/(kg \cdot K)$. *The specific heat capacities of water and oil are about* 4200 $J/(kg \cdot K)$ *and* 2000 $J/(kg \cdot K)$ *respectively, and the saturated water sandstone is* 1055 $J/(kg \cdot K)$.

Internal energy: *it is the sum of the kinetic energy of random thermal motion and the potential energy of intermolecular interaction of the molecules that make up an object. The unit is J.*

Heat exchange coefficient: *it is the heat exchanged by convection between solid surface and fluid per unit temperature difference, per unit time and per unit area. The symbol is* h_n, *and the unit is* $J/(m^2 \cdot s \cdot K)$.

Thermal diffusivity: *it is a measure of the rate at which a temperature disturbance at one point in a body passes to another. The equation is:*

$$E_h = \frac{\lambda_h}{\rho C_h} \tag{5.32}$$

where E_h *is the thermal diffusion coefficient,* m^2/s. *The thermal diffusion coefficient of saturated water sandstone is* 4.13×10^4 m^2/s. C_h *is the specific heat capacity,* $J/(kg \cdot K)$.

Entropy: *it is a state function that represents the energy of a material system, defined here as the energy per unit mass of material, symbol H, unit J/kg.*

5.2.3 *Analysis*

5.2.3.1 *Heat transfer in flow*

(1) Thermal conduction

As long as there is a temperature gradient inside the fluid and inside the rock, there will be thermal conduction, and its law is Fourier basic law of thermal conduction, namely:

$$J_h = -\lambda_h gradT \tag{5.33}$$

It describes the linear relationship between the thermal velocity of heat in the medium and the temperature gradient.

The thermal conductivity of the rock is an order of magnitude greater than that of the fluid, so in thermal oil recovery, while the energy injected heats the rock as well as the crude oil, most of the energy is used to heat the rock.

(2) Thermal diffusion

Similar to the diffusion phenomenon caused by the difference of component concentration, diffusion also exists in heat, which is divided into molecular thermal diffusion and convective thermal diffusion. Generally, the comprehensive thermal diffusion coefficient is expressed as:

$$E_h = \frac{\lambda_h}{\rho C_h} \tag{5.34}$$

(3) Heat exchange

If the heat is injected into the pore through the fluid, there is a temperature difference between the liquid in the pore and the rock. In this case, there is heat exchange or heat loss between the two media. Its equation is:

$$J_{hn} = h_n(T_f - T_l) \tag{5.35}$$

where J_{hn} is the heat velocity generated by interfacial heat exchange, $\mathrm{J/(m^2 \cdot s)}$; T_f is the temperature of the rock, K; and T_l is the temperature of the fluid, K.

(4) Thermal radiation

When pores are full of gas, thermal radiation occurs between solid particles, which is not considered in flow.

5.2.3.2 *Governing equation of flow*

Flow conforms to Darcy's law, then:

$$v_{li} = -\frac{KK_{rl}}{\mu_l}\frac{\partial p_l}{\partial x_i}(l = o, w) \tag{5.36}$$

According to the mass conservation law, the basic flow differential equation is established as:

$$\frac{\partial \rho_l}{\partial t} + \frac{\partial}{\partial x_i}(\rho_l v_{li}) = 0 \tag{5.37}$$

5.2.3.3 *Energy equation*

(1) Energy equation of fluid

The internal energy of liquid can be expressed by $U = C_h T$, then the energy equation can be established as follows:

$$\rho_l C_{hl}\left(\frac{\partial T_l}{\partial t} + v_i\frac{\partial T_l}{\partial x_i}\right) = -\frac{\partial J_h}{\partial x_i} - \frac{\partial}{\partial x_i}\left(E_{hi}\frac{\partial T_l}{\partial x_i}\right) + h_n(T_f - T_l) - F \tag{5.38}$$

where C_{hl} is the specific heat capacity of liquid; T_l is the initial temperature of the liquid; T_f is the final temperature of the liquid; x_i is the direction in x, y and z; v_i is the flow velocity in a certain direction; E_{hi} is the diffusion coefficient in one direction; F is the heat loss of rock at the top and bottom of reservoir; ρ_l is greatly affected by temperature, and its equation of state is $\rho_l = \rho_0[1 - C_{TL}(T_l - T_{l0})]$.

In the above equation, the first term on the left indicates the change of internal energy of the liquid; the second term on the left represents the change of energy carried by the flow; the first term on the right represents the change of energy generated by thermal conduction of the liquid; the second term on the right represents the change of energy generated by the thermal diffusion of the liquid; the third term on the right represents the change of energy generated by the heat exchange between rock and liquid; and the fourth item on the right represents the energy loss generated by the rock at the top and bottom of the reservoir.

(2) Energy equation of rock

$$\rho_f C_{hf}\frac{\partial T_f}{\partial t} = -\frac{\partial J_{hs}}{\partial x_i} - h_n(T_f - T_l) \tag{5.39}$$

where $J_{hs} = -\lambda_{hf}\frac{\partial T}{\partial x_i}$ is the velocity of heat transfer from liquid to solid rock.

5.2.3.4 *Comprehensive equations*

Assuming that the temperature balance between the rock and the liquid can be reached instantaneously, then the liquid energy equation and the rock energy equation can be combined.

The comprehensive energy equation is:

$$\begin{cases} [\phi\rho_l C_{hl} + (1-\phi)\rho_f C_{hf}]\dfrac{\partial T}{\partial t} + \rho_l C_{hl} v_i \dfrac{\partial T}{\partial x_i} \\ = \dfrac{\partial}{\partial x_i}[\phi\lambda_{hl} + (1-\phi)\lambda_{hf}]\dfrac{\partial T}{\partial x_i} - \dfrac{\partial}{\partial x_i}\left(\phi E_{hi}\dfrac{\partial T}{\partial x_i}\right) - F \end{cases} \tag{5.40}$$

The flow equation is:

$$\frac{\partial \rho_l \phi S_l}{\partial t} + \frac{\partial}{\partial x_i}(\rho_l v_{li}) = 0 \tag{5.41}$$

The equation of motion is:

$$v_{li} = -\frac{K K_{rl}}{\mu_l}\frac{\partial p_l}{\partial x_i}(l = o, w) \tag{5.42}$$

The viscosity of liquid is generally a function of temperature, that is, $\mu_l = \mu_l(T_l)$.

The equation of state is:

$$\rho_l = \rho_0[1 - C_{TL}(T_l - T_{l0})] \tag{5.43}$$

$$\phi = \phi_0[1 - C_{Tf}(T_f - T_{f0})] \tag{5.44}$$

The auxiliary equation is:

$$S_o + S_w = 1 \tag{5.45}$$

$$p_{cow} = p_o - p_w \tag{5.46}$$

Solve the above equation to obtain the pressure and temperature.

In the case of one-dimensional flow, if heat diffusion, heat loss and injection at the top and bottom of the reservoir are ignored, then:

$$[\phi\rho_l C_{hl} + (1-\phi)\rho_f C_{hf}]\frac{\partial T}{\partial t} = \frac{\partial}{\partial x_i}[\phi\lambda_{hl} + (1-\phi)\lambda_{hf}]\frac{\partial T}{\partial x_i} - \rho_l C_{hl} v_i \frac{\partial T}{\partial x_i} \tag{5.47}$$

The formation comprehensive specific heat C_{ht} and formation comprehensive thermal conductivity λ_{ht} are defined as:

$$C_{\text{ht}} = \phi\rho_l C_{hl} + (1-\phi)\rho_f C_{hf} \tag{5.48}$$

$$\lambda_{ht} = \phi\lambda_{hl} + (1-\phi)\lambda_{hf} \tag{5.49}$$

Then the basic differential equation of one-dimensional single-phase flow is:

$$C_{ht}\frac{\partial T}{\partial t} = \lambda_{ht}\frac{\partial^2 T}{\partial x^2} - \rho_l C_{hl} v \frac{\partial T}{\partial x} \tag{5.50}$$

If it is steady state, then:

$$\frac{\partial^2 T}{\partial x^2} = \frac{\rho_l C_{hl} v}{\lambda_{ht}}\frac{\partial T}{\partial x} \tag{5.51}$$

If the boundary condition is $x = 0$, $T = t_0$; $x = L$, $T = T_L$, then the solution is:

$$\frac{T - T_0}{T_L - T_0} = \frac{e^{\frac{ax}{L}} - 1}{e^a - 1} \tag{5.52}$$

where $a = \frac{C_{hl}\rho_l}{\lambda_{hl}} vL$ is a dimensionless heat transfer parameter. The flow velocity v can be obtained according to the flow equation, and it is a constant, then

$$a = \frac{KC_{hl}\rho_l}{\mu\lambda_{hl}}(p_e - p_w) \tag{5.53}$$

5.2.3.5 *Non-isothermal flow based on entropy*

The energy equation can also be derived based on entropy.

(1) Energy continuity equation

As shown in Figure 5.4, the energy exchange between a certain part of the rock and the surrounding rock mass in the reservoir includes heat transfer generated by fluid movement (flow), heat transfer generated by temperature difference, heat source E and heat loss F at the top and bottom of the reservoir.

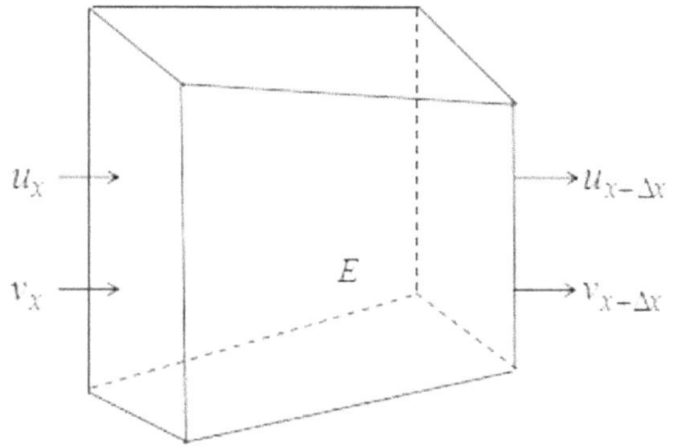

Figure 5.4. One-dimensional non-isothermal flow for REV

The energy entering the REV is:

$$(\mathrm{A}u)|_x\Delta t + (\mathrm{A}\rho_{\mathrm{w}}H_w v_w)|_x\Delta t + (\mathrm{A}\rho_{\mathrm{o}}H_o v_o)|_x\Delta t + E_{in}\mathrm{A}\Delta x\Delta t \tag{5.54}$$

The energy conducted out of the REV is:

$$\begin{aligned}&(\mathrm{A}u)|_{x+\Delta x}\Delta t + (\mathrm{A}\rho_{\mathrm{w}}H_w v_w)|_{x+\Delta x}\Delta t\\ &+ (\mathrm{A}\rho_{\mathrm{o}}H_o v_o)|_{x+\Delta x}\Delta t + F\mathrm{A}\Delta x\Delta t\end{aligned} \tag{5.55}$$

Energy changes in pore rocks include fluid energy changes and rock energy changes, so the increment of energy in the REV is:

$$\mathrm{A}\Delta x\left\{\begin{array}{l}[\varphi(\rho_w H_w S_w + \rho_o H_o S_o) + (1-\varphi)\rho_f H_R]|_{t+\Delta t}\\ -[\varphi(\rho_w H_w S_w + \rho_o H_o S_o) + (1-\varphi)\rho_f H_R]|_t\end{array}\right\} \tag{5.56}$$

where u is the thermal velocity and it is the same as J_h; H_w, H_o and H_R are the entropies of water, oil and rock respectively.

As can be seen from the increment of energy in the REV, in thermal oil recovery, the energy injected is used to heat the rock as well as the crude oil, and a large part of the energy is used to heat the rock. Therefore, if the energy of thermal fluid injected is not sufficient, the heating range will be very small.

According to the law of energy conservation, the equation of energy continuity is obtained as:

$$\begin{aligned}&-\frac{\partial u_x}{\partial x} - \frac{\partial(\rho_w H_w v_w)}{\partial x} - \frac{\partial(\rho_o H_o v_o)}{\partial x} + E_{in} - F\\ &= \frac{\partial}{\partial t}[\varphi(\rho_w H_w S_w + \rho_o H_o S_o) + (1-\varphi)\rho_f H_R]\end{aligned} \tag{5.57}$$

(2) Fluid motion equation and thermal conduction equation

According to Darcy's law, the fluid motion equation is:

$$v_l = -\frac{KK_{rl}}{\mu_l}\frac{\partial p_l}{\partial x}(l = o, w) \tag{5.58}$$

The thermal conductivity phenomenon satisfies Fourier's law, and the thermal conductivity equation is:

$$u = -\lambda_h gradT \tag{5.59}$$

(3) Mathematical model of non-isothermal flow

The equation of motion and the equation of thermal conduction are substituted into the energy equation as:

$$\lambda_h \frac{\partial^2 T}{\partial x^2} + \frac{\partial}{\partial x}\left(\rho_w H_w \frac{KK_{rw}}{\mu_w}\frac{\partial p_w}{\partial x}\right) + \frac{\partial}{\partial x}\left(\rho_o H_o \frac{KK_{ro}}{\mu_o}\frac{\partial p_o}{\partial x}\right) + E_{in} - F = \frac{\partial}{\partial t}[\phi(\rho_w H_w S_w + \rho_o H_o S_o) + (1-\phi)\rho_f H_R] \tag{5.60}$$

In the process of non-isothermal flow, in addition to satisfying the energy conservation relationship shown in the above equation, the fluid flow also satisfies the conservation of mass, that is,

$$\frac{\partial}{\partial x}\left(\rho_w \frac{KK_{rw}}{\mu_w}\frac{\partial p_w}{\partial x}\right) = \frac{\partial}{\partial t}(\phi \rho_w S_w) \tag{5.61}$$

$$\frac{\partial}{\partial x}\left(\rho_o \frac{KK_{ro}}{\mu_o}\frac{\partial p_o}{\partial x}\right) = \frac{\partial}{\partial t}(\phi \rho_o S_o) \tag{5.62}$$

Similarly, the mathematical model of three-dimensional non-isothermal flow can be written as:

$$\nabla(\lambda_h \nabla T) + \nabla\left(\rho_w H_w \frac{KK_{rw}}{\mu_w}\nabla p_w\right) + \nabla\left(\rho_o H_o \frac{KK_{ro}}{\mu_o}\nabla p_o\right) + E_{in} - F = \frac{\partial}{\partial t}[\phi(\rho_w H_w S_w + \rho_o H_o S_o) + (1-\phi)\rho_f H_R] \tag{5.63}$$

$$\nabla\left(\rho_w \frac{KK_{rw}}{\mu_w}\nabla p_w\right) = \frac{\partial}{\partial t}(\phi\rho_w S_w) \tag{5.64}$$

$$\nabla\left(\rho_o \frac{KK_{ro}}{\mu_o}\nabla p_o\right) = \frac{\partial}{\partial t}(\phi\rho_o S_o) \tag{5.65}$$

The auxiliary equation is:

$$S_o + S_w = 1 \tag{5.66}$$

$$p_{cow} = p_o - p_w \tag{5.67}$$

As can be seen from the mathematical model, temperature is a function of time and space, and fluid viscosity is a function of temperature. Therefore, the mathematical model of non-isothermal flow requires the combination of fluid physical property models to be solved.

5.3 Mass transfer by diffusion

5.3.1 *Introduction*

In the classical flow mechanics in porous media, the components in the fluid move according to Darcy's law, but when the injected fluid is with oil displacing agent, tracer, miscible gas, etc., the flow of these injectors in the reservoir is restricted by the so-called dispersion phenomenon in addition to Darcy's law. Under such circumstances, the phenomenon of "hydraulic dispersion" is often encountered in the study of flow.

The most easily understood phenomenon of hydraulic dispersion is "Brownian motion" or "molecular thermal motion", that is, when the fluid is not flowing, when an miscible tracer is injected, the diffusion is formed under the influence of the difference in concentration, which is also called molecular diffusion.

Generally speaking, hydraulic dispersion is caused by the following reasons: (1) molecular diffusion caused by concentration difference; (2) external forces acting on the fluid; (3) complex microstructure of pore space; (4) changes in flow caused by differences in fluid parameters such as density and viscosity; (5) interaction between liquid and solid phases, etc.

Due to the existence of hydraulic dispersion, the mass transfer in the flow process is composed of three aspects: average flow caused by Darcy's law, molecular diffusion caused by concentration gradient and convective diffusion caused by mechanical dispersion.

5.3.2 *Concepts*

Brownian motion: *it is the constant random motion of the molecules of all matter.*

Hydraulic dispersion: *it is the mass transfer in the dilution process of solute tracers. There are two main ways: molecular diffusion and convection diffusion.*

Molecular diffusion: *it is the movement of matter through space due to the thermal motion of molecules, atoms, etc., under differential concentration or other driving forces.*

Convective diffusion: *when fluid moves, solute tracer moves in different directions in the pores, which causes continuous dispersion in the pores and occupies more and more pore space. It is also known as mechanical diffusion.*

Longitudinal diffusion: *the diffusion direction in the medium is consistent with the flow direction.*

Transverse diffusion: *the diffusion direction in the medium is perpendicular to the flow direction. When the flow velocity is 0, transverse diffusion also exists.*

Fick diffusion law: *it describes the relationship between diffusion velocity and concentration gradient during molecular diffusion when mass transfer occurs independently of macroscopic mixing.*

Gaussian error function: *it can be defined as*

$$erf(\xi) = \frac{2}{\sqrt{\pi}} \int_0^{\xi} e^{-\xi^2} d\xi \qquad (5.68)$$

Adsorption: *when a fluid is in contact with a porous solid, one or more components of the fluid accumulate on the solid surface. This* *phenomenon is called adsorption. The reverse process of adsorption is called desorption.*

5.3.3 Analysis

5.3.3.1 Diffusion equation

The diffusion velocity of the ideal diffused material along the flow direction can be expressed by Fick's diffusion law, which indicates that the mass flow rate of tracer on unit time and unit area caused only by the diffusion phenomenon can be expressed as:

$$u = -D^* \frac{\partial C}{\partial x} \tag{5.69}$$

where D^* is the diffusion coefficient, cm^2/s; C is the tracer concentration, $\mathrm{g/cm}^3$; and u is the mass flow rate, $\mathrm{g/(cm^2 s)}$.

5.3.3.2 Continuity equation

As shown in Figure 5.5, in the REV, the mass velocity of the diffused material at point M is u_i, and the mass velocity of the component at point M' is:

$$u_i - \frac{\partial u_i}{\partial x}\frac{dx}{2} \tag{5.70}$$

After dt time, the mass flow rate of the inflow surface $a'b'$ is:

$$\left(u_i - \frac{\partial u_i}{\partial x}\frac{dx}{2}\right) dydzdt \tag{5.71}$$

Similarly, the mass velocity of the component at point M'' is:

$$\left(u_i + \frac{\partial u_i}{\partial x}\frac{dx}{2}\right) \tag{5.72}$$

The mass flow rate of the outflow surface $a''b''$ after dt time is:

$$\left(u_i + \frac{\partial u_i}{\partial x}\frac{dx}{2}\right) dydzdt \tag{5.73}$$

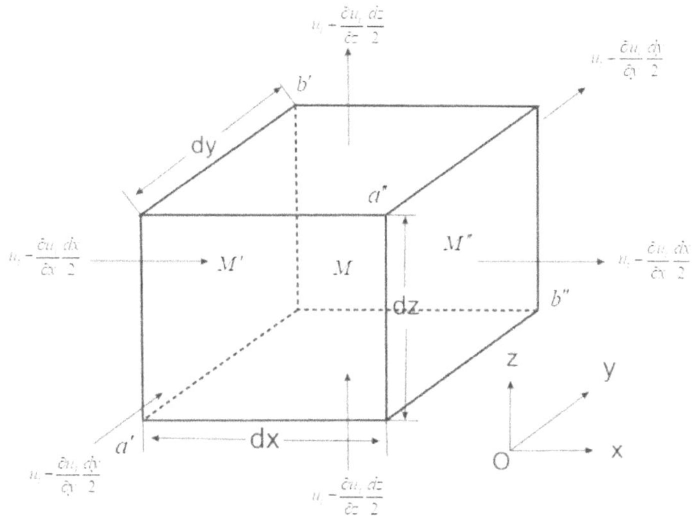

Figure 5.5. REV used to derive the continuity equation of tracer diffusion

In the x direction, the mass difference between the inflow and outflow in dt time is:

$$-\frac{\partial u_i}{\partial x} dxdydzdt \tag{5.74}$$

Similarly, the mass difference between inflow and outflow in the direction of y and z is:

y direction:

$$-\frac{\partial u_i}{\partial y} dxdydzdt \tag{5.75}$$

z direction:

$$-\frac{\partial u_i}{\partial z} dxdydzdt \tag{5.76}$$

The mass flow difference in dt time is:

$$-\left(\frac{\partial u_i}{\partial x} + \frac{\partial u_i}{\partial y} + \frac{\partial u_i}{\partial z}\right) dxdydzdt \tag{5.77}$$

At t time, the fluid mass concentration in the cell is C, and the concentration changes $(\partial C/\partial t)dt$ during dt time. The pore volume of the cell is $\phi\ dxdydz$, and the mass change caused by the change in mass concentration is:

$$\phi\frac{\partial C}{\partial t}dxdydzdt \tag{5.78}$$

According to the mass conservation law, the mass changes calculated by the two methods should be equal, then we have:

$$-\left(\frac{\partial u_i}{\partial x}+\frac{\partial u_i}{\partial y}+\frac{\partial u_i}{\partial z}\right)=\frac{\partial(\phi C)}{\partial t} \tag{5.79}$$

If the fluid flow is considered, the concentration change caused by the flow needs to be added, namely:

$$-\left(\frac{\partial u_i}{\partial x}+\frac{\partial u_i}{\partial y}+\frac{\partial u_i}{\partial z}\right)-\left(v_x\frac{\partial C}{\partial x}+v_y\frac{\partial C}{\partial y}+v_z\frac{\partial C}{\partial z}\right)=\frac{\partial(\phi C)}{\partial t} \tag{5.80}$$

Assume that the rock is incompressible. Divide the diffusion velocity u_i and the flow velocity v by the porosity ϕ, and obtain the true diffusion velocity u and the true flow velocity v_ϕ of the pore. Then the equation becomes:

$$-\left(\frac{\partial u}{\partial x}+\frac{\partial u}{\partial y}+\frac{\partial u}{\partial z}\right)-\left(v_{\phi x}\frac{\partial C}{\partial x}+v_{\phi y}\frac{\partial C}{\partial y}+v_{\phi z}\frac{\partial C}{\partial z}\right)=\frac{\partial C}{\partial t} \tag{5.81}$$

The above equation is the continuity equation of mass transfer by diffusion.

5.3.3.3 *One-dimensional flow equation with diffusion*

From the continuity equation, the one-dimensional flow equation is obtained as:

$$-\frac{\partial u}{\partial x}-v_\phi\frac{\partial C}{\partial x}=\frac{\partial C}{\partial t} \tag{5.82}$$

Substitute into the diffusion equation:

$$D^* \frac{\partial^2 C}{\partial x^2} - v_\phi \frac{\partial C}{\partial x} = \frac{\partial C}{\partial t} \tag{5.83}$$

The first term on the left is the concentration change caused by diffusion; the second term on the left is the concentration change caused by fluid flow; and the right side term is the accumulation term.

If the diffusion effect is not taken into account, the change in solute concentration is only caused by the flow. The equation is:

$$-v_\phi \frac{\partial C}{\partial x} = \frac{\partial C}{\partial t} \tag{5.84}$$

This is a one-dimensional wave equation. The concentration distribution equation can be solved as:

$$C(x,t) = \frac{C_1}{2}\left[1 - erf\left(\frac{x - v_\phi t}{2\sqrt{D^* t}}\right)\right] \tag{5.85}$$

where C_1 is the concentration at the beginning of the diffusion. The Gaussian error function is:

$$erf(\xi) = \frac{2}{\sqrt{\pi}} \int_0^\xi e^{-\xi^2} d\xi \tag{5.86}$$

5.3.3.4 *One-dimensional flow equation with adsorption and diffusion*

Chemical agents in many oil reservoirs not only have diffusion effect, but also adsorb on the surface of rock solid particles due to the action of molecular force or electrostatic field, and the adsorption will affect the change of concentration.

Let C_r be the adsorption concentration, and C_r^* be the critical concentration for adsorption, that is, no adsorption is allowed after reaching this threshold value. K_1 and K_2 are the adsorption constant and desorption constant respectively, unit s^{-1}. The adsorption velocity can be written as:

$$\left(\frac{dC_r}{dt}\right)_r = K_1\left(1-\frac{C_r}{C_r^*}\right)C \tag{5.87}$$

Desorption corresponds to adsorption, which is a dynamic equilibrium process. The velocity can be written as:

$$\left(\frac{dC_r}{dt}\right)_d = -K_2\frac{C_r}{C_r^*} \tag{5.88}$$

Then, the relationship between the total adsorption concentration and time is:

$$\frac{dC_r}{dt} = K_1\left(1-\frac{C_r}{C_r^*}\right)C - K_2\frac{C_r}{C_r^*} \tag{5.89}$$

If time approaches infinity, the above equation can be rewritten as:

$$C_r(\infty) = \frac{aC}{1+bC} \tag{5.90}$$

where $a = \frac{K_1 C_r^*}{K_2}$, $b = \frac{K_1}{K_2}$.

This formula is the true equilibrium adsorption concentration formula, also known as Langmuir isothermal adsorption formula.

In the unit volume rock with porosity φ, the volume of the particles is $1-\varphi$, and part of the volume of the particles S_r is the adsorption area. Thus, in the rock of unit volume with porosity, the adsorption volume should be $(1-\varphi)S_r/\varphi$, and the concentration equation with adsorption and diffusion is:

$$D^*\frac{\partial^2 C}{\partial x^2} - v_\phi\frac{\partial C}{\partial x} = \frac{\partial C}{\partial t} + \frac{1-\phi}{\phi}S_r\frac{\partial C_r}{\partial t} \tag{5.91}$$

According to the Langmuir equation, there is:

$$\frac{\partial C_r}{\partial t} = \frac{dC_r}{dC}\frac{\partial C}{\partial t} = \frac{a}{(1+bC)^2} \tag{5.92}$$

Substitute into the equation:

$$D^*\frac{\partial^2 C}{\partial x^2} - v_\phi\frac{\partial C}{\partial x} = \left[1 + \frac{(1-\phi)\,S_r a}{\phi(1+bC)^2}\right]\frac{\partial C}{\partial t} \tag{5.93}$$

The equation is a second order non-linear partial differential equation with variable coefficients. You can solve it numerically to get an approximate solution, or you can solve it analytically under special conditions.

5.4 Multiphase and multicomponent flow

5.4.1 *Introduction*

Traditional flow mechanics in porous media are based on the "phase" as the object of the flow model, such as liquid phase, gas phase, oil phase or water phase. The famous black oil model assumes that the gas component exists in the gas phase and the oil phase, the oil phase only and the water phase only. In addition to oil, gas and water components, sometimes there are other chemicals in the reservoir, such as polymers, surfactants, etc., and it is necessary to study the concentration distribution of chemical components in the reservoir. In this case, multiphase and multicomponent flow plays an important role.

Multiphase and multicomponent flow is the key model used for chemical flooding, heavy reservoir development and gas hydrate development.

Chemical flooding is an important branch of enhanced oil recovery. The chemical agents used for chemical flooding include polymer, surfactant, alkali, etc. These chemicals interact with residual oil or formation to improve oil recovery. High pressure gas injection and rich gas injection are often used in heavy reservoirs to form miscible development mode. The recent development of combustible ice is also a problem of multiphase and multicomponent flow.

Multiphase multicomponent flow refers to the flow of a mixture of components in subsurface pore media, which may exist in liquid state or gaseous state. In the process of flow, the phases and components exchange with each other, and the phase transition process is very complex. The theory of multiphase and multicomponent flow needs to be studied extensively to solve such problems.

Multiphase multicomponent flow process includes a series of complicated physical and chemical changes. At present, the process of multiphase multicomponent flow associated with physical and chemical changes is extremely difficult to establish an accurate mathematical model, or even to establish basic differential equations. In addition, the equation is also very difficult to solve, and generally needs to be solved numerically under some specific conditions. In this section, we introduce the establishment process of multiphase and multicomponent flow model under isothermal conditions.

5.4.2 *Concepts*

Black oil model: *it is also known as the low volatile oil two-component model, and it is a mathematical model that describes the movement rules of two fluid systems in the reservoir. The two fluids are black oil with non-volatile components and dissolved gas with volatile components.*

Phase state: *it is the state that a macroscopic physical system has. The most common states are solid, liquid and gas.*

Phase diagram: *it is also known as phase equilibrium diagram. It is used to show the composition of the phase equilibrium system and the relationship of some parameters (such as temperature and pressure) in a diagram. There are generally p-T phase diagrams, p-V phase diagrams and triangular phase diagrams.*

Phase equilibrium: *under certain conditions, a system is said to be in phase equilibrium when the properties and number of phases in the system do not change with time. In phase equilibrium, the content of each component in phase is determined.*

Phase equilibrium constant: *it refers to the ratio of the molar fraction of a certain component in the gas phase to the molar fraction of the same component in the liquid phase when the gas-liquid phase reaches an equilibrium state at a certain temperature and pressure, also known as the equilibrium ratio. The symbol is* K_i.

Mass fraction: *it is the ratio of the mass of a component in a phase to the total mass of all components in that phase, symbol* C_i.

5.4.3 *Analysis*

5.4.3.1 *Phase state of multicomponent systems*

In a multiphase multicomponent system, each component can exist in a certain equilibrium proportion in the gas phase or liquid phase, and its mass fraction in each phase should be determined according to the phase diagram. Take the triangular diagram of methane (CH_4), n-butane (n-C_4H_{10}) and decane ($C_{10}H_{22}$) as an example, and the phase diagram of these components are as shown in Figure 5.6 with the temperature as 344.26 K.

Assuming that an equilibrium system is composed of n components, C_{ig} represents the mass fraction of the ith component in the gas phase, C_{io} represents the mass fraction of the ith component in the oil phase, then:

$$\sum_{i=1}^{n} C_{ig} = 1, \quad \sum_{i=1}^{n} C_{io} = 1 \tag{5.94}$$

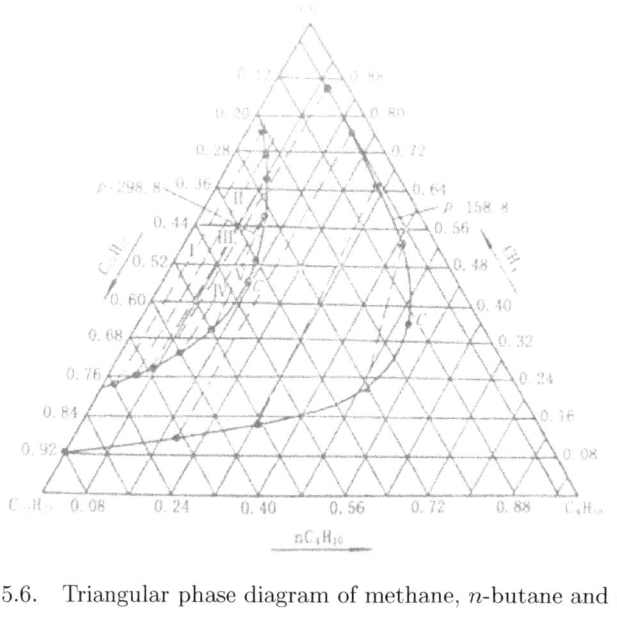

Figure 5.6. Triangular phase diagram of methane, n-butane and decane

In addition, under a certain temperature and pressure, the phase equilibrium constant of the components is determined, that is,

$$K_i = \frac{C_{ig}}{C_{io}} \tag{5.95}$$

where K_i is the phase equilibrium constant of the ith component; C_{ig} is the mass fraction of the ith component in the gas phase; and C_{io} is the mass fraction of the ith component in the oil phase.

5.4.3.2 *Differential equations of multiphase and multicomponent flow*

The assumptions of the model are: (1) the flow in reservoir is isothermal flow; (2) the fluid flow in reservoir is linear and accords with Darcy's law; (3) there are N components in the reservoir fluid; (4) the phase state reaches equilibrium instantly in the reservoir; (5) the effects of gravity and capillary force are not considered; and (6) the porous media are rigid and incompressible.

(1) Continuity equation

Take the single component i as an example, and suppose that its mass fractions in three phases of oil, gas and water are C_{io}, C_{ig} and C_{iw} respectively. Then:

The mass difference of component i in the gas phase flowing in and out of REV during dt time in the REV is:

$$-\mathrm{div}(\vec{v}_g\, \rho_g C_{ig})dxdydzdt \tag{5.96}$$

The mass difference of component i in the oil phase flowing in and out of REV during dt time is:

$$-\mathrm{div}(\vec{v}_o\, \rho_o C_{io})dxdydzdt \tag{5.97}$$

The mass difference of component i in the water phase flowing in and out in the REV during dt time is:

$$-\mathrm{div}(\vec{v}_w\, \rho_w C_{iw})dxdydzdt \tag{5.98}$$

The pore volume of REV is:

$$\phi dxdydz \tag{5.99}$$

The oil phase mass of component i in the REV is:

$$C_{io}\phi\rho_o S_o dxdydz \tag{5.100}$$

The mass of component i in the gas phase in the REV:

$$C_{ig}\phi\rho_g S_g dxdydz \tag{5.101}$$

The mass of component i in the liquid phase in the REV is:

$$C_{iw}\phi\rho_w S_w dxdydz \tag{5.102}$$

The change rate of fluid mass in dt time is:

$$\frac{\partial(C_{io}\phi\rho_o S_o + C_{ig}\phi\rho_g S_g + C_{iw}\phi\rho_w S_w)}{\partial t} dxdydzdt \tag{5.103}$$

Fluid mass change rate within dt time is:

$$\frac{\partial(C_{io}\phi\rho_o S_o + C_{ig}\phi\rho_g S_g + C_{iw}\phi\rho_w S_w)}{\partial t} dxdydzdt \tag{5.104}$$

According to the law of mass conservation, there is:

$$\begin{aligned} &\operatorname{div}(\vec{v}_g\,\rho_g C_{ig} + \vec{v}_o\,\rho_o C_{io} + \vec{v}_w\,\rho_w C_{iw}) \\ &\quad = -\phi\frac{\partial(C_{io}\rho_o S_o + C_{ig}\rho_g S_g + C_{iw}\rho_w S_w)}{\partial t} \end{aligned} \tag{5.105}$$

(2) Equation of motion

According to Darcy's law, the equations of motion for oil, gas and water are respectively:

$$\begin{aligned} &\vec{v}_o = -\frac{KK_{ro}}{\mu_o}\operatorname{grad}(p); \quad \vec{v}_g = -\frac{KK_{rg}}{\mu_g}\operatorname{grad}(p); \\ &\vec{v}_w = -\frac{KK_{rw}}{\mu_w}\operatorname{grad}(p) \end{aligned} \tag{5.106}$$

(3) Basic differential equations

The equation of motion is substituted into the continuity equation:

$$\operatorname{div}\left[\left(\frac{C_{io}K_{ro}\rho_o}{\mu_o}+\frac{C_{ig}K_{rg}\rho_g}{\mu_g}+\frac{C_{iw}K_{rw}\rho_w}{\mu_w}\right)grad(p)\right] = \frac{\varphi}{K}\frac{\partial(C_{io}\rho_oS_o+C_{ig}\rho_gS_g+C_{iw}\rho_wS_w)}{\partial t} \tag{5.107}$$

(4) Auxiliary equations

The saturations satisfy:

$$S_o+S_g+S_w=1 \tag{5.108}$$

The components satisfy:

$$\sum_{i=1}^{n}C_{io}=1,\quad \sum_{i=1}^{n}C_{ig}=1,\quad \sum_{i=1}^{n}C_{iw}=1 \tag{5.109}$$

The phase equilibrium condition is:

$$k_{igo}=\frac{C_{ig}}{C_{io}},\quad k_{iow}=\frac{C_{io}}{C_{iw}} \tag{5.110}$$

5.4.3.3 *Multiphase and multicomponent flow in special cases*

In the above model, the flow models in special cases are considered.

(1) Two-phase flow of oil and gas

When there are only two phases of oil and gas, the oil component is only in the oil phase, while the gas component is in the oil phase and gas phase, that is, $C_{oo}=1$, $C_{og}=0$, $C_{go}+C_{gg}=1$.

For the oil component, it has:

$$\operatorname{div}\left[\left(\frac{K_{ro}\rho_o}{\mu_o}\right)grad(p)\right]=\frac{\phi}{K}\frac{\partial(\rho_oS_o)}{\partial t} \tag{5.111}$$

For the gas component, it has:

$$\operatorname{div}\left[\left(\frac{C_{go}K_{ro}\rho_o}{\mu_o}+\frac{C_{gg}K_{rg}\rho_g}{\mu_g}\right)grad(p)\right]=\frac{\phi}{K}\frac{\partial(C_{go}\rho_oS_o+C_{gg}\rho_gS_g)}{\partial t} \tag{5.112}$$

(2) Black oil model

The oil component is only in the oil phase, the water component is only in the water phase, but the gas component is in both the oil phase and the gas phase, that is, $C_{oo} = 1$, $C_{ww} = 1$, $C_{gw} = 0$, $C_{go} + C_{gg} = 1$.

For the oil component, it has:

$$\text{div}\left[\left(\frac{K_{ro}\rho_o}{\mu_o}\right) grad(p)\right] = \frac{\phi}{K}\frac{\partial(\rho_o S_o)}{\partial t} \tag{5.113}$$

For the water component, it has:

$$\text{div}\left[\left(\frac{K_{rw}\rho_w}{\mu_w}\right) grad(p)\right] = \frac{\phi}{K}\frac{\partial(\rho_w S_w)}{\partial t} \tag{5.114}$$

For the gas component, it has:

$$\text{div}\left[\left(\frac{C_{go}K_{ro}\rho_o}{\mu_o} + \frac{C_{gg}K_{rg}\rho_g}{\mu_g}\right) grad(p)\right] = \frac{\phi}{K}\frac{\partial(C_{go}\rho_o S_o + C_{gg}\rho_g S_g)}{\partial t} \tag{5.115}$$

(3) Condensate gas reservoir

The condensate components are in the oil phase and gas phase, the gas component is in the gas phase and oil phase, the heavy oil component is only in the oil phase, and the water component is only in the water phase, namely $C_{ww} = 1$, $C_{Ho} = 1$, $C_{Lo} + C_{Lg} = 1$, $C_{Hw} = C_{Lw} = C_{gw} = 0$, $C_{go} + C_{gg} = 1$. C_{Hi} is the mass fraction of heavy oil component in each phase, and C_{Li} is the mass fraction of condensate component in each phase.

For the condensate component, it has:

$$\text{div}\left[\left(\frac{C_{Lo}K_{ro}\rho_o}{\mu_o} + \frac{C_{Lg}K_{rg}\rho_g}{\mu_g}\right) grad(p)\right] = \frac{\phi}{K}\frac{\partial(C_{Lo}\rho_o S_o + C_{Lg}\rho_g S_g)}{\partial t} \tag{5.116}$$

For the heavy oil component, it has:

$$\text{div}\left[\left(\frac{K_{ro}\rho_o}{\mu_o}\right) grad(p)\right] = \frac{\phi}{K}\frac{\partial(\rho_o S_o)}{\partial t} \tag{5.117}$$

For the gas component, it has:

$$\text{div}\left[\left(\frac{C_{go}K_{ro}\rho_o}{\mu_o}+\frac{C_{gg}K_{rg}\rho_g}{\mu_g}\right)grad(p)\right]=\frac{\phi}{K}\frac{\partial(C_{go}\rho_o S_o+C_{gg}\rho_g S_g)}{\partial t} \tag{5.118}$$

For the water component, it has:

$$\text{div}\left[\left(\frac{K_{rw}\rho_w}{\mu_w}\right)grad(p)\right]=\frac{\phi}{K}\frac{\partial(\rho_w S_w)}{\partial t} \tag{5.119}$$

5.5 Oil displacement with nanofluids

5.5.1 *Introduction*

Nanotechnology is to study the composition of mesoscopic substances at the nanometer scale (0.1 $\sim$ 100 nm), and create new substances by directly manipulating and arranging atoms and molecules. Due to the special surface effect, small size effect and quantum size effect, nano-materials show excellent optical, thermal, electrical, magnetic, mechanical and chemical properties. They are substantially different from common solid materials, and have huge potentials in aerospace, medicine, communications, industrial manufacturing, environment, energy and many other fields.

In June 2010, Saudi Aramco injected 250 bbl (barrel) of diluted nano-robots in Arab-D formation. After putting into production, the fluorescence spectrum analysis was used to detect the sample in the production wells. It has been confirmed that nano-robots have higher recovery rate, stability and flowing capability, which shows that nano-robotic research has made revolutionary progress. The application of nano-robots helps us improve the understanding on the formation fluid and pore structure, and makes the "tracking" of fluid flow in the formation and the identification of flow environment come true.

As early as 1966, nano-materials were applied in micro-emulsions for oil displacement. Compared with traditional chemical flooding to enhance oil recovery, the modified nano-particle dispersion system has good stimulation and injection effect in chemical flooding.

5.5.2 *Concepts*

Nanotechnology: *it refers to the study of the composition of mesoscopic materials at the nanometer scale* (0.1 ~ 100 nm), *the creation of new materials through direct manipulation and arrangement of atoms and molecules, and the application of nanometer materials and equipment in various fields.*

Nano-particles: *it refers to particles with size at the nano-scale, which is defined as less than 100 nm in at least one dimension.*

Capillary number: *it is the ratio of the driving force to the resistance of the remaining oil droplets in the pores of the oil reservoir, and the symbol is* N_c. *It has various expressions, one of which is*

$$N_c = \frac{v\mu}{\sigma} \tag{5.120}$$

where v *is the displacement velocity;* μ *is the viscosity; and* σ *is the interfacial tension.*

Interfacial tension: *it can be regarded as the contraction force acting on the liquid interface per unit length. The symbol is* σ, *and the common unit is mN/m. The gas-water interfacial tension under normal pressure is 72 mN/m, and the oil-water interfacial tension under reservoir conditions is about 20 mN/m.*

Viscosity-concentration curve: *it describes the relationship between the viscosity and concentration of a chemical.*

Injection pore volume: *it is the ratio of the volume of the injected liquid solution to the pore volume, and the symbol is PV.*

Mass concentration: *it is the ratio of the mass of a chemical in a solution to the volume of the solution, and the unit is usually in mg/L or ppm.*

5.5.3 *Analysis*

5.5.3.1 *Mechanism of nanotechnology in oil displacement*

(1) Reduction of interfacial tension

The main function of the oil displacement agent is to reduce the interfacial tension of residual oil droplets.

As shown in Figure 5.7, the relationship between the mass concentration of a nano-emulsion and the oil-water interfacial tension is described. In formation water system, with the increase in concentration of nano-emulsion, the oil-water interfacial tension gradually decreases and eventually reaches the order of 10^{-3} mN/m, which is the ultra-low interfacial tension.

Most of the capillary number in the water flooding of an actual oilfield is in the range of $10^{-7} \sim 10^{-5}$. In order to reduce the residual oil, the capillary number needs to be increased by $10^2 \sim 10^4$. In actual production, it is impossible to increase the production pressure difference, and thus the interfacial tension between oil and water needs to be reduced. At this time, the oil-water interfacial tension is reduced to $10^{-2} \sim 10^{-3}$ orders of magnitude.

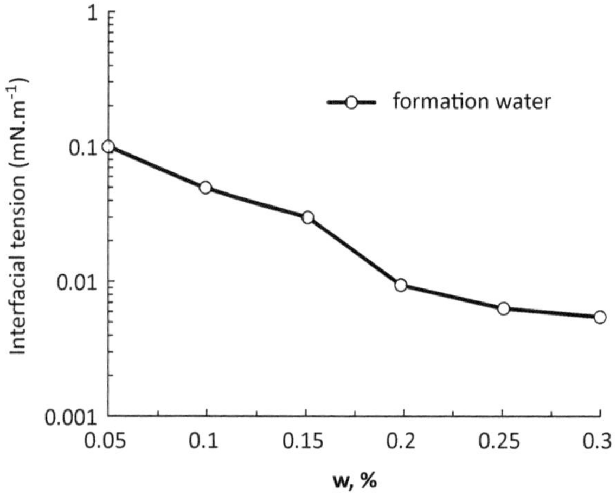

Figure 5.7. Influence of nano-emulsion concentration on oil-water interfacial tension

Nano-emulsions can reduce the interfacial tension between oil and water, thus producing a better oil displacement effect. In principle, the flow mechanism lies in oil droplet deformation and emulsification.

After water flooding, the residual oil disperses into pores and contacts with the nano-emulsion. As a result, the oil-water interfacial tension can be reduced, oil beads are deformed, the Garmin effect and adhesion to rocks are reduced, and the oil can flow with the displacing fluid.

During the oil droplet deformation of the residual oil, the oil droplet elongates or breaks under the action of the shear stress of the displacing fluid flow, the emulsification takes effect and it forms oil-water droplets, which can pass through smaller pores.

(2) Reduction of the mobility ratio

The silicone based nano-material used for the oil displacement can reduce the mobility ratio. The main composition of such nano-material is SiO_2 particles, and the particle diameter is generally $5 \sim 10\,\text{nm}$. It has good stability and optimal viscosity.

Due to its stable viscosity, it can continuously and effectively increase the viscosity of displacing phase, reduce mobility ratio and improve sweep efficiency in the process of oil displacement. In addition to its main function, the nano-particles can change the wettability of rock, increase permeability of the oil phase and improve the oil displacement efficiency.

5.5.3.2 *Mathematical model of nano-particle flow*

The multiphase and multicomponent flow model is applied to describe the flow of nano-particles.

The oil and water equations are consistent with the compositional model:

$$\operatorname{div}\left(\frac{K_{ro}\rho_o}{\mu_o} grad(p)\right) = \frac{\phi}{K}\frac{\partial(\rho_o S_o)}{\partial t} \tag{5.121}$$

$$\operatorname{div}\left(\frac{K_{rw}\rho_w}{\mu_w} grad(p)\right) = \frac{\phi}{K}\frac{\partial(\rho_w S_w)}{\partial t} \tag{5.122}$$

The equation of the displacing agent considering adsorption is:

$$\operatorname{div}\left(C\frac{KK_{rw}}{\mu_w}\,grad(p)\right) = \frac{\partial}{\partial t}(\phi C S_w + C_r) \tag{5.123}$$

where C_r is the adsorption concentration, which can be obtained by using the Langmuir isothermal adsorption equation. That is:

$$C_r = \frac{aC}{1 + bC} \tag{5.124}$$

where b is the constant related to temperature and pressure.

In the basic differential equation, the viscosity of water needs to be determined by the viscosity-concentration curve of the nano-particles. The relationship between the viscosity and concentration in nano-particle solutions is:

$$\mu_{we} = \mu_{w0}e^{\lambda C} \tag{5.125}$$

where μ_{w0} and μ_{we} are respectively the viscosity of the solution before and after the addition of nano-particles; λ is the viscosity growth index; and C is the mass concentration of the solution, mg/L.

The viscosity-concentration curve of a nano-particle solution is shown in Figure 5.8.

The effect of interfacial tension reduction in the flow mathematical model is to modify the relative permeability curve.

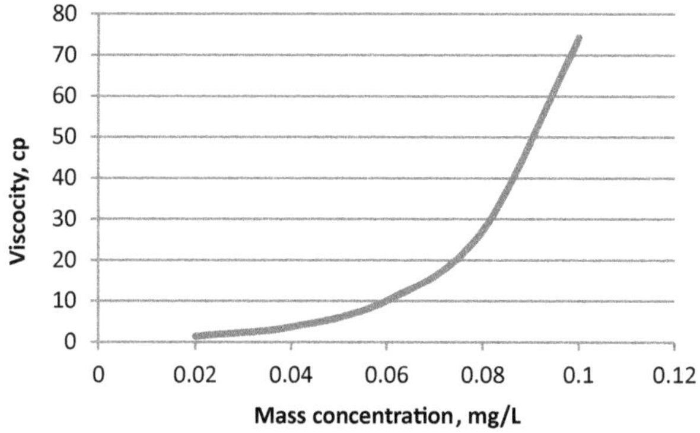

Figure 5.8. Viscosity-concentration curve of a nano-particle solution

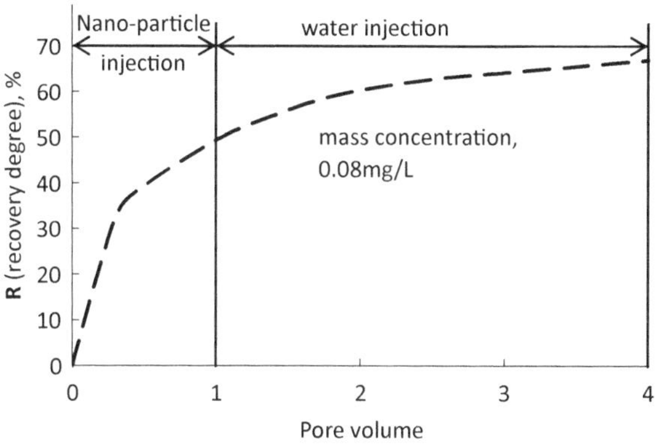

Figure 5.9. Relationship between injection pore volume and recovery degree

The displacing agent reduces the interfacial tension of the oil drop, and makes the residual oil saturation on the relative permeability curve move to the right. The correction steps are as follows:

According to the experiment results, the curve relating the concentration C of displacing fluid to the interfacial tension σ is plotted, and the concentration C of displacing fluid and the capillary number N_c curve are analyzed.

Combined with displacement efficiency, the relationship between residual oil saturation S_{or} and displacing fluid concentration is analyzed to determine the lowest residual oil saturation.

Compared with the water flooding results, the endpoint value of oil relative permeability curve is determined.

For the shape of the oil-water relative permeability curve, it is processed in two ways: in the immiscible phase, index n is reduced according to the normalized relative permeability curve; and in the miscible phase, similar to piston-like oil displacement characteristic, exponent n is equal to 1.

When evaluating the oil displacement performance, the relationship between the pore volume of the injected solution and the recovery degree is often used to reflect the effect of oil displacement. Figure 5.9 shows the displacement effect of a solution.

Chapter 6

Flow in unconventional reservoirs

Unconventional reservoirs and fluids highlight the complex situation of flow. In order to adapt to the development trend of unconventional resources, this chapter introduces the flow problems in unconventional reservoirs briefly, which mainly include shale gas, tight oil, coal bed methane, shale oil, natural gas hydrate, geothermal, water-soluble gas, etc. Each part focuses on its basic geological and resource characteristics, flow characteristics, key parameters and factors considered in the establishment of the mathematical model, so as to provide the basic understanding on these problems.

6.1 Flow in shale gas reservoirs

6.1.1 *Introduction*

The breakthrough of gas energy in the United States introduces a new important member, which is shale gas. The United States produced $599 \times 10^8\,m^3$ of shale gas in 2008, accounting for 10.5% of the country's total natural gas production. By 2015, its shale gas production and proportion reached $4296 \times 10^8\,m^3$ and 56.1%.

China's gas energy development has new momentum in recent years. Since the breakthrough of shale gas development in 2010, China's shale gas production has reached $100 \times 10^8\,m^3$ in 2017. According to China's shale gas development plan, production is expected to reach $300 \times 10^8\,m^3$ in 2020 and $1000 \times 10^8\,m^3$ in 2030. It can be seen that the development of shale gas takes an important strategic position in China's energy policy.

Shale gas is an unconventional gas resource. The key characteristics of the shale gas reservoir are the self-generation and self-storage gas reservoir, the nanoscale pores in shale matrix, the existence of the natural and hydraulic fractures, the multi-scale flow, the adsorption and desorption of gas, the low production rate, and the long production period. The characteristic parameters of shale gas reservoirs are summarized in Table 6.1.

6.1.2 *Concepts*

Shale rock: *it usually refers to a sedimentary rock composed of fine particles smaller than 3.9 μm, clay, organic matter, etc.*

Shale gas: *it is mainly located in dark muddy shale or high carbon muddy shale. It exists as adsorbed or free gas, and occurs in pores or natural fractures in the shale matrix.*

Shale gas content: *it refers to the total amount of natural gas per ton of rock at surface temperature and pressure (101.325 kPa, 25°C), including free gas, adsorbed gas, dissolved gas, etc. Most of the shale gas is adsorbed gas and free gas.*

Slippage effect: *it is also known as the Klinkenberg effect. The low-speed flow characteristic of gas in rock pore media is different from that of liquid. Gas does not produce thin adsorption layer at the pore wall of rock, and the velocity of gas molecules has no obvious difference at the pore center and near the pore wall, which is called slippage effect.*

Volumetric hydraulic fracturing: *it refers to the process of hydraulic fracturing in which the fractures keep propagating and form the complex fracture network between natural and hydraulic fractures.*

Fractured horizontal wells: *they are horizontal wells affected by the hydraulic fracturing to form a stimulated reservoir volume (SRV) around the well. Figure 6.1 shows a typical fractured horizontal well.*

Poisson's ratio of rock: *it refers to the ratio of lateral strain to axial strain at linearly elastic regions when the rock is under uniaxial*

Table 6.1. Characteristic parameters and range of shale gas reservoirs

Characteristic parameter	Key evaluation parameters	Data range
Organic matter abundance	Total organic carbon content (TOC)	TOC > 2%
	Thermal maturity	$R_o = 1.3$–1.7%
Inorganic mineral	Quartz or calcite content	> 40%
	Clay mineral content	< 30%
	Clay mineral composition	The content of clay minerals, such as smectite, is low
	Brittleness coefficient	> 35%
Physical properties	Matrix pore size	$> 1\,\mu$m
	Matrix permeability	$K < 0.1\,\mu$D
	Matrix porosity	< 5%
	Fracture permeability	It closely relates to the fracture aperture, and it's thousands of times larger than the matrix
	Fracture porosity	< 0.45%
	Water saturation	$S_w < 40\%$
	Oil saturation	$S_o < 30\%$
Rock mechanics	Poisson's ratio	$\upsilon < 0.25$
	Modulus of elasticity	$E > 20{,}000$ MPa
	Rock density	1.89–3.02 t/m^3, on average 2.47 t/m^3
Thickness	High gamma shale thickness	> 30 m
Gas content	Total gas content	$0.1903 \times \text{TOC} + 0.1185 \times \Phi$, 0.1–3 m^3/t
	Free gas content	0.12–0.82 m^3/t in standard conditions
	Adsorbed gas content	20–80% is the content of adsorbed gas, which changes little under high pressure
	Langmuir pressure P_L	10 MPa
	Langmuir volume V_L	2.83–8 m^3/t
Hydraulic fracturing parameters	Stage spacing	40–100 m
	Cluster spacing	20–30 m
	Pump rate	5–20 m^3/min
	Proppant concentrations	About 25%
	Half-bandwidth of fracturing	About 15 m
	Half-length of fracture	About 150 m

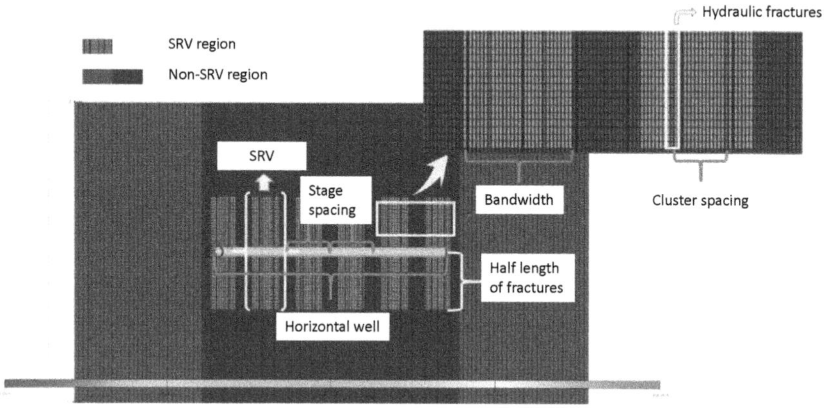

Figure 6.1. Schematic diagram of horizontal well with hydraulic fracturing

tension or compression. It is generally less than 0.5. The smaller the value, the easier it is to be broken.

Elastic modulus of rock: *in the elastic deformation stage of rock, its stress and strain has a linear proportional relationship (i.e., complying with Hooke's law), and the proportional coefficient is called elastic modulus. The unit is MPa and the value is usually on the magnitude of* 10^4 *MPa.*

6.1.3 *Analysis*

6.1.3.1 *Shale gas flow*

Shale gas tends to be produced with different characteristics in different stages:

① Due to the high permeability of hydraulic fracture and the gas-water two phase flow, the initial gas production is relatively large, and the pressure drops rapidly. The free gas in the small reserves of hydraulic fracture is depleted quickly, resulting in a rapid decline in the production, with a declining rate of up to 20%.

② The hydraulic fracture flow area increases, the pressure decreases slowly, the water production decreases, and the gas in the contact surface with matrix and natural fractures is rapidly depleted, so the gas production trend is stable.

③ Gas production in natural fracture and matrix is gradually declining due to slower pressure drop and lower permeability.

④ With the pressure drops in the natural fracture flow reservoir, more and more free gas in the matrix in contact with the fracture network slowly flows into the fracture. At the same time, adsorbed gas in the matrix reservoir is desorbed to supplement the free gas production, and the production is stable.

⑤ The pressure drop of the matrix caused by the the overall pressure drop in the reservoir increases, the desorption capability increases, and the gas supply increases correpondingly.

⑥ The flow area inside the matrix gradually expands, and the free gas and desorption from the matrix become the main source of gas supply to increase the gas production.

⑦ When the pressure drop spreads to the whole area of the matrix, the reservoir energy appears to be exhausted and the gas production decreases slowly.

6.1.3.2 *Shale gas desorption and adsorption*

The main difference between shale gas and conventional natural gas reservoirs is that shale gas adheres to the surface of organic matter in the adsorbed state. For shale gas, it is difficult to achieve commercial development by relying only on the free gas in the matrix pores and fractured pores, and the existence of desorption and adsorption phenomenon provides a continuous gas source for the successful development of the reservoir.

Langmuir isothermal adsorption curve can quantitatively describe the equilibrium relationship between shale gas desorption and adsorption under the condition of constant temperature. The equation is:

$$V_E = \frac{V_L p_m}{p_L + p_m} \tag{6.1}$$

where V_E is the total adsorption capacity; V_L is the Langmuir volume; p_L is the Langmuir pressure; and p_m is the reservoir pressure.

Figure 6.2 shows the adsorption characteristic curve of shale gas. As can be found from the figure, if the formation pressure

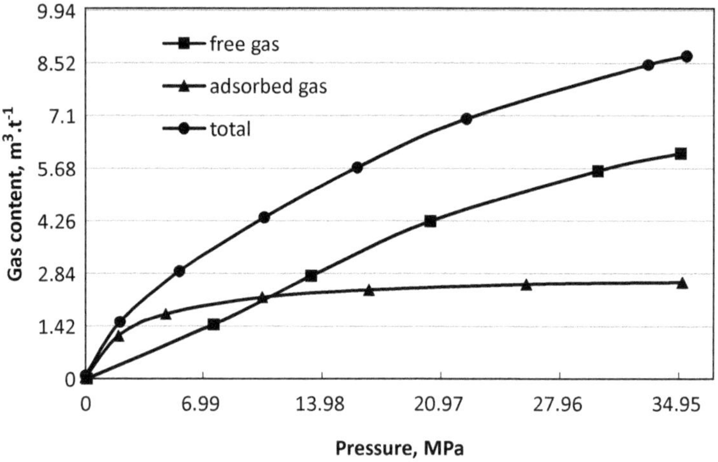

Figure 6.2. Adsorbed gas content in Barnett shale gas reservoir

coefficient is 1.0, the absorbed gas content in the shallow layer increases significantly with the depth of the formation. As for free gas content, it always increases with increase in pressure. Therefore, deep shale gas reservoirs are generally dominated by free gas.

In the United States, some shale gas content can reach $8\,m^3/t$. Generally, shale reservoirs have low gas content. In China, most of them are about $3\,m^3/t$ or even $1\text{–}2\,m^3/t$.

6.1.3.3 *Shale gas flow model*

If there is free gas and adsorbed gas in the reservoir, and there is no oil or water phase, the flow conforms to Darcy's law, and desorption and adsorption are considered. The Warren-Root dual media model is used to establish the flow model.

(1) Matrix flow model

The matrix flow equation considering desorption and adsorption is:

$$\nabla\left[\frac{K_m}{\mu}\rho_g\nabla p_m\right] - q_{mf} = \frac{\partial}{\partial t}(\phi_m\rho_g + \rho_{gsc}V_E) \tag{6.2}$$

In the equation, all the compacted terms can be written as:

$$V_E = \frac{V_L p_m}{p_L + p_m} \tag{6.3}$$

$$\rho_g = \frac{p_m M}{ZRT} \tag{6.4}$$

$$\rho_{gsc} = \frac{p_{sc} M}{Z_{sc} RT_{sc}} \tag{6.5}$$

Substitute into the flow equation and obtain:

$$\frac{M}{RT}\nabla\left[\frac{K_m}{\mu}\frac{p_m}{Z}\nabla p_m\right] - q_{mf} = \frac{\partial}{\partial t}\left(\phi_m \frac{p_m M}{ZRT} + \rho_{gsc}\frac{V_L p_m}{p_L + p_m}\right) \tag{6.6}$$

The first term on the right side of the conservation equation is:

$$\begin{aligned}
\frac{\partial}{\partial t}\left(\phi_m \frac{p_m M}{ZRT}\right) &= \phi_m \frac{M}{RT}\frac{\partial}{\partial t}\left(\frac{p_m}{Z}\right) \\
&= \phi_m \frac{M}{RT}\left[-\frac{p_m}{Z^2}\frac{\partial}{\partial t}\left(\frac{1}{Z}\right) + \frac{1}{Z}\frac{\partial p_m}{\partial t}\right] \\
&= \phi_m \frac{M}{RT}\left[-\frac{p_m}{Z^2}\frac{\partial Z}{\partial P_m}\frac{\partial p_m}{\partial t} + \frac{1}{Z}\frac{\partial p_m}{\partial t}\right] \\
&= \phi_m \frac{M}{RT}\frac{p_m}{Z}\left[\frac{1}{p_m} - \frac{1}{Z}\frac{\partial Z}{\partial p_m}\right]\frac{\partial p_m}{\partial t} \\
&= \varphi_m \frac{M}{RT}\frac{p_m}{Z} C_{gm}\frac{\partial p_m}{\partial t}
\end{aligned} \tag{6.7}$$

where $C_{gm} = \frac{1}{p_m} - \frac{1}{Z}\frac{\partial Z}{\partial p_m}$.

The second term on the right side of the conservation equation is:

$$\begin{aligned}
\frac{\partial}{\partial t}\left(\rho_{gsc}\frac{V_L p_m}{p_L + p_m}\right) &= \rho_{gsc} V_L \frac{(p_L + p_m)\frac{\partial p_m}{\partial t} - p_m \frac{\partial p_m}{\partial t}}{(p_L + p_m)^2} \\
&= \rho_{gsc}\frac{V_L p_L}{(p_L + p_m)^2}\frac{\partial p_m}{\partial t}
\end{aligned} \tag{6.8}$$

Therefore, the right-hand side of the equation can be reduced to:

$$\begin{aligned}
&\varphi_m \frac{M}{RT}\frac{p_m}{Z} C_{gm} \frac{\partial p_m}{\partial t} + \rho_{gsc} V_L \frac{p_L}{(P_m + P_L)^2}\frac{\partial p_m}{\partial t} \\
&= \varphi_m \frac{M}{RT}\frac{p_m}{Z}\left[C_{gm} + \frac{\rho_{gsc} V_L p_L}{\varphi_m \frac{M}{RT}\frac{p_m}{Z}(p_m + p_L)^2}\right]\frac{\partial p_m}{\partial t} \\
&= \phi_m \frac{M}{RT}\frac{p_m}{Z}\left[C_{gm} + \frac{\rho_{gsc} V_L p_L}{\phi_m \rho_{gm}(p_L + p_m)^2}\right]\frac{\partial p_m}{\partial t} \qquad (6.9)
\end{aligned}$$

The matrix comprehensive compressibility considering desorption and adsorption is defined as:

$$\overline{C_m} = C_{gm} + \frac{\rho_{gsc} V_L p_L}{\phi_m \rho_m (p_L + p_m)^2} \qquad (6.10)$$

Thus, the right side of the matrix flow equation is:

$$\frac{M}{RT}\phi_m \overline{C}_{tm} \frac{p_m}{Z}\frac{\partial p_m}{\partial t} \qquad (6.11)$$

Then the matrix flow equation can be expressed as:

$$\nabla\left[\frac{k_m}{\mu}\frac{p_m}{Z}\nabla p_m\right] - \frac{RT}{M} q_{mf} = \phi_m \overline{C}_{tm} \frac{p_m}{z}\frac{\partial p_m}{\partial t} \qquad (6.12)$$

The pseudo-pressure function is defined as: $\psi_m = 2\int_0^{p_m} \frac{p}{\mu Z} dp$.

The matrix flow equation can be written as:

$$\nabla[K_m \nabla \psi_m] - \frac{RT}{M} q_{mf} = \phi_m \overline{C}_{tm} \mu \frac{\partial \psi_m}{\partial t} \qquad (6.13)$$

The formula of cross flow is defined as:

$$\begin{aligned}
q_{mf} &= \alpha \frac{K_m}{\mu} \rho_g (p_m - p_f) \\
&= \alpha \frac{K_m}{\mu}\frac{p_m M}{ZTR}(p_m - p_f) \\
&= \alpha \frac{K_m M}{RT}(\psi_m - \psi_f) \qquad (6.14)
\end{aligned}$$

where α is the matrix shape factor, m^{-2}.

Then the matrix flow equation can be written as:

$$\nabla^2 \cdot (\psi_m) - \alpha(\psi_m - \psi_f) = \frac{\phi_m \overline{C}_{tm} \mu}{K_m} \frac{\partial \psi_m}{\partial t} \tag{6.15}$$

(2) Fracture flow model

The flow of shale gas molecules in the fracture can be regarded as being consistent with Darcy's law, and its flow equation is:

$$\nabla \left[\rho_f \frac{K_f}{\mu} \nabla p_f \right] + q_{mf} = \frac{\partial}{\partial t}(\phi_f \rho_f) \tag{6.16}$$

In the equation,

$$\rho_f = \frac{p_f M}{ZRT} \tag{6.17}$$

Substitute into the flow equation and simplify it to get:

$$\nabla \left[\frac{p_f}{Z} \frac{k_f}{\mu} \nabla p_f \right] + \frac{RT}{M} q_{mf} = \phi_f C_{fg} \frac{p_f}{Z} \frac{\partial p_f}{\partial t} \tag{6.18}$$

In this equation,

$$C_{fg} = \frac{1}{p_f} - \frac{1}{Z} \frac{\partial Z}{\partial p_f} \tag{6.19}$$

The pseudo-pressure function is defined as:

$$\psi_f = 2 \int_0^{p_f} \frac{p}{\mu Z} dp \tag{6.20}$$

Combined with the cross flow equation, the following equation can be obtained:

$$\nabla [K_f \nabla \psi_f] + \alpha K_m (\psi_m - \psi_f) = \phi_f C_{fg} \mu \frac{\partial \psi_f}{\partial t} \tag{6.21}$$

Let $\alpha_{mf} = \alpha k_m$, then:

$$\nabla [K_f \nabla \psi_f] + \alpha_{mf} (\psi_m - \psi_f) = \phi_f C_{fg} \mu \frac{\partial \psi_f}{\partial t} \tag{6.22}$$

6.1.3.4 *Multi-scale flow of shale gas*

(1) Diffusion

With the increase of pressure gradient, a large number of gas molecules desorb from the surface of organic matter into the nano pores in the shale matrix. It is difficult to describe the flow by using Darcy's law due to the extremely low matrix permeability, and shale gas molecular diffusion process from high to low concentration area dominates the flow, especially when pore diameter is small and the pressure is low.

From the point of view of molecular motion theory, the essence of gas diffusion is the result of random thermal motion of gas molecules. According to the study on the diffusion mechanism of gas in porous media, Knudsen number can be used to evaluate the size of diffusion effect. Knudsen number K_n is defined as:

$$K_n = \frac{\lambda}{d} \tag{6.23}$$

$$\lambda = \frac{k_B T}{\sqrt{2}\pi\delta^2 P} \times 10^{21} \tag{6.24}$$

where K_n is Knudsen number; d is the average pore diameter, nm; λ is the gas molecular mean free path, nm; k_B is the Boltzmann constant, 1.3805×10^{-23} J/K; and δ is the gas molecule diameter, nm.

If the pressure remains unchanged, when the pore diameter increases, K_n becomes smaller and the diffusion effect is weakened.

If the pore diameter remains unchanged, when the pressure increases, K_n decreases and the diffusion effect weakens.

If the pore diameter is small and the pressure is high or the pore diameter is large and the pressure is small, the diffusion effect may be weak and may only appear as Darcy flow. When $K_n > 0.01$, Darcy flow is obvious. Therefore, $K_n < 0.01$ is set as the applicable range of diffusion effect.

The free path of gas molecules ($\delta = 0.38\,\text{nm}$, $T = 338\,\text{K}$) under different pressures is calculated by taking methane, the main component of shale gas, as shown in Figure 6.3.

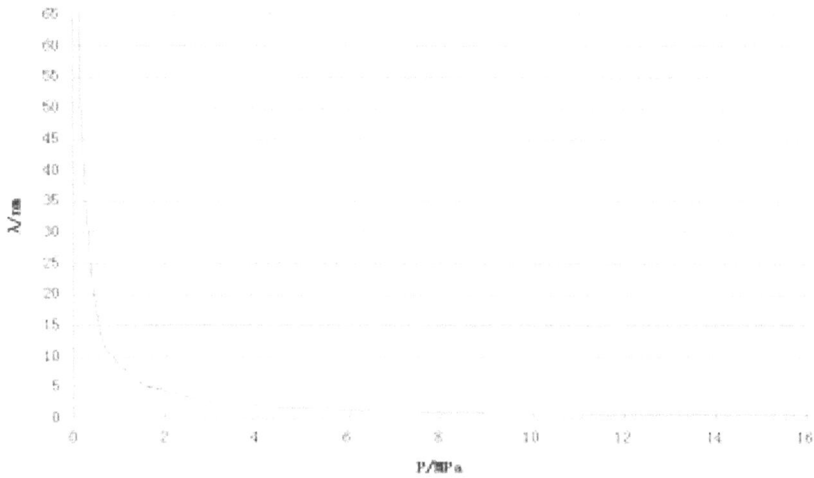

Figure 6.3. Relationship between molecular free path and pressure

At room temperature and pressure (20°C, 0.1 MPa), the average free path of a methane molecule is about 63 nm. In the actual shale gas reservoirs, the pressure of methane molecules is much higher than normal pressure, so the average free path of methane molecules is much shorter. The overall appearance is λ decreases with increase in pressure, but the pressure at 0.5–2 MPa has a significant inflection point, then the average free path slowly decreases. When the pressure is greater than 8 MPa, the λ value is less than 1 nm, and becomes stable at 0.5 nm.

When the gas reservoir temperature is constant ($T = 338\,\text{K}$), the relationship between Knudsen number K_n and pressure under different pore diameters is shown in Figure 6.4.

As can be seen from the figure, in shale gas reservoirs composed of pores with diameter less than 100 nm, the bottom hole pressure is generally lower than 10 MPa or even 5 MPa in the actual production. Therefore, the flow process of the matrix is affected by diffusion, and the smaller the pore radius, the more significant the diffusion effect will be with the decrease in bottom hole pressure.

In 2007, Florence *et al.* proposed an expression for permeability of non-Darcy diffusion flow suitable for tight gas and shale gas by

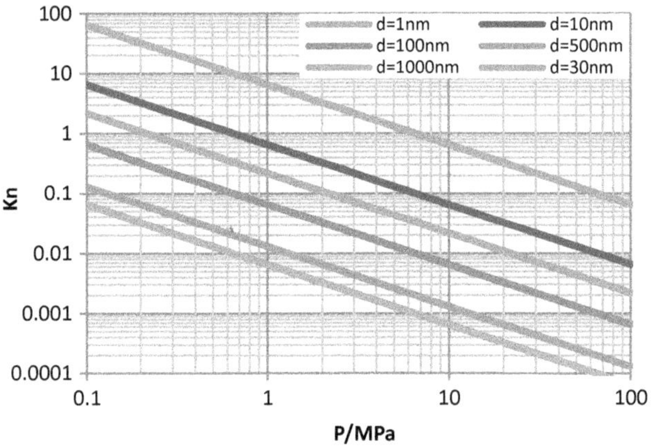

Figure 6.4. Relationship between K_n and pressure under different pore diameters

introducing Knudsen number, which converted microscopic flow into macroscopic flow:

$$K = K_0 \left[1 + \frac{128}{15\pi^2} \tan^{-1}(4K_n^{0.4})K_n\right] \left[1 + \frac{4K_n}{1 + K_n}\right] \tag{6.25}$$

where K_0 is the original permeability.

Because shale gas generally flows under high pressure in the reservoir (greater than 5 MPa), the molecular free path changes very little with pressure. Therefore, Knudsen number can be considered as a constant, and the matrix permeability equation can be applied to the flow model.

(2) Slip effect

For the quantitative analysis of slippage effect, Klinkenberg gave its relation to rock permeability as early as 1941, and it has been applied till now. The formula is:

$$K = K_0 \left(1 + \frac{b}{p}\right) \tag{6.26}$$

where b is the slippage coefficient, $b = 0.04275K_0^{-0.34}$; p is pressure, MPa; and K is the permeability, mD.

In Poiseuille equation, the formula for pore diameter d can be written as:

$$d = 5.66 \times 10^3 \sqrt{\frac{K_0}{\phi}} \tag{6.27}$$

where d is the pore diameter, nm; and ϕ is porosity.

Therefore, the variation amplitude of permeability under different pore diameters can be calculated. For example, for a shale gas reservoir with a formation pressure of 20 MPa and a matrix porosity of 0.045, the variation amplitude of permeability of different pore sizes affected by slippage effect is shown in Figure 6.5. Permeability increment rate is defined as $(K - K_0)/K_0$, and its value decreases with increase in pore diameter. When the pore diameter is 50 nm, the increment rate of permeability is about 15%; when the pore size is 300 nm, the increment rate of permeability by slippage effect is less than 5%.

The pressure in the slippage effect equation to correct the permeability can be taken as the average formation pressure and considered as a constant, and the corrected permeability value can be used in the flow model.

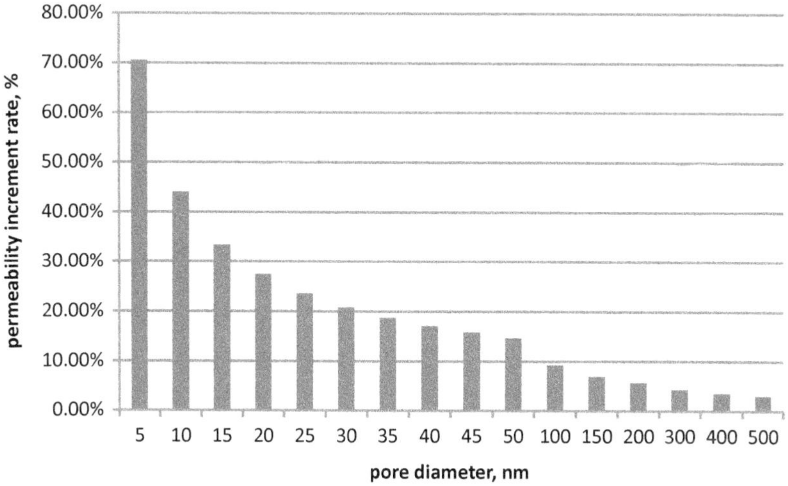

Figure 6.5. Change of the slippage effect with pore diameter

(3) Stress sensitive effect

Because the pore radius of shale gas reservoirs is very small, when the confining pressure increases, the pore radius decreases. At the same time, the content of clay particles in shale reservoirs is relatively high, and the increasing confining pressure will cause clay particles to fall off, and then the pore throat will be blocked, resulting in great changes in the permeability of shale gas reservoirs.

When the shale reservoir has natural fractures, with the continuous increase in confining pressure, the fracture will close and the shale permeability will decrease sharply.

Therefore, in the flow process of shale gas reservoirs, a significant increase in the production pressure difference may result in a sharp decrease in reservoir permeability, which is not conducive to improving the productivity of shale gas wells.

With the increase in stress sensitive effect of fracture network and matrix, the stress sensitive effect of fracture network is far greater than that of matrix. Therefore, the stress sensitive effect in fracture is only considered when establishing the flow model.

The change of shale reservoir fracture permeability K_f with confining pressure can be fitted with an exponential relationship, namely:

$$K_f/K_{f0} = ae^{-cp_{\mathit{eff}}} \tag{6.28}$$

where K_{f0} is the initial fracture permeability, mD; p_{eff} is the effective stress, MPa; a is the stress sensitive fitting constant term; and c is the stress sensitive effect coefficient.

In the actual production process of shale gas, the overlying rock pressure of the reservoir is constant, and the actual pressure change is the pore pressure change of the reservoir. Therefore, effective stress p_{eff} should be the difference between overlying rock pressure and reservoir pressure. Therefore, reservoir pressure is used to express the change relationship between the overlying rock pressure and permeability:

$$K_f/K_{0i} = e^{-c(P_i - P_w)} \tag{6.29}$$

The change of fracture permeability caused by stress sensitive effect needs to be obtained by fitting the experimental data. If the production pressure difference is constant, this value can be used to replace the fracture permeability in the flow model.

6.2 Flow in tight oil reservoirs

6.2.1 *Introduction*

Tight sandstone and carbonate oil reservoir both belong to the category of tight oil. In order to distinguish it from shale oil, tight oil reservoir is widely considered to be caught in or close to the high-quality source rock layers of tight sandstone or carbonate. In this book, tight oil reservoir refers to tight sandstone oil reservoir. Its characteristic parameters are shown in Table 6.2.

Tight oil generally has no natural productivity. Similar to shale gas and shale oil, it requires hydraulic fracturing to achieve the

Table 6.2. Characteristic parameters of tight oil reservoirs

Characteristic parameter	Key evaluation parameters	Data range
Organic matter abundance	Total organic carbon (TOC) content	TOC $> 2\%$
	Thermal maturity	$R_o = 0.6 \sim 1.3\%$
Geological structure	Trap	Slope, simple structure or no trap
	Area	Continuous and widely distributed
Rock property	Brittleness coefficient	$> 30\%$
Physical property	Matrix permeability	$K \leq 0.1\,\text{mD}$, air permeability $< 1\,\text{mD}$
	Matrix porosity	$8 \sim 12\%$
	Water saturation	$30 \sim 40\%$
	Pore throat	Pore diameter $40 \sim 900\,\text{nm}$
	Pore type	Organic pore, inorganic pore and fracture
	Fluid	Light, density is less than $0.825\,\text{g/cm}^3$
Reservoir property	Thickness	$5 \sim 80\,\text{m}$
	Pressure	Relatively high

industrial production. Flow in tight oil is characterized by non-linear flow due to a high start-up pressure gradient. In the production process, the initial production capacity is high, but the production capacity will decrease rapidly, and the water content will rise sharply after water breakthrough. Production rate in the late stage is low, and the injection pressure of wells is high. It is difficult to maintain the reservoir energy, and water injection does not have an obvious effect on the production.

6.2.2 *Concepts*

Tight oil: *it is accumulated oil trapped in or adjacent to a high-quality source rock or carbonate reservoir without the large-scale long-distance migration. The physical properties of the tight reservoir are defined as the reservoir with air permeability measured on the ground surface less than 1 mD and air permeability measured under overburden pressure conditions less than 0.1 mD.*

Start-up pressure gradient: *when the pressure gradient reaches a certain value, the fluid can start to flow. This threshold value is called start-up pressure gradient. The symbol is G, and the common unit is MPa/m.*

Confining pressure: *it is the pressure exerted on the rock by the surrounding rock mass. The confining pressure of rock in the deep underground is mainly caused by the weight of overlying rock. The unit is MPa, and the equation is:*

$$p_z = 10^{-6}\rho_f g D \tag{6.30}$$

where p_z *is the confining pressure, MPa;* ρ_f *is the average density of overlying rock,* $10^3\,kg/m^3$*;* g *is the gravitational acceleration, 9.8 N/kg ; and D is the depth of the rock, m.*

Effective confining pressure: *the existence of fluid pressure in the rock partially counteracts the effect of confining pressure, and the remaining confining pressure is called effective confining pressure.*

Stress: *when the object is deformed due to external factors (stress, humidity, temperature field change, etc.), the internal force of interaction is generated between the various parts of the object to resist the action of such external factors and tries to make the object recover from the deformation. The internal force per unit area at a point in the section is called the stress. The symbol is σ, and the unit is MPa.*

Effective stress: *it is the difference between rock mass internal stress and pore pressure.*

6.2.3 *Analysis*

6.2.3.1 *Start-up pressure gradient of the tight oil reservoir*

In tight oil flow, in addition to overcoming viscous resistance, the interaction between solid and liquid interfaces in the boundary layer should also be overcome. Only when the displacement pressure is greater than a certain value, can the fluid flow. At this time, the displacement pressure is called start-up pressure gradient. The start-up pressure gradient is affected by crude oil viscosity, effective confining pressure and rock wettability.

(1) Typical curve of the low velocity non-linear flow

The flow characteristics in tight oil reservoirs is shown in Figure 6.6. In the low-velocity flow stage, when the displacement pressure is small, the flow curve shows non-linear characteristics (ad). As the pressure increases, the curve transits to a linear shape and shows linear characteristics (de). Pressure gradient at point a represents the real start-up pressure gradient, that is the pressure gradient in the largest pore, pressure gradient at point c is the pseudo start-up pressure gradient, and b is the required minimum pressure gradient for linear flow.

(2) Influencing factors of start-up pressure gradient

① Crude oil viscosity and permeability

As shown in Figure 6.7, when the permeability is very small, the start-up pressure gradient increases dramatically with decrease

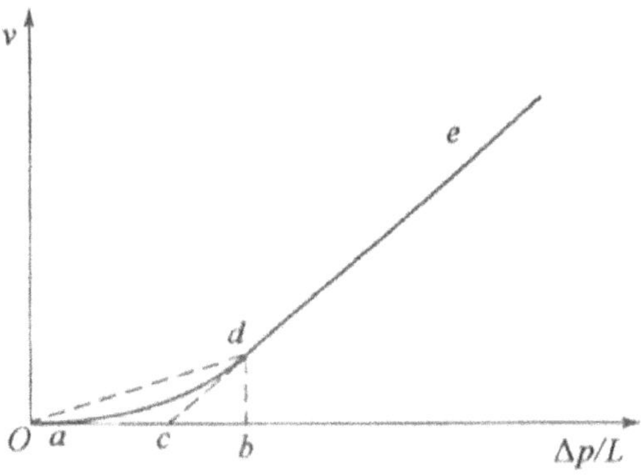

Figure 6.6. Typical curve of flow in tight oil reservoirs

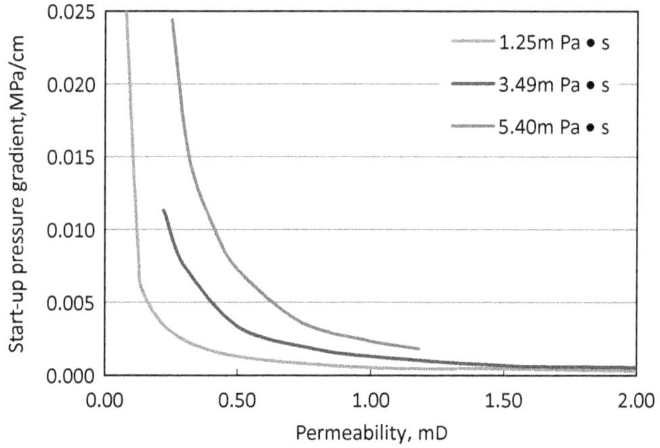

Figure 6.7. Comparison of start-up pressure gradient under different viscosities

in permeability. When the core permeability is larger than a certain value, the start-up pressure gradient is small and changes gently. When the permeability continues to increase, the start-up pressure gradient tends to be very small, and the flow satisfies the linear flow characteristic described by Darcy's law.

With a certain permeability, the higher the crude oil viscosity, the larger the start-up pressure gradient. The start-up pressure gradient of crude oil is positively correlated with the viscosity. With the increase in permeability, the difference in start-up pressure gradient of crude oil with different viscosities will decrease.

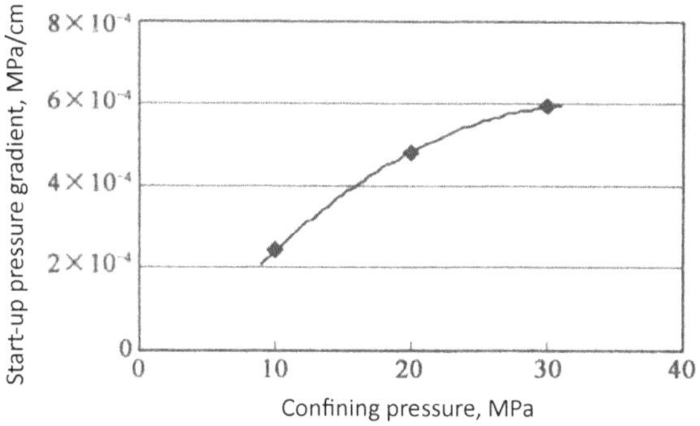

Figure 6.8. Relationship between start-up pressure gradient and confining pressure

② Effective confining pressure

As shown in Figure 6.8, with the increase in confining pressure, the start-up pressure gradient increases. The increasing rate of the start-up pressure gradient decreases, and is mainly due to the fact that the confining pressure make the cement between the particles compressed first, then the pore volume and throat radius decrease. When the confining pressure continues to increase, the elastic deformation of the rock particles themselves occurs, but the decreasing rate of the pore volume tends to become less obvious.

③ Rock wettability

When the wettability of rock sample changes from water-wet to oil-wet, the start-up pressure gradient increases gradually. The reason is that when the wettability changes to oil-wet, the adsorption effect of the polar molecules in oil phase on the particle surface increases. Under the same driving pressure gradient, the thickness of boundary layer within the pore walls increases, the flow resistance increases, and the start-up pressure gradient increases accordingly.

(3) Description of start-up pressure gradient

The description of the start-up pressure gradient is accounted for in the fluid motion equation, namely:

$$v_l = \frac{K}{\mu}\left(1 - \frac{G}{\operatorname{grad} p}\right)\operatorname{grad} p \tag{6.31}$$

where G is start-up pressure gradient, Pa/m.

When $\nabla p \leq G$, the driving force of the fluid cannot overcome the flow resistance and the fluid cannot flow. Only when $\nabla p \geq G$, can the fluid flow. With the increase in driving pressure gradient, the influence of the start-up pressure gradient gradually decreases.

6.2.3.2 *Stress sensitive effect of tight oil reservoir*

During the development process of tight oil reservoirs, it is hard to maintain the formation energy by providing external energy sources. With the production of reservoir fluid, the formation pressure keeps declining, the effective confining pressure increases, and the elastic-plastic deformation of the reservoir rock will occur. It will result in the closure of natural fractures, deformation of the reservoir rock can be partial or total irreversible deformation, which will lead to the redistribution of formation porosity and permeability.

Due to the poor physical properties of the tight oil reservoir, it is difficult to inject external fluids to maintain the reservoir pressure. When the flow rate is too low to keep continuous production, generally it needs to shut-in the well to restore formation pressure near the wellbore area. Therefore, reservoir pressure near the wellbore will fluctuate, and this pressure fluctuation will create an irreversible loss of the reservoir properties, such as porosity and permeability.

The relationship between permeability and stress presents a power relationship. Permeability and effective stress are dimensionless, and the relationship is obtained as:

$$\frac{K}{K_0} = a\left(\frac{\sigma}{\sigma_0}\right)^{-\mathrm{b}} \tag{6.32}$$

where σ is the current stress, MPa; σ_0 is the initial stress, MPa; a is correlation index; and b is the stress sensitive effect coefficient.

When $\sigma = \sigma_0$, $K = K_0$ and then $a = 1$. Take the logarithm of both sides of the above equation and get:

$$\lg\frac{K}{K_0} = -b\lg\frac{\sigma}{\sigma_0} \tag{6.33}$$

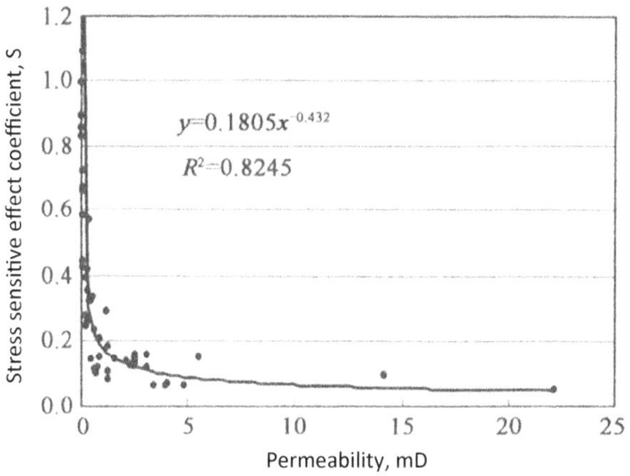

Figure 6.9. Relationship between initial permeability and stress sensitive effect coefficient

As can be seen from the above equation, $\frac{K}{K_0} : \frac{\sigma}{\sigma_0}$ is a line with slope $-b$ that passes the point $(1,1)$ in the double logarithmic coordinate system.

Therefore, the stress sensitive effect coefficient is defined as:

$$b = -\frac{\lg \frac{K}{K_0}}{\lg \frac{\sigma}{\sigma_0}} \tag{6.34}$$

The relationship between stress sensitive effect coefficient b and initial permeability cores is shown in Figure 6.9. The smaller the initial permeability of core, the larger the corresponding stress sensitive coefficient. When the permeability is small, the stress sensitive effect coefficient increases sharply. When the permeability is large, the stress sensitive effect coefficient tends to level off. Due to the fact that the tight oil reservoir permeability is generally less than 0.1 mD, stress sensitive effect has a significant impact on its development.

The stress sensitive effect coefficient of the reservoir is in a power law relationship with the initial permeability of the formation, which can be expressed as:

$$b = c(K_0)^{-m} \tag{6.35}$$

For different reservoirs, c and m coefficients have different values, which can be measured by experiments.

6.2.3.3 *Non-linear flow model with start-up pressure gradient*

If a production well is located in the center of a homogeneous, equal-thickness, horizontal circular reservoir and it is assumed that the outer circular boundary is the supply boundary, then the fluid motion equation can be written as:

$$v = -\frac{K_0}{\mu}\left(\frac{\mathrm{d}p}{\mathrm{d}r} - G\right) \tag{6.36}$$

The basic differential equation can be converted into:

$$\frac{\mathrm{d}p}{\mathrm{d}r} + \frac{v}{r} = 0 \tag{6.37}$$

Inner and outer boundary conditions are:

$$r = R_w, \quad p = p_w; \quad r = R_e, \quad p = p_e \tag{6.38}$$

The pressure distribution equation of low velocity non-linear steady state flow in tight oil under the consideration of start-up pressure gradient can be obtained as:

$$\begin{aligned} p &= p_w + \frac{(p_e - p_w) - G(R_e - R_w)}{\ln \frac{R_e}{R_w}} \ln \frac{r}{R_w} + G(r - R_w) \\ &= p_e - \frac{(p_e - p_w) - G(R_e - R_w)}{\ln \frac{R_e}{r}} \ln \frac{R_e}{r} - G(R_e - r) \end{aligned} \tag{6.39}$$

Thus, the production rate can be obtained as:

$$Q = \frac{2\pi K_0 h}{\mu B} \cdot \frac{(p_e - p_w) - G(R_e - R_w)}{\ln \frac{R_e}{R_w}} \tag{6.40}$$

6.2.3.4 *Non-linear flow model with stress-sensitive effect*

The equation of motion is:

$$v = -\frac{K_0}{\mu}\left(\frac{p_Z - p}{\sigma_0}\right)^{-b} \cdot \frac{dp}{dr} \tag{6.41}$$

The continuity equation and boundary conditions are the same as (6.37) and (6.38).

According to the above conditions, the productivity equation of radial steady state flow considering stress sensitive effect can be solved:

$$Q = \frac{2\pi K_0 h}{\mu B \sigma_0{}^{-b}} \cdot \frac{(p_Z - p_w)^{1-b} - (p_Z - p_e)^{1-b}}{(1-b)\ln \frac{R_e}{R_w}} \tag{6.42}$$

The pressure distribution is:

$$\begin{aligned} p &= p_Z - \left[\frac{Q\mu B(1-b)\sigma_0{}^{-b} \ln \frac{R_e}{r}}{2\pi K_0 h} + (p_Z - p_e)^{1-b}\right]^{\frac{1}{1-b}} \\ &= p_Z - \left[\frac{(p_Z - p_w)^{1-b} - (p_Z - p_e)^{1-b}}{\ln \frac{R_e}{R_w}} \ln \frac{R_e}{r} + (p_Z - p_e)^{1-b}\right]^{\frac{1}{1-b}} \end{aligned} \tag{6.43}$$

6.3 Flow in shale oil reservoirs

6.3.1 *Introduction*

Shale oil and shale gas are similar. They exist in the organic matter of muddy shale reservoirs, and have hydrocarbon in the adsorption state. Shale oil has permeability an order of magnitude lower than tight oil, which is less than $0.1 \sim 0.01$ mD. Shale oil, similar to shale gas and tight oil and gas, requires hydraulic fracturing to produce effectively. The produced oil from the shale oil reservoir is mainly free hydrocarbon, which is different from shale gas reservoir. In shale gas reservoir, the produced gas contains both free gas and adsorbed gas.

The breakthrough in massive hydraulic fracturing led to the revolution of shale gas, and then shale oil. Since 2000, US shale oil has been produced in large quantities (about 4 million tons per year). By 2015, it reached more than 200 million tons, accounting for more than 50% of the total crude oil.

The world oil industry is in a stage of transition to unconventional oil and gas. Shale oil has become another hot spot in the field of oil and gas development and a potential important source of oil and gas supply. Shale oil is widely distributed in the nano-scale pores in shale oil reservoirs. The pore diameter is mainly $50 \sim 300\,\text{nm}$, the porosity is generally less than 10%, and the permeability is less than $0.01 \times 10^{-3}\,\mu\text{m}^2$.

6.3.2 *Concepts*

Abundance of organic matter: *refers to the amount of organic matter contained in sedimentary rocks, often expressed by the content of residual organic carbon, extracts and pyrolytic hydrocarbons, symbol TOC. Shale oil and gas standards are around 2%.*

Brittleness coefficient: *it is proportional to the brittle minerals (mainly silica and calcium) in the rock. The effective hydraulic fracturing generally requires that the brittleness coefficient is more than 30%.*

Microscopic scale: *it corresponds to the scale with obvious quantum effect, such as pore structure and organic matter distribution studied by scanning electron microscope (SEM).*

Mesoscopic scale: *it refers to the scale between macroscopic and microscopic, generally considered between nanometer and millimeter.*

Macroscopic scale: *investigate the momentum, heat and mass transfer caused by the fluid's overall movement, and based on the conservation principle, conduct overall calculation on a certain range, and establish the relevant mathematical model.*

6.3.3 *Analysis*

According to the different flow characteristics in shale oil reservoir around the well, the flow region can be divided into two zones, and the mathematical model can be established by coupling the two flow processes in the two-zone shale oil reservoir.

6.3.3.1 *Equation of motion in shale oil reservoirs*

The core flooding experiment is used to measure the pressure gradient of fluid passing through the core under different flow conditions, and the relationship curve between the start-up pressure gradient and permeability can be drawn, as shown in Figure 6.10 and Figure 6.11.

In the process of shale oil development, the intersection point between the extension of the straight line segment and the pressure gradient axis does not pass through the origin of the coordinate, and the curve has obvious non-linear characteristics. Fluid flow no longer follows Darcy's law, and low-speed non-Darcy flow occurs, which has a pseudo-start-up pressure gradient. The smaller the core permeability, the greater the start-up pressure gradient. The start-up pressure gradient for shale oil can be as high as 5–20 MPa/m, much higher than that for low permeability or even tight reservoirs (generally less than 1 MPa/m).

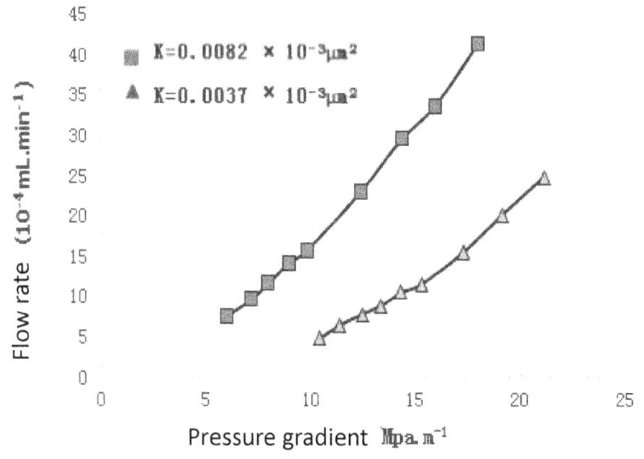

Figure 6.10. Non-Darcy flow curve

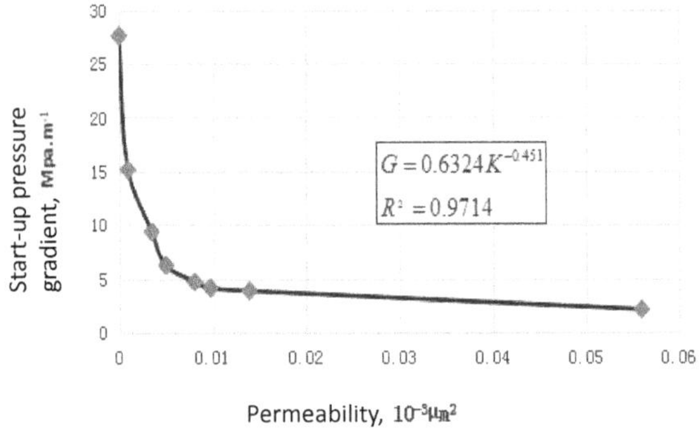

Figure 6.11. Relationship between start-up pressure gradient and permeability

When fluid flows in shale oil reservoirs, the start-up pressure gradient is closely related to the formation permeability. With the decrease in permeability, the start-up pressure gradient of formation fluid increases sharply, and there is a power function relationship between the start-up pressure gradient and permeability, namely:

$$G = 0.6324K^{-0.451} \tag{6.44}$$

where G is the start-up pressure gradient, MPa/m; and K is the permeability, $10^{-3}\,\mu\text{m}^2$.

Therefore, the flow motion equation in the nano-micron scale pore throat in the shale oil reservoir is expressed as:

$$v = -\frac{K}{\mu}\nabla p\left(1 - \frac{G}{|\nabla p|}\right) \tag{6.45}$$

6.3.3.2 *Equivalent permeability of SRV area*

In order to study the flow velocity, an ideal model is established: the formation thickness in cylindrical radial flow reservoir is h, the reservoir radius is r, the well diameter is R_w, the bottomhole pressure is p_w and flow rate is Q, as shown in Figure 6.12.

According to the radial Darcy flow formula, the following equation can be obtained:

$$Q = \frac{2\pi K_n h}{\mu}\frac{dp}{dr} \tag{6.46}$$

where K_n is the effective permeability of fracture.

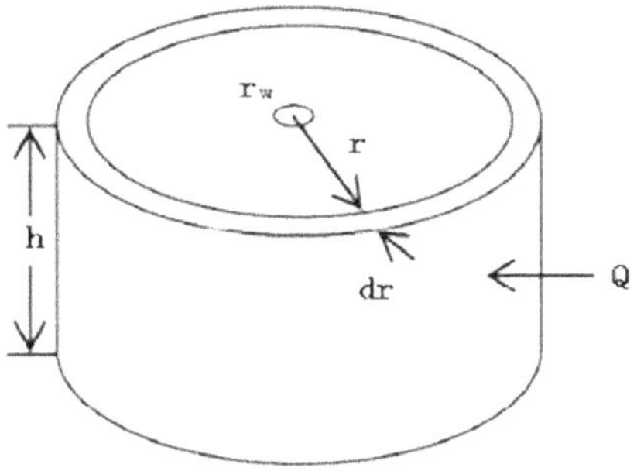

Figure 6.12. Schematic diagram of radial flow

The fracture volumetric coefficient f_n is the ratio between the fracture volume in the cylinder region V_f and the total volume of the cylinder V_r, that is:

$$f_n = \frac{V_f}{V_r} \tag{6.47}$$

Then, the equivalent permeability of the SRV fracture zone K_e is:

$$K_e = K_m(1 - f_n) + K_n f_n \tag{6.48}$$

6.3.3.3 *Mathematical model of productivity*

As shown in Figure 6.13, assume that the SRV is a $2 \times x_f \times b$ oval cylinder symmetrical along the main fracture. According to the flow characteristics of shale oil reservoir after hydraulic fracturing, the reservoir fluid flow can be divided into two zones:

In zone 1, the low-speed non-Darcy flow in the elliptic fracture network area formed by hydraulic fracturing;

In zone 2, linear Darcy flow within the main hydraulic fractures.

According to the fluid flow characteristics during production of hydraulic fractured wells in shale oil reservoirs and the non-linear flow characteristics of shale oil reservoirs, a productivity prediction model for hydraulic fractured vertical wells in shale oil reservoirs is established. The assumptions are: ① the upper and lower boundaries of shale oil reservoir are closed and the formation is infinite; ② perform hydraulic fracturing on the vertical well and form an elliptical fracture network. The short semi-axial length is b, and

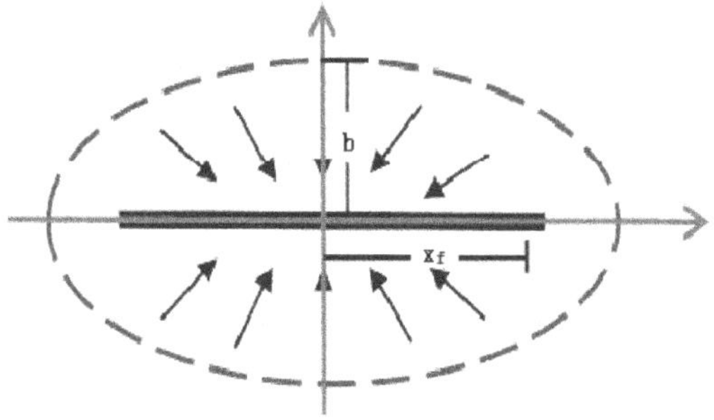

Figure 6.13. Main hydraulic fractures and SRV area

the long semi-axial length is the half-length of the main fractures; ③ the fluid in reservoir and fracture is single-phase incompressible fluid, and the flow is isothermal and steady state, without considering the influence of gravity; ④ the impact of start-up pressure gradient in the process of flow is considered.

(1) Linear flow of the main hydraulic fractures

The fluid flow in the hydraulic fractures obeys Darcy's law and belongs to linear flow. Its motion equation is:

$$v = -\frac{K(p)}{\mu}\nabla p \tag{6.49}$$

It can be re-written as:

$$-\frac{\partial p}{\partial x} = \frac{\mu}{K_f}v \tag{6.50}$$

where K_f is the permeability of hydraulic fracture.

By separating variables and integrating from (x_f, p_{xf}) to (R_w, p_w), the relationship between fluid flow rate and pressure difference in the fracture can be obtained as:

$$p_{xf} - p_w = \frac{\mu}{K_f}\frac{x_f Q}{2w_f h} \tag{6.51}$$

where p_{xf} is the pressure at both ends of the hydraulic fracture; w_f is the aperture of hydraulic fracture; and x_f is the half-length of hydraulic fracture.

(2) Elliptic flow in the SRV area

After hydraulic fracturing, the relationship between the Cartesian coordinate system (x, y) and the elliptic coordinate system (η, ξ) in the elliptical flow zone of the reservoir is as follows:

$$\begin{aligned} x &= a\cos\eta \quad y = b\sin\eta \\ a &= x_f\cosh\xi, \quad b = x_f\sinh\xi \end{aligned} \tag{6.52}$$

where a and b are the long and short axis of the ellipse respectively.

For the hydraulic fractured shale oil reservoir, its equivalent permeability K_e is still not high, the flow in this area is non-linear flow, and there is a start-up pressure gradient, so the rate of steady state flow in the elliptic region is:

$$\frac{QB}{4x_f h\cosh\xi} = \frac{K_e}{\mu}\left(\frac{\partial p}{\partial y} - G\right) \tag{6.53}$$

where B is the volume factor.

The relationship between flow rate and pressure difference is obtained as:

$$p_e - p_{xf} = \frac{\mu BQ}{2\pi K_e h}(\xi_i - \xi_w) + \frac{2x_i G}{\pi}(\sinh\xi_i - \sinh\xi_w) \tag{6.54}$$

(3) Coupling of the hydraulic fracture and SRV area

The pressure at the interface of the two zones is equal, and productivity after fracturing of the vertical well is obtained as:

$$\begin{aligned} p_e - p_w = {} & \frac{\mu BQ}{2\pi K_e h}(\xi_i - \xi_w) + \frac{2x_i G}{\pi}(\sinh\xi_i - \sinh\xi_w) \\ & + \frac{\mu}{K_f}\frac{x_f Q}{2w_f h} \end{aligned} \tag{6.55}$$

6.4 Flow in coal bed methane reservoirs

6.4.1 *Introduction*

Coal bed methane (CBM) is a kind of unconventional natural gas that is generated in the process of coal formation and is mainly stored in the self-storage coal seam in the form of free and adsorbed states.

Its main composition is methane (about 90%), and a small amount of CO_2, N_2, etc.

When CBM is produced, the matrix gas penetrates into the fracture through desorption and diffusion, and then flows into the well. Some CBM reservoirs also need fracturing to improve the flow capacity of fractures.

Pressure drop in coal seam can release CBM from the matrix and enter the cleavage (coal seam is a naturally fractured reservoir, and the fracture is called cleavage). The initial gas saturation is very low, then a small amount of CBM begins to flow, and the initial production of CBM well is very low.

With the continuous depletion of coal seam, saturation of gas content in the cleavage gradually increases, resulting in the increase of gas production. The production of CBM increases gradually with decrease in pressure of the coal reservoir. The depletion process may take months or years, during which gas production may be increased 2–10 times. After a CBM well or field reaches its peak and shows an obvious decline curve, dynamic analysis of the production well and numerical simulation of the CBM reservoir can be used to predict future dynamics and remaining reserves of CBM.

6.4.2 *Concepts*

Coalification: *it is the geological process by which plants are converted into coal under a certain temperature and pressure.*

Coal rank: *it represents the degree of maturity that can be achieved during carbonization.*

Coal cleat: *cleats are natural fractures in coal beds. They occur as two orthogonal sets. The dominant continuously distributed cleat is the "face cleat", and the secondary cleat is the "butt cleat".*

6.4.3 *Analysis*

6.4.3.1 *Flow mechanism of coal bed methane*

The main path of coal bed methane (CBM) flow in reservoirs is the fracture system in coal. The flow of the CBM molecules diffused into the fracture and the water molecules in the fracture is driven by the pressure gradient, and it can be considered that this process conforms to Darcy's law.

CBM is produced through pumping out the water and reducing the pressure of the CBM reservoir, which is obviously different from conventional oil and gas extraction. The process of CBM flowing from coal matrix into production well can be divided into three stages.

The first stage is single-phase water flow. Before the CBM reservoir pressure reaches critical desorption pressure, the decrease in pressure around the wellbore is not so obvious, CBM has not been desorbed, and only water is produced near the wellbore.

The second stage is water flow with gas bubbles. When the coal reservoir pressure drops slightly lower than the critical desorption pressure, there is a certain amount of coal seam gas near the wellbore desorbed from micro-pores in the coal matrix to fractures in the coal system. There is no continuous gas flow in the fracture system, and the gas exists as bubbles. The existence of gas bubbles reduces the flow capability of water, and the relative permeability of water phase declines.

The third stage is gas and water two-phase flow. When the coal reservoir pressure drops below the critical desorption pressure, more gas will be desorbed forming a continuous gas flow with pressure reduction. The relative permeability of gas phase gets larger, and the relative permeability of water phase constantly decreases. The gas production rate will increase and form a two-phase flow.

For the same CBM reservoir, these three stages occur step-by-step in the process of pressure drop.

The main flow path of CBM in coal reservoirs is coal cleats or fractures. Generally, the coal reservoir is shallow buried, the formation pressure is low, the permeability is low, and the gas diffusion cannot be ignored. Therefore, the flow mechanisms include diffusion and Darcy flow, which follow Fick's law and Darcy's law respectively. Apart from CBM, water also exists in the coal cleats. Driven by pressure gradient, water flow is in the direction of pressure reduction, and the flow agrees with Darcy's law. Gas and water flow are two-phase immiscible flow, and the velocity is proportional to the relative permeability.

6.4.3.2 *Mathematical equation of coal bed methane flow*

Assuming that the initial water saturation of the coal reservoir is connate water saturation, the continuity equation of gas flow in the cleat can be written as:

$$\rho F + \nabla(\rho \nu) = -\Phi_f \frac{\partial \rho}{\partial t} \tag{6.56}$$

where ρ is the CBM density; F is the rate of CBM flowing from matrix to cleats; and Φ_f is the cleat porosity.

According to Darcy's law, the relationship between CBM velocity and pressure gradient is established as:

$$v = -\frac{K_g}{\mu_g} \nabla p \tag{6.57}$$

If the coal seam is full of real gas, then:

$$\rho = \frac{M}{RT}\left(\frac{p}{Z}\right) \tag{6.58}$$

By using Langmuir's isothermal adsorption equation, the amount of adsorbed gas per unit rock volume can be expressed by:

$$V = V_L \frac{p}{p + p_L} \rho_B \tag{6.59}$$

where V is the CBM content; V_L is the Langmuir volume constant; p_L is the Langmuir pressure constant; and ρ_B is the coal density.

According to the real gas state equation:

$$p v_g = nZRT \tag{6.60}$$

The relationship between CBM flow in coal matrix under reservoir conditions and standard conditions is:

$$\frac{pF}{ZT} = \frac{p_{sc} F_{sc}}{Z_{sc} T_{sc}} \tag{6.61}$$

If the above equation is substituted into the continuity equation of gas flow and the coal seam is isothermal, the basic differential equation can be simplified and obtained as:

$$k_g \nabla \left(\frac{p}{\mu_9 Z} \nabla P \right) = \Phi_f \frac{\partial}{\partial t} \left(\frac{p}{z} \right) + \frac{p_{sc} T}{Z_{sc} T_{sc}} \rho_B V_{Lis} \frac{p_L}{p + p_L} \frac{\partial p}{\partial t} \tag{6.62}$$

Assuming that CBM has the characteristics of real gas, the following equation can be obtained:

$$c_g = \frac{1}{\frac{p}{z}} \frac{\partial}{\partial p} \left(\frac{p}{Z} \right) \tag{6.63}$$

The pseudo-pressure function is defined as:

$$m(p) = 2 \int_{p_b}^{p} \frac{p}{\mu_g Z} dp \tag{6.64}$$

Based on the compressibility of CBM under isothermal conditions and the introduction of real gas pseudo pressure function, the basic differential equation is:

$$\nabla^2 m(p) = \frac{\Phi_f \mu_g}{K_g} \left[c_g + \frac{B_g \rho_B V_L}{\Phi_f} \frac{p_L}{(p + p_L)} \right] \frac{\partial m(p)}{\partial t} \tag{6.65}$$

where $B_g = \frac{p_{sc} ZT}{p Z_{sc} T_{sc}}$.

When stable production occurs, the basic differential equation is the Laplace equation, which is similar to the steady state flow of single-phase incompressible liquid in classical flow mechanics in porous media. Then, the equation of CBM production is:

$$q_g = \frac{K K_{rg} h [m(p_R) - m(p_w)]}{1.310 T (In\, R_e / R_w - 0.75 + S)} \tag{6.66}$$

where q_g is the CBM production rate, $10^3\ \text{m}^3/\text{d}$; h is the thickness of coal seam, m; p_R is the average reservoir pressure, MPa; p_w is the bottom hole flow pressure, MPa; T is the coal seam temperature, K; R_e is the supply boundary radius, m; R_w is the wellbore radius, m; and S is the skin coefficient.

In the case of transient production, the basic differential equation is the Fourier equation, which is similar to the elastic transient flow of the single phase slightly-compressible liquid in classical flow mechanics in porous media. The pressure and flow rate relationship can be obtained, and then the transient well test can be carried out.

6.5 Flow in gas hydrate reservoirs

6.5.1 *Introduction*

Gas hydrate, also known as "combustible ice," is a solid-like "ice" in the formation. When the pressure is lowered or heated, the "ice" melts into water and methane gas. Ideally, the amount of natural gas released by $1\,m^3$ "ice" can reach $160 \sim 180\,m^3$ under standard conditions. It is estimated that the total amount of methane in natural gas hydrate stored in underground reservoirs is about $2.0 \times 10^{16}\,m^3$, which is equal to twice the total carbon content of traditional fossil energy (coal, oil and natural gas).

The geological and production characteristics of gas hydrate are summarized as follows: ① gas hydrate is mainly distributed in oceans with a depth of more than 300 m and in the permafrost on the land where there is no cap rock; ② the gas source is from the microbial effect to sediments in the shallow layer or the pyrolysis of organic matter in the deep layer; ③ gas hydrate exists in formations with high pressure and low temperature; ④ the proper temperature for gas hydrate under formation pressure is about 25°C; ⑤ the decomposition of gas hydrate can lead to an increase in permeability and porosity; ⑥ gas hydrate can be in gas, water and solid phases, and phase transition can occur among all the three phases; ⑦ gas hydrate decomposition involves the heat transfer process.

Gas hydrate flow is a multiphase multicomponent non-isothermal physicochemical flow process. This process includes phase transition, energy change and the change of physical properties of reservoir media.

The extraction methods of natural gas hydrate mainly include: pressure reduction, heating and carbon dioxide injection.

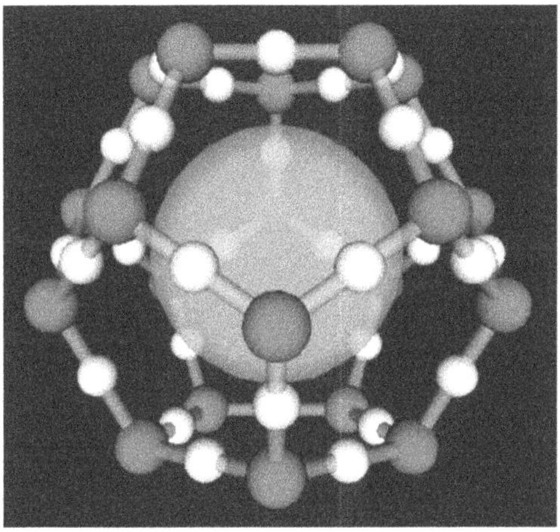

Figure 6.14. Molecular structure of gas hydrate

6.5.2 *Concepts*

Gas hydrate: *its molecular structure is shown in Figure 6.14. A methane molecule with one carbon and four hydrogen atoms is located at the middle of the structure and the methane molecule is trapped by water molecules. Water and gas molecules attract each other through van der Waals force.*

Porosity of gas hydrate reservoir: *it is the ratio of the volume of water, gas and gas hydrate in the rock to the total volume of the rock, symbol* ϕ_H.

Saturation of gas hydrate reservoir: *it is the ratio of gas hydrate volume* V_H *in the rock to the total pore volume* V_φ *of the rock. The symbol is* S_H*, and the equation is*

$$S_H = \frac{V_H}{V_\varphi} \tag{6.67}$$

where S_H *is the gas hydrate saturation;* V_H *is the volume of gas hydrate; and* V_φ *is the total pore volume of the rock.*

Permeability of of gas hydrate reservoir: *it is the capacity of porous media containing gas hydrate to allow the fluid to pass* *through, which decreases exponentially with the increase of gas hydrate saturation* S_H*, the symbol is* K_H*, which satisfies*

$$K_H = K_{D0}(1 - S_H)^N \tag{6.68}$$

where K_H *is the natural gas hydrate permeability;* K_{D0} *is the permeability of porous media when* $S_H = 0$*; and* N *is the permeability decline index.*

6.5.3 *Analysis*

6.5.3.1 *Flow characteristics of gas hydrate*

The production process of gas hydrate mainly includes three flow processes.

(1) Hydrate decomposition process

During the extraction of gas hydrate, the stable state of gas hydrate is broken down by means of lowering pressure, heating or injecting chemicals to decompose it. The chemical process of decomposition of gas hydrate can be described as:

$$CH_4 \cdot (H_2O)_n \rightarrow nH_2O + CH_4 \tag{6.69}$$

where n is the hydration index.

According to the kinetic characteristics of gas hydrate decomposition, the decomposition rate of gas hydrate is expressed as:

$$m_g = K_d A_s (p_{HC} - p) \tag{6.70}$$

where K_d is the reaction velocity constant, $g/(cm^2 \cdot Pa \cdot s)$; p_{HC} is the reaction critical pressure, Pa; and A_s can be expressed as:

$$A_s = \left(\frac{\varphi_H^3}{2K_H}\right)^{\frac{1}{2}} \tag{6.71}$$

According to the chemical equation of hydrate decomposition, 1 mol natural gas hydrate can generate 1 mol methane gas and n mol water after decomposition, so:

$$m_H = -\frac{nM_w + M_g}{M_g} \tag{6.72}$$

$$m_w = m_g \frac{nM_w}{M_g} \tag{6.73}$$

where m_g, m_w and m_H are respectively the masses of gas, water and hydrate produced by the decomposition of methane hydrate in rocks per unit time and per unit volume, $\text{g}/(\text{s} \cdot \text{cm}^3)$;

M_g, M_w and M_H are respectively the molar masses of gas, water and natural gas hydrate, g/mol.

The reason the mass formula of hydrate has a negative sign is that in the decomposition process, the mass of hydrate decreases continuously while the mass of its product increases constantly, so the negative sign is added to distinguish it.

(2) Multiphase flow process after decomposition

In the process of natural gas hydrate extraction, with the decomposition of hydrate, the parameters, such as saturation of each phase, reservoir porosity and permeability, are constantly changing. At the same time, when the pressure decreases, the loose solid hydrate and solid particles in the reservoir may partially fall off and flow with it.

If the solid rock particles do not fall off after hydrate decomposition, reservoir porosity is only related to compressibility, and has nothing to do with the amount of hydrate. Then its change is:

$$\varphi_H = \varphi_0 + C_f(p - p_0) \tag{6.74}$$

In the natural gas hydrate model, the hydrate is a solid phase and the gas and water phases flow in pores other than the solid phase. During the decomposition process of the hydrate, the hydrate phase keeps decreasing, the flow space and the permeability keep increasing, and the permeability changes exponentially with the

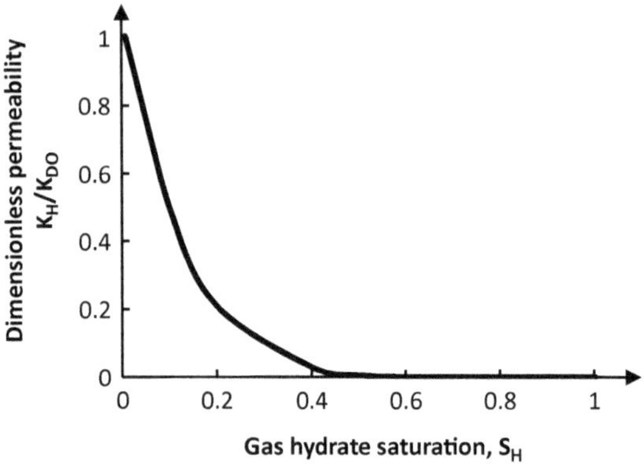

Figure 6.15. Curve of dimensionless permeability change with hydrate saturation

change in saturation of the natural gas hydrate. The equation is:

$$K_H = K_{D0}(1 - S_H)^N \tag{6.75}$$

When $N = 7$, the variation of dimensionless permeability with gas hydrate saturation is shown in Figure 6.15.

The equations of gas and water relative permeability are:

$$K_{rg} = K_{rg0} S_{gD}^{1/2} (1 - S_{HD}^{1/m})^{2m} \tag{6.76}$$

$$K_{rw} = K_{rw0} S_{wD}^{1/2} [1 - (1 - S_{wD}^{1/m})^m]^2 \tag{6.77}$$

where the normalized saturation is:

$$S_{wD} = \frac{S_w - S_{wc}}{1 - S_{wc} - S_{gr}} \quad S_{HD} = \frac{S_w + S_H - S_{wc}}{1 - S_{wc} - S_{gr}} \tag{6.78}$$

$$S_{gD} = \frac{1 - S_w - S_H - S_{gr}}{1 - S_{wc} - S_{gr}} \tag{6.79}$$

where K_{rw0} and K_{rg0} are the endpoint values of the water-gas relative permeability curve. When $m = 0.45$, $S_{wc} = 0.3$, $S_{gr} = 0.05$, $K_{rw0} = 0.5$ and $K_{rg0} = 1$.

The product of relative permeability and K_H gives the phase permeability of each phase.

The flow of natural gas hydrate is a very complicated process, involving three phases of solid, liquid and gas. Multiple components, such as hydrate, water, light hydrocarbon, heavy hydrocarbon and inhibitor, are involved and the formation temperature may change in the process of flow. In addition, the formation of hydrate has a "memory effect", that is, the water generated by the decomposition of hydrate is easier to form hydrate than free water, so the regeneration of hydrate will also have an important impact on the flow process.

(3) Flow process of the injected fluid

The decomposition of natural gas hydrate is an endothermic process, and some of the hydrate development needs to inject a certain amount of fluid to transfer heat or inhibitors to the reservoir. This process is a dual process of injection fluid flow and hydrate decomposition, involving many physical and chemical changes. According to the conservation of material energy and by means of numerical simulation, the variation of the injection fluid's advance front distribution, temperature and pressure propagation velocity in the reservoir can be obtained.

6.5.3.2 *Establishment of flow mathematical model*

The multiphase and multicomponent model is used to establish the conservation equations of mass and energy.

4 phases: gas phase, water phase, solid hydrate phase, solid ice phase;

9 components: hydrate, water, methane, decomposed methane, non-methane hydrocarbon gases, decomposed non-methane hydrocarbon gases, salts, water-soluble inhibitors and pseudo-enthalpy components.

(1) Governing equation

The non-isothermal hydrate system can be described by mass balance equation and energy conservation equation, and the conservation of mass and energy of component k can be expressed as:

$$\frac{d}{dt}\int_{V_n} M^k dV = \int_{\Gamma_n} F^k \cdot n d\Gamma + \int_{V_n} q^k dV \tag{6.80}$$

where M^k is the mass and energy increment of component k in unit volume; F^k is the flow term of component k and energy; q^k is the source/sink term of component k; $K = 1, \ldots, 8$, N_{K+1}, represents 9 material components, $k = N_{K+1}$ is the energy component; V_n is the domain volume; and Γ_n is the boundary.

(2) Accumulation item

When $k = 1, \ldots, 8$, the governing equation is mass conservation equation, and the left side represents the mass increment of k component in the REV per unit time, and its expression is:

$$M^k = \sum_{\beta=A,G,I,H} \phi_H S_\beta \rho_\beta X_\beta^k \tag{6.81}$$

where β is an index, and there are 4 phases in the model, including: water phase ($\beta = A$), gas phase ($\beta = G$), solid hydrate phase ($\beta = H$) and solid ice phase ($\beta = I$); ϕ_H is the porosity; S_β is the phase saturation; ρ_β is the phase density; and X_β^k is the mass fraction of k components in β phase.

Obviously, the summation of saturation in each phase is equal to 1, and the sum of the mass fraction of each component in each phase is equal to 1, that is:

$$\sum_{\beta=A,G,I,H} S_\beta = 1, \quad \sum_k X_\beta^k = 1 \tag{6.82}$$

When $k = N_{K+1}$, it is the energy conservation equation. The left side represents the heat increment, including the contribution of rock and all phases, which can be expressed as:

$$M^{N_{K+1}} = (1 - \phi_H)\rho_R C_R T + \sum_{\beta=A,G,I,H} \phi S_\beta \rho_\beta u_\beta \tag{6.83}$$

where ρ_R and C_R are solid phase density and specific heat capacity of rock respectively; and u_β is the internal energy of the β phase.

(3) Flowing phase

In the mass conservation equation, the contribution of the two solid phases ($\beta = H, I$) to the flow is zero. The flow term of water phase and gas phase is:

$$F = \sum_{\beta=A,G} (F_\beta^k + J_\beta^k) \tag{6.84}$$

where $J_\beta^k = -\varphi S \tau_G D_\beta^k \rho_\beta \nabla X_\beta^k$ and $\tau_G = \varphi^{1/3} S_G^{7/3}$; τ_G is the tortuosity; and D_β^k is the hydrodynamic diffusion tensor.

For the water phase,

$$F_A^k = -K_H \frac{K_{rA}\rho_A}{\mu_A}(\nabla p_A - \rho_A g) X_A^k \tag{6.85}$$

For the gas phase, the mass flow rate considering Darcy flow and diffusion is expressed as:

$$F_G^k = -K_H \left(1 + \frac{b}{p_G}\right) \frac{K_{rG}\rho_G}{\mu_G}(\nabla p_G - \rho_G g) X_G^k + J_G^k \tag{6.86}$$

The heat flow including conduction and convection in the energy conservation equation is:

$$\begin{aligned} F^{N_{K+1}} = -\{&(1-\varphi_H)K_{tR} + \varphi_H[S_H K_{tH} + S_I K_{tI} \\ &+ S_A K_{tA} + S_G K_G]\}\nabla T + \sum_{\beta=A,G} h_\beta F_\beta \end{aligned} \tag{6.87}$$

where the phase of enthalpy is $h_A = \sum_k X_A^k h_A^k$, $h_G = \sum_k X_G^k h_G^k$; $K_{t\beta}$ is the thermal conductivity of components; and K_{tR} is the thermal conductivity of rock.

(4) Source and sink terms

In the mass conservation equation, the injection velocity of each component is expressed as:

$$q^k = \sum_{\beta=A,G} X_\beta^k q_\beta \tag{6.88}$$

In the dynamic model, the additional source and sink terms due to hydrate decomposition, hydrocarbon gas and water release must be considered. Therefore, the source sink term is changed to $q^k + m^k$, where the calculation method of each phase m^k has been mentioned in the flow characteristics of hydrate, and k represents water, methane and non-methane hydrocarbons.

In the energy balance equation, the heat in the decomposition process should be considered, so the heat flow velocity is:

$$q^{N_{K+1}} = q_d + \sum_{\beta=A,G} h_\beta q_\beta + m_H \Delta H^0 \tag{6.89}$$

where q_d is the energy absorbed or released; ΔH^0 is the heat of hydrate decomposition.

(5) Phase equilibrium and thermo-physical properties

According to Kamath's regression equation, the relationship between equilibrium pressure and temperature can be calculated as:

$$p_E = \exp\left(e_1 + \frac{e_2}{T}\right) \tag{6.90}$$

where

$$\begin{cases} e_1 = 37.980, e_2 = -8533.80, & 0°\text{C} < \text{T} \leq 25°\text{C} \\ e_1 = 14.717, e_2 = -1886.79, & -25°\text{C} \leq \text{T} \leq 0°\text{C} \end{cases} \tag{6.91}$$

Numerical solution method can be applied to solve the established mathematical model.

6.6 Flow in geothermal reservoirs

6.6.1 *Introduction*

In addition to traditional fossil energy, the application of geothermal energy attracts more and more attention. Its heat source comes from the heat generated by the decay of radioactive elements contained in the earth's internal materials. In other words, the earth itself is a heat source. According to the form of heat storage, geothermal energy is divided into shallow geothermal energy, hydrothermal energy and dry hot rock geothermal energy. The latter two are the objects of flow mechanics in porous media in this section.

The utilizations of geothermal energy include the extraction of medium-low temperature geothermal water, the development of high-temperature geothermal fields used for electricity generation, and the creation of the enhanced geothermal energy system. The flow and heat transfer processes in fractured rock have important influences on geothermal energy extraction and economic benefits. The former two extraction methods are similar to those of conventional oil and gas reservoirs. The enhanced geothermal systems (EGS) for hot dry rock reservoirs require hydraulic fracturing, which is similar to unconventional oil and gas reservoirs.

As shown in Figure 6.16, EGS focuses mainly on the heat energy in the hot dry rock. Hydraulic fracturing of the hot dry rock reservoir creates a large number of fractures. Therefore, the region out of the fractured zone can provide a stable heat supply source, and within the fractured zone, cold water is injected. After the flow and heat transfer in heat storage, production wells can produce high temperature steam or hot water. These high temperature steam or hot water is exported to power plants, and the remaining water is injected back to the reservoir system.

When creating an artificial thermal reservoir, cold water and high pressure are injected into the dry hot rock, which causes cooling and contraction effect of the high-temperature rock mass.

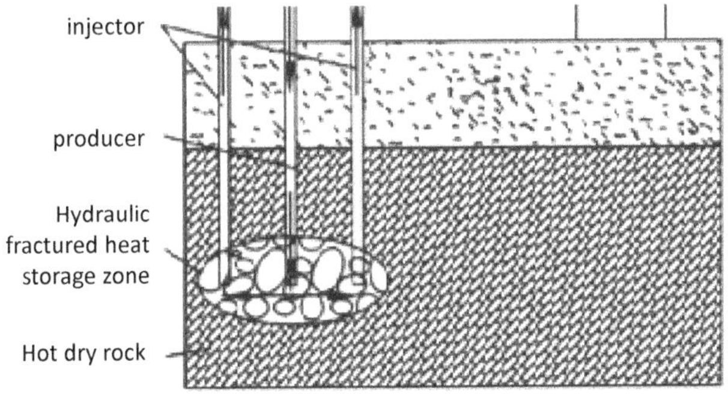

Figure 6.16. Hydraulic fractured geothermal storage and extraction method of dry hot rock

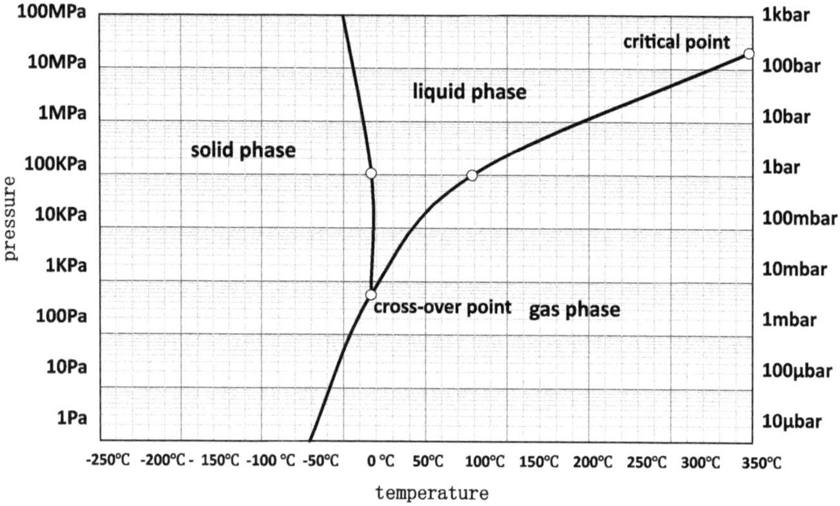

Figure 6.17. Phase diagram of water

The thermal stress changes generated cause shear failure of the rock during fracturing, and the generated fracture will not close when the pressure decreases, which is different from the mechanism of unconventional oil and gas reservoir fracturing.

Whether it is the natural heat storage or hydraulic fractured heat storage, fractures become the main medium of water flow and heat exchange. But the phase state of water can change at different temperatures and pressures. It can be single-phase liquid water, gas-liquid phase, or gas phase. It depends on the phase equilibrium of water. Figure 6.17 shows the phase diagram of water.

The geothermal reservoir is 3000 m deep, with a pressure of about 30 MPa, a temperature gradient of 5°C/100 m, and a temperature of about 150°C. During the flow process, the pressure maintains the original formation pressure, so flow in the reservoir is a single liquid phase. Even if the temperature reaches 300°C, the flow is still a single liquid phase.

Therefore, the mathematical model of geothermal flow is established based on the assumption of the single liquid water phase in the fractured medium.

6.6.2 *Concepts*

Thermal stress: *the stress that occurs when a body is unable to expand and contract freely due to external constraints and mutual constraints between the internal parts when the temperature changes.*

Geothermal resources: *it refers to the geothermal fluid and its useful components in the interior parts of the earth and can be economically utilized by human beings. According to the form of heat storage, it can be divided into shallow geothermal, hydrothermal and hot dry rock geothermal. According to temperature, a low temperature geothermal reservoir has a temperature below 150° C, otherwise it is a high temperature geothermal reservoir. A shallow geothermal reservoir has a depth of 200–2000 m, and it is a deep geothermal reservoir if the depth is greater than 2000 m.*

Geothermal reservoir: *it is a reservoir located in a favorable geological structure and contains hot water and steam with certain physical and chemical properties. Generally, it includes four elements, namely, heat storage space, cap rock, hot fluid pathway and heat source.*

Geothermal storage: *it refers to the strata, rocks or tectonic zones buried underground with effective pores and permeability, in which the geothermal fluid stored can be extracted.*

Dry hot rock: *it is a deep geothermal reservoir rock with a temperature of more than 200° C and a depth of more than 2000 m. There is no fluid or only a small amount of fluid inside. Heat energy is stored in the rock, and there is a thermal barrier layer composed of sedimentary rock.*

Geothermal fluid: *geothermal fluid is underground hot water, geothermal steam and heat-carrying gas, which are stored in the ground and the temperature is higher than normal. The common heat-carrying gases include carbon dioxide, hydrogen sulfide, hydrogen, oxygen, nitrogen and methane. Geothermal fluid is usually dominated by geothermal hot water and steam.*

Storage coefficient: *it is the volume of hot water discharged from the rock per unit volume in the aquifer with a unit pressure drop. The symbol is C_s, and the equation is:*

$$C_s = \frac{1}{V_f}\frac{dV_{hw}}{dp} \tag{6.92}$$

where C_s is the storage coefficient, MPa^{-1}; V_f is the rock volume, m^3; and V_{hw} is the volume of the discharged hot water from rock, m^3.

Enhanced geothermal system (EGS): *it is a hydraulic fractured geothermal reservoir, where the deep thermal energy can be economically extracted from the dry hot rock through the well system.*

6.6.3 *Analysis*

6.6.3.1 *Single fracture model*

As shown in Figure 6.18, there is only one injection well and one production well in the well system, so the flow between the two wells can be approximately regarded as a unidirectional flow in a vertical fracture. The assumptions are: ① the water temperature in the fracture remains the same along the y-axis, that is, the water temperature in the fracture is only related to the x-coordinate; ② the aperture of the fracture is $2d_f$, the porosity in the fracture is 1, the permeability is K_f, and is steady state flow; ③ the initial temperature of fracture and matrix is T_0, and the temperature of injected water is T_{in}; ④ the matrix is infinitely large with the temperature T_0, and the matrix only stores heat not water; and ⑤ the heat conduction is parallel to the direction of the fracture in the matrix and the heat conduction in the fracture is ignored.

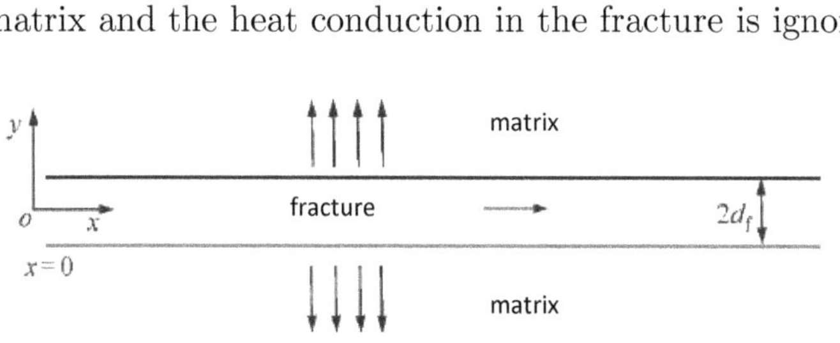

Figure 6.18. Diagram of flow in the single fracture of the well system

The flow equation of water in the fracture is:

$$\frac{d^2 p_f}{\partial x^2} = 0 \tag{6.93}$$

It is known that the pressure p_f distribution is uniform, and the flow velocity v_w is constant.

The energy equation in the fracture is:

$$d_f \rho_w C_{hw} \frac{\partial T_w}{\partial t} + d_f \rho_w C_{hw} v_w \frac{\partial T_w}{\partial x} - J_h = 0 \tag{6.94}$$

The energy equation in the matrix is:

$$\rho_m C_{hm} \frac{\partial T_m}{\partial t} + \lambda_{hm} \frac{\partial^2 T_m}{\partial y^2} + J_h = 0 \tag{6.95}$$

where m and w represent the matrix and water respectively; C_h is the specific heat; λ_h is the thermal conductivity; and J_h is the thermal velocity between matrix and fractures.

J_h is expressed as Newton's heat conduction law, namely:

$$J_h = \lambda_{hm} \left. \frac{\partial T_m}{\partial y} \right|_{y=d_f} \tag{6.96}$$

The initial conditions are:

$$\begin{cases} T_w(x, t = 0) = T_0 \\ T_m(x, y, t = 0) = T_0 \end{cases} \tag{6.97}$$

The outer boundary conditions are:

$$T_m(x \to \infty, y \to \infty) = T_0 \tag{6.98}$$

The inner boundary conditions are:

$$T_m(x = 0, t > 0) = T_{in} \tag{6.99}$$

Therefore, the temperature distribution along the x-axis in the fracture can be obtained as:

$$T_w = T_0 + (T_{in} - T_0) erfc \left[\frac{\lambda_{hm} x}{2 \rho_w C_{hw} d_f \sqrt{E_{hm} v_w (v_w t - x)}} \right] \cdot U\left(t - \frac{x}{v_w}\right) \tag{6.100}$$

where *erfc(X)* is the error function; $U(X)$ is the step function, $X > 0$, $U(X) = 1$, $X < 0$, $U(X) = 0$, $X = 0$, its value is different; and E_{hm} is the thermal diffusion coefficient of rock, where the equation is:

$$E_{hm} = \frac{\lambda_{hm}}{\rho_m C_{hm}} \tag{6.101}$$

6.6.3.2 *Discrete network model*

According to the discrete network model, the system is composed of two parts: matrix and fracture. Then the flow mathematical model is established and the discrete network model is solved by numerical simulation.

The flow equation of matrix is:

$$C_{tm}\frac{\partial p_m}{\partial t} + \nabla\left(-\frac{K_m}{\mu}\nabla p_m\right) - Q_c = 0 \tag{6.102}$$

The energy equation of matrix is:

$$C_{htm}\frac{dT_m}{dt} - C_{hw}\left(\frac{K_m}{\mu}\nabla p_m\right)\nabla T_m - \nabla(\lambda_{htm}\nabla T_m) + J_h = 0 \tag{6.103}$$

The flow equilibrium equation of the fracture is:

$$d_f C_{tf}\frac{\partial p_f}{\partial t} + \nabla\left(-\frac{K_f}{\mu}d_f\nabla p_f\right) + Q_c = 0 \tag{6.104}$$

The energy conservation equation of the fracture is:

$$d_f C_{thf}\frac{\partial T_w}{\partial t} - d_f\rho_w C_{hw}\left(\frac{K_f}{\mu}\nabla p_f\right)\nabla T_w - d_f\nabla\left(\lambda_{thf}\nabla T_w\right) - J_h = 0 \tag{6.105}$$

where C_t is the comprehensive compressibility; C_{ht} is the comprehensive specific heat; λ_{ht} is the comprehensive thermal conductivity; and Q_c is the cross flow.

The above differential equations are established and can be solved by numerical method.

6.7 Flow in water-soluble gas reservoirs

6.7.1 *Introduction*

In gas-water two-phase flow, it is assumed that gas and water are mutually immiscible, but in the actual reservoir, some natural gas will be dissolved in water. Under the condition of conventional reservoir, the solubility of natural gas in water is only 1–10 $\mathrm{m}^3/\mathrm{m}^3$. If it is in the deep high-pressure reservoir, its solubility can reach 20 $\mathrm{m}^3/\mathrm{m}^3$, and the reserves are still considerable.

The world's water-soluble gas resources are estimated to be on the magnitude of $1 \times 10^{16} \sim 1 \times 10^{18}\,\mathrm{m}^3$, which could be tens or hundreds of times larger than the total reserves of conventional natural gas.

Water-soluble gas is a kind of special natural gas resource, which is mainly methane dissolved in formation water under conventional reservoir conditions. Generally speaking, natural gas is concentrated in underground water storage reservoirs in the way of dissolution. When the water is pumped to the surface, the dissolved natural gas will be precipitated out of the water due to the decrease of water-soluble gas saturation, and will become fossil energy like conventional natural gas.

Water-soluble gas is also an important component of coal bed methane. Coaled methane exists in coal seam in three states (adsorbed state, dissolved state and free state), and the amount of water-soluble gas is far less than that of adsorbed state. This is because the amount of underground water in most coal bed reservoirs is limited, but the dissolved coal bed methane may escape from the water and form reservoirs at any time.

Water-soluble gas is also related to the loss of natural gas in gas storage. The long-term contact between natural gas and water can assume that the dissolved natural gas in groundwater reaches its maximum saturation. Because of the enormous volume of water, the loss caused by dissolution is considerable.

6.7.2 *Concepts*

Water-soluble gas: *it is a common form of natural gas that dissolves in formation water and is mainly dissolved gas.*

Dissolved gas-water ratio: *it refers to the dissolved gas volume per unit volume of water under standard temperature and pressure conditions.*

6.7.3 *Analysis*

6.7.3.1 *Characteristics of water-soluble gas*

(1) Compositional characteristics of water-soluble gas

Compared with other types of gases, the methane content in water-soluble gas is the highest, other hydrocarbons with higher molecular weight, such as ethane, are rarely present, so the relative density of water-soluble gas is very low, lower than the oil-dissolved gas and free gas. In addition, with an increase in water-soluble natural gas buried deep, under normal circumstances the methane content increases and the carbon dioxide content decreases. Since methane is the main component of water-soluble gas, the properties of pure methane can be used to represent the properties of water-soluble gas.

(2) Geological characteristics of the water-soluble gas reservoir

Water-soluble gas reservoirs are widely distributed in marine and continental sedimentary basins in the Mesozoic and Cenozoic ages. It is not restricted by structure and can exist in anticline, syncline, monocline and fault tectonic belt. Water-soluble gas reservoirs can be divided into open and closed. When the mineral extraction in the formation water is not considered, water-soluble gas reservoirs with closed boundary and high pressure should be favorable. In many cases, there are a large number of fractures, especially water-soluble gas reservoirs with abnormally high pressure. Therefore, it can be considered as a reservoir with fracture and pore dual media in the flow simulation.

6.7.3.2 *Mathematical model of flow*

The assumptions to establish the flow model are: ① only gas and water exist in the gas reservoir; ② the gas reservoir medium is a dual porosity medium; ③ the capillary force and viscous force are considered in the model; ④ the reservoir is under isothermal flow condition; ⑤ the salinity remains unchanged in the process of flow; and ⑥ the rock and fluid are slightly compressible.

When the flow velocity is relatively small, the flow can satisfy Darcy's law, and the flow motion equation can be obtained as:

For the water component, the equation of motion is:

$$\overrightarrow{V_w} = -\frac{KK_{rw}}{\mu_w} grad(p_w + \gamma_w D) \tag{6.106}$$

For the gas component, the equation of motion is:

$$\overrightarrow{V_g} = -\frac{KK_{rg}}{\mu_g B_g} grad(p_g + \gamma_g D) \tag{6.107}$$

The potential function is defined as:

$$\Phi_k = \mathrm{p}_k + \gamma_k D \tag{6.108}$$

For the water component, it is:

$$\overrightarrow{V_w} = -\frac{KK_{rw}}{\mu_w} grad\Phi_w \tag{6.109}$$

For the gas component, it is:

$$\overrightarrow{V_g} = -\frac{KK_{rg}}{\mu_g B_g} grad\Phi_g \tag{6.110}$$

For the matrix system, the amount of cross flow rate should be equal to the accumulation rate. For the fracture system, the fracture accumulation rate should be equal to the sum of the difference between the fracture inflow and outflow rates, the production rate and cross flow rate. Therefore:

For the matrix, it has:

$$-\tau_{mf} = \frac{\partial}{\partial t}\left(\frac{\phi_m S_m}{B_m}\right) \tag{6.111}$$

For the fracture, it has:

$$-\frac{\partial(\mathrm{v_x})}{\partial x} - \frac{\partial(\mathrm{v_y})}{\partial y} - \frac{\partial(\mathrm{v_z})}{\partial z} + q + \tau_{mf} = \frac{\partial}{\partial t}\left(\frac{\phi S}{B}\right) \tag{6.112}$$

For the water component in the fracture system, it has

$$\mathrm{div}\frac{KK_{rw}}{\mu_w B_w} grad\,\Phi_w + \tau_{wmf} + q_w = \frac{\partial}{\partial t}\left(\frac{\phi S_w}{B_w}\right) \tag{6.113}$$

For the gas component in the fracture system, it has

$$\mathrm{div}\frac{KK_{rg}}{\mu_g B_g} grad\,\Phi_g + \mathrm{div}\frac{R_s KK_{rw}}{\mu_w B_w} grad\,\Phi_w + \tau_{gmf} + q_g$$
$$= \frac{\partial}{\partial t0}\left(\frac{\phi S_g}{B_g} + \frac{\phi R_s S_w}{B_w}\right) \tag{6.114}$$

For the water component in the matrix system, it has

$$-\tau_{wmf} = \frac{\partial}{\partial t}\left(\frac{\phi_m S_{wm}}{B_{wm}}\right) \tag{6.115}$$

For the gas component in the matrix system, it has

$$-\tau_{gmf} = \frac{\partial}{\partial t}\left(\frac{\phi_m S_{gm}}{B_{gm}} + \frac{\phi_m R_{xm} S_{wm}}{B_{wm}}\right) \tag{6.116}$$

The auxiliary equations are as follows:

$$S_g + S_w = 1 \tag{6.117}$$

$$S_{gm} + S_{wm} = 1 \tag{6.118}$$

$$p_w = p_g - p_{cgw} \tag{6.119}$$

$$p_{wm} = p_{gm} - p_{cgwm} \tag{6.120}$$

$$\Phi_w = p_w + \gamma_w D \tag{6.121}$$

$$\Phi_g = p_w - p_{cgw} + \gamma_w D \tag{6.122}$$

$$\Phi_{wm} = p_{wm} + \gamma_{wm} D \tag{6.123}$$

$$\Phi_{gm} = p_{wm} - p_{cgwm} + \gamma_{wm} D \tag{6.124}$$

where R_s and R_{sm} are respectively the dissolved gas-water ratio in the fracture and matrix; q_w and q_g are respectively the water and gas production rates, with production as the positive sign and injection as the negative sign; and τ_{jmf} is the fluid exchange between rock and fracture for the 'j' phase.

The initial conditions are:

$$P(x, y, z, t) = |_{t=0} = p_0(x, y, z) \tag{6.125}$$

$$S(x, y, z, t) = |_{t=0} = S_0(x, y, z) \tag{6.126}$$

where p_0 and S_0 are respectively the initial value of any point in the medium system of the gas reservoir.

The bottom hole constant pressure is:

$$P|_{r=R_w} = C \tag{6.127}$$

The constant production is:

$$Q|_{r=R_w} = C \tag{6.128}$$

The outer boundary condition is:

The constant pressure on the boundary is:

$$P_e = f_P(x, y, z, t) \tag{6.129}$$

The constant flow rate on the boundary is:

$$\left.\frac{\partial P}{\partial n}\right|_E = f_q(x, y, z, t) \tag{6.130}$$

The mixed boundary condition can be:

$$\left(\frac{\partial P}{\partial n} + aP\right)\bigg|_E = f_q(x, y, z, t) \tag{6.131}$$

where $\frac{\partial P}{\partial n}\big|_E$ is the outward normal direction on the boundary E.

The above model needs to be solved by numerical simulation.

Chapter 7

Theoretical application

In order to reflect the importance of the basic theory of flow mechanics in porous media, this chapter briefly introduces the application of the basic theory to reservoir numerical simulation, modern well testing, production engineering, reservoir engineering, and enhanced oil recovery. Each part focuses on the interactions with flow mechanics in porous media.

7.1 Application to reservoir numerical simulation

7.1.1 *Introduction*

The mathematical models established in flow mechanics are solved by analytical and numerical methods. The analytical solutions of many mathematical models have been introduced. In this section, the numerical methods are briefly introduced.

The key to reservoir numerical simulation is to discretize the reservoir into a finite grid and write the differential equation in its numerical discretization form. In the finite difference method, the continuous differential equation of flow is discretized into the discrete finite difference grid, and the discretized equations of all grids are solved to obtain the distribution of pressure and saturation on each grid.

The advantage of numerical simulation is that it can effectively solve flow problems under complex conditions, such as heterogeneous, non-uniform thickness, non-isothermal, multiphase and multicomponent, and physicochemical flow. Because a large number of calculations are needed to solve equations, numerical simulation is generally realized by computer programming.

7.1.2 *Concepts*

Reservoir numerical simulation: *it is a numerical calculation method to study the flow law of oil, gas, water and injecting fluids in the development process of oil and gas reservoirs.*

Finite difference method: *it is a method used for finding numerical solutions of partial differential equations. In other words, the continuous partial differential equation is transformed into a discrete finite grid.*

First order difference quotient: *it is the differential treatment of first order partial derivatives.*

Second order difference quotient: *it is the differential treatment of second order partial derivatives.*

Difference equation: *it is an equation obtained through replacing the differential equation with difference quotients.*

7.1.3 *Analysis*

7.1.3.1 *Numerical grid*

The reservoir is divided into a finite number of grids, as shown in Figure 7.1.

Before the establishment of the discretized equation, the grid needs to be numbered, and the way of numbering may affect the complexity of the discretized equation. An example to number the one-dimensional grid is shown in Figure 7.2.

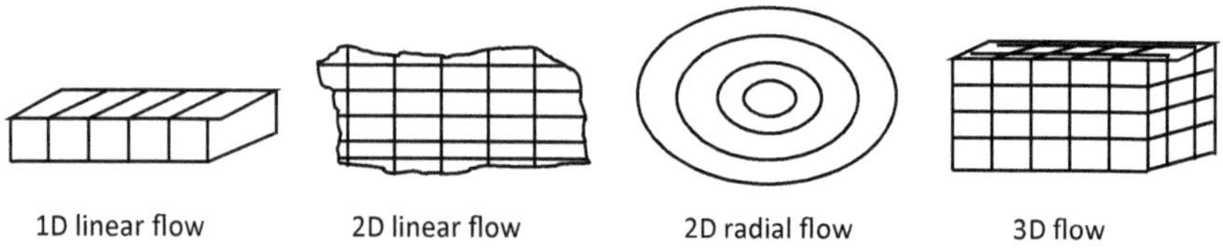

Figure 7.1. Schematic diagram of the discretized grid for different flow regimes

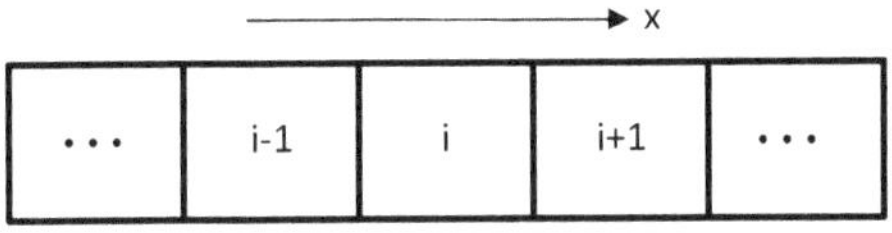

Figure 7.2. Numbering of the one-dimensional linear flow grid

7.1.3.2 *Difference quotient*

In the process of establishing the flow continuity equation, the derivative is approximated by the difference quotient. Combined with the Taylor series method, the truncation error is ignored and the distance between adjacent grids is set as Δx, then the difference quotients are as follows.

First order forward spatial difference quotient:

$$\left.\frac{\partial p}{\partial x}\right|_{x_i,t_n} = \frac{p_{i+1}^n - p_i^n}{\Delta x} \tag{7.1}$$

First order backward spatial difference quotient:

$$\left.\frac{\partial p}{\partial x}\right|_{x_i,t_n} = \frac{p_i^n - p_{i-1}^n}{\Delta x} \tag{7.2}$$

First order central spatial difference quotient:

$$\left.\frac{\partial p}{\partial x}\right|_{x_i,t_n} = \frac{p_{i+1}^n - p_{i-1}^n}{\Delta x} \tag{7.3}$$

Forward temporal difference quotient:

$$\left.\frac{\partial p}{\partial t}\right|_{x_i,t_n} = \frac{p_i^{n+1} - p_i^n}{\Delta t} \tag{7.4}$$

Backward temporal difference quotient:

$$\left.\frac{\partial p}{\partial t}\right|_{x_i,t_n} = \frac{p_i^n - p_i^{n-1}}{\Delta t} \tag{7.5}$$

Central temporal difference quotient:

$$\left.\frac{\partial p}{\partial t}\right|_{x_i,t_n} = \frac{p_i^{n+1} - p_i^{n-1}}{\Delta t} \tag{7.6}$$

Second order spatial difference quotient:

$$\left.\frac{\partial^2 p}{\partial x^2}\right|_{x_i,t_n} = \frac{p_{i+1}^n - 2p_i^n + p_{i-1}^n}{(\Delta x)^2} \tag{7.7}$$

where x_i is the ith grid in the x direction; t_n is the nth time step; p_i^n is the pressure at the nth time step in the ith grid, and it is generally considered to be known; p_i^{n+1} is the pressure at the $n{+}1^{th}$ time step in the ith grid, which is generally regarded as unknown and needs to be solved; Δx is the grid size; and Δt is the length of time step.

7.1.3.3 *Establishment of discretized equation*

The discretized equation is established by taking one-dimensional single-phase transient flow as an example. The basic differential equation is:

$$\frac{\partial^2 p}{\partial x^2} = \frac{1}{\eta}\frac{\partial p}{\partial t} \tag{7.8}$$

(1) Explicit difference scheme

For grid node (x_i, t_n), the spatial difference quotient uses the values at time n, the temporal difference quotient uses the forward difference quotient, then the discretized equation is:

$$\left.\frac{\partial^2 p}{\partial x^2}\right|_{x_i,t_n} = \frac{1}{\eta}\left.\frac{\partial p}{\partial t}\right|_{x_i,t_n} \tag{7.9}$$

$$\frac{p_{i+1}^n - 2p_i^n + p_{i-1}^n}{(\Delta x)^2} = \frac{1}{\eta}\frac{p_i^{n+1} - p_i^n}{\Delta t} \tag{7.10}$$

The difference scheme at any point is:

$$p_i^{n+1} = (1-2\delta)p_i^n + \delta(p_{i+1}^n + p_{i-1}^n) \tag{7.11}$$

$$\delta = \frac{\eta \Delta t}{\Delta x^2} \tag{7.12}$$

If the pressure value of each mesh node at time n is known, the pressure value at time $n{+}1$ after a time step Δt can be calculated.

(2) Implicit difference scheme

For the grid node $(x_i,\ t_{n+1})$, the spatial difference quotient uses the value at time step $n+1$, and the temporal difference quotient uses the backward difference quotient, then the discretized equation is:

$$\left.\frac{\partial^2 p}{\partial x^2}\right|_{x_i,t_{n+1}} = \frac{1}{\eta}\left.\frac{\partial p}{\partial t}\right|_{x_i,t_{n+1}} \tag{7.13}$$

$$\frac{p_{i+1}^{n+1} - 2p_i^{n+1} + p_{i-1}^{n+1}}{(\Delta x)^2} = \frac{1}{\eta}\frac{p_i^{n+1} - p_i^n}{\Delta t} \tag{7.14}$$

The difference scheme at any point is:

$$-\delta p_{i-1}^{n+1} + (1-2\delta)p_i^{n+1} - \delta p_{i+1}^{n+1} = p_i^n \tag{7.15}$$

This equation contains three unknowns. In this case, the discretized equations of all grids need to be set up together to form a system of equations, and then the pressure value at the moment $n+1$ can be obtained by solving the system of equations.

Then the implicit discretized equation is established by taking two-dimensional single-phase fluid transient flow as an example. The basic differential equation is:

$$\frac{\partial^2 p}{\partial x^2} + \frac{\partial^2 p}{\partial y^2} = \frac{1}{\eta}\frac{\partial p}{\partial t} \tag{7.16}$$

For grid nodes $(x_i,\ y_j,\ t_{n+1})$, the spatial difference quotient uses the values at the time step $n+1$, and the time difference quotient uses the backward difference quotient, then the discretized equation is:

$$\left.\frac{\partial^2 p}{\partial x^2}\right|_{x_i,y_i,t_{n+1}} + \left.\frac{\partial^2 p}{\partial y^2}\right|_{x_i,y_i,t_{n+1}} = \frac{1}{\eta}\left.\frac{\partial p}{\partial t}\right|_{x_i,t_{n+1}} \tag{7.17}$$

$$\frac{p_{i+1,j}^n - 2p_{i,j}^n + p_{i-1,j}^n}{(\Delta x)^2} + \frac{p_{i,j+1}^n - 2p_{i,j}^n + p_{i,j-1}^n}{(\Delta y)^2} = \frac{1}{\eta}\frac{p_{i,j}^{n+1} - p_{i,j}^n}{\Delta t} \tag{7.18}$$

The difference scheme at any point is:

$$-\beta p_{i,j-1}^{n+1} - \alpha p_{i-1,j}^{n+1} + (2\alpha + 2\beta + 1)p_{i,j}^{n+1} - \alpha p_{i+1,j}^{n+1} - \beta p_{i,j+1}^{n+1} = p_{i,j}^{n} \tag{7.19}$$

$$\alpha = \frac{\eta \Delta t}{(\Delta x)^2}, \quad \beta = \frac{\eta \Delta t}{(\Delta y)^2} \tag{7.20}$$

This equation contains 5 unknowns. In this case, the discretized equations of all grids need to be set up together to form a system of equations, and then the pressure value at the moment $n+1$ can be obtained by solving the system of equations..

If it is the two-dimensional oil-water two-phase flow, the discretized equation needs to be established for oil and water phases, namely:

For the oil phase, it has:

$$c_{oi,j}P_{oi,j-1}^{n+1} + a_{oi,j}P_{oi-1,j}^{n+1} + e_{oi,j}P_{oi,j}^{\mathrm{n}+1} + b_{oi,j}P_{oi+1,j}^{n+1} + d_{oi,j1}P_{oi,j+1}^{n+1}$$
$$= f_{oi,j} + V_{ij}(\varphi\rho_o)_{ij}\frac{S_{oi,j}^{n+1} - S_{oi,j}^{n}}{\Delta t^n} \tag{7.21}$$

For the water phase, it has:

$$c_{wi,j}P_{wi,j-1}^{n+1} + a_{wi,j}P_{wi-1,j}^{n+1} + e_{wi,j}P_{wi,j}^{n+1} + b_{wi,j}P_{wi+1,j}^{n+1} + d_{wi,j+1}P_{wi,j+1}^{n+1}$$
$$= f_{wi,j} + V_{ij}(\varphi\rho_w)_{ij}\frac{S_{wi,j}^{n+1} - S_{wi,j}^{n}}{\Delta t^n} \tag{7.22}$$

There are 2 equations and 12 unknowns (5 for oil phase, 5 for water phase pressure and 2 for oil and water saturations) on a grid node, and the equation system on all the grids needs to be solved.

7.2 Application to the modern well test

7.2.1 *Introduction*

Steady state well test is used to find the relationship between pressure difference and production rate and to determine the productivity, formation parameters and the proper working schedule.

Transient well test is used to find the relationship between the bottom hole pressure and time, and to calculate the formation parameters, stimulation effect, interwell connectivity, etc.

In the oilfield, the steady state well test and transient well test are usually carried out after the new well is drilled, in order to determine oil production capability and obtain initial formation parameters. When the well is producing oil under normal conditions, the pressure buildup test can be used which requires the transient well test. Modern well testing is based on the theory of transient well test.

This section briefly introduces the basic principles of modern well testing in order to further understand the application of flow mechanics in porous media.

7.2.2 *Concepts*

Modern well test: *it is on the basis of flow mathematical model, and the well test interpretation method is established through dimensionless transformation, and then the measured curve is fitted to obtain the formation parameters.*

Well test interpretation: *it refers to the solution of all kinds of well test models commonly used in dimensionless flow mathematical models, which can be used to build sample curves with different parameters.*

Dimension: *it refers to the basic properties of physical quantities, and is also a function related to various physical quantities, generally expressed in units. The four basic dimensions commonly used in flow mechanics are: [L] represents length dimension, unit m; [M] represents mass dimension, unit g; [T] represents time dimension, unit s; [K] represents temperature dimension, unit K. All the key parameters can be represented by these four basic dimensions, e.g., the pressure dimension is* $ML^{-1}T^{-2}$ *and the viscosity dimension is* $ML^{-1}T^{-1}$.

7.2.3 *Analysis*

Taking the establishment of well test model for infinite homogeneous reservoir in vertical well as an example, the process of modern well test is briefly introduced.

7.2.3.1 *Establishment of dimensionless flow model*

The basic differential equation of radial flow in elastic transient flow of single-phase liquid is:

$$\frac{\partial^2 p}{\partial r^2}+\frac{1}{r}\frac{\partial p}{\partial r}=\frac{1}{\eta}\frac{\partial p}{\partial t} \tag{7.23}$$

The dimensionless pressure is defined as:

$$p_D=\frac{Kh(p_i-p)}{1.842\times 10^{-3}Q\mu B} \tag{7.24}$$

The dimensionless time is defined as:

$$t_D=\frac{3.6Kt}{\phi\mu C_t R_w^2} \tag{7.25}$$

The dimensionless wellbore storage factor is defined as:

$$C_D=\frac{0.159C}{\phi C_t h R_w^2} \tag{7.26}$$

The dimensionless radius is defined as:

$$r_D=\frac{r}{R_w} \tag{7.27}$$

Then, the dimensionless basic differential equation can be written as:

$$\frac{\partial^2 p_D}{\partial r_D^2}+\frac{1}{r_D}\frac{\partial p_D}{\partial r_D}=\frac{\partial p_D}{\partial t_D} \tag{7.28}$$

The initial condition is:

$$p_D(r_D,0)=0 \tag{7.29}$$

The inner boundary conditions are:

$$P_{wD}=\left[P_D-S\left(\frac{\partial P_D}{\partial r_D}\right)\right]_{r_D=1} \tag{7.30}$$

$$C_D\frac{dP_{wD}}{dt_D}-\left[r\frac{\partial P_D}{\partial r_D}\right]_{r_D=1}=1 \tag{7.31}$$

The outer boundary condition is:

$$\lim_{r_D-\infty}[P_D(r_D,t_D)]=0 \tag{7.32}$$

7.2.3.2 *Analytical solution of the well test model*

The Laplace transform is written as:

$$\overline{p}_D=\int_0^\infty p_D e^{-ut_D}dt_D \tag{7.33}$$

Through the transformation of the well test mathematical model, the well bottom pressure $\overline{P}_{wD}$ in Laplace space is expressed as:

$$\overline{P}_{wD}=\frac{K_0(\sqrt{u})+S\sqrt{u}K_1(\sqrt{u})}{u\{\sqrt{u}K_1(\sqrt{u})+C_Du[K_0(\sqrt{u})+S\sqrt{u}K_1(\sqrt{u})]\}} \tag{7.34}$$

where $K_0(x)$ and $K_1(x)$ are the zero and first order Bessel functions respectively; and u is the Laplace operator.

This expression is difficult to apply and needs Laplace inversion and simplification.

(1) Simplification in early stage

In the early stage, time t_D is small, then the dimensionless bottom hole pressure can be expressed as:

$$p_{wD}(S,C_D,t_D)=\frac{1}{C_D}\left(t_D-\frac{t_D^2}{2C_DS}+\frac{8t_D^{5/2}}{15\sqrt{\pi}C_DS^2}\right) \tag{7.35}$$

(2) Simplification in the late stage

In the late stage, time t_D is relatively large, then the dimensionless bottom hole pressure can be expressed as:

$$p_{wD}(S,t_D)=0.5(\ln t_D+0.80907+2S) \tag{7.36}$$

7.2.3.3 *Characteristics of well test curve*

On the log-log plot, take log (t_D/C_D) as the x-axis and log (p_D) as the y-axis. As shown in Figure 7.3, the figure is a typical log-log

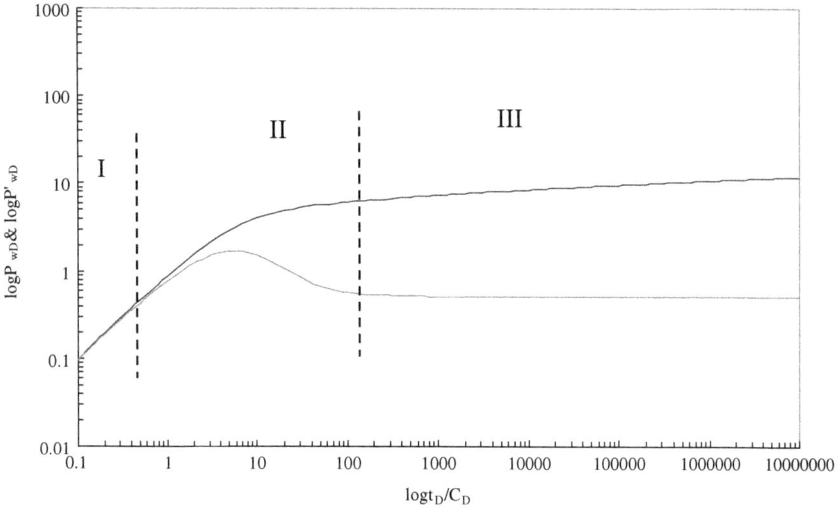

Figure 7.3. Pressure and pressure derivative on the log-log plot

characteristic for identification of homogeneous infinite reservoir. It can be analyzed in three sections.

① In section I, the pressure and derivative curves have the slope with the straight line of 45°, which indicates that this stage is influenced by the wellbore storage.

② In section II, it is the transition section, the pressure derivative curve peaks and then shows the downward slope. The height of the peak depends on the size of the parameter C_De^{2S}. Because the skin factor S is in an exponential position, the curve shape is affected more by the S value. When C_De^{2S} is large, the peak value is higher, the downward slope is steeper, and the appearance of the peak value is relatively late.

③ In section III, a horizontal segment shows up, which is the characteristics of radial flow.

7.2.3.4 *Well test fitting method*

In the process of fitting, the reservoir model should be selected, the relevant parameters and well test data should be input. On the log-log plot, the well test curve can be fitted with the samples. The formation parameters can be obtained by applying "the characteristic points based pressure derivative fitting method".

This method is to obtain the best matching characteristic point values of the measured pressure derivative curve by fitting characteristic points of the theoretical pressure derivative curve with those on the measured pressure derivative curve.

The procedures are: ① the measured and theoretical pressure and pressure derivative curves are made in the coordinate system of the same size. The theoretical curve can be calculated by taking any set of parameters; ② the characteristic points are identified on the theoretical pressure derivative curve; ③ the characteristic points of the theoretical curve is moved so that the theoretical curve and the measured curve get the best match, then read the horizontal and vertical coordinates of the matched point values on the theoretical and measured curves, that is $(t_D/C_D)_M, (p_D)_M$; ④ the peak value of the pressure derivative p'_{max} is recorded; ⑤ the fitting values can be calculated by using the following equations.

The fitting value of time can be computed as:

$$t_M = \left(\frac{t_D/C_D}{t}\right)_M \tag{7.37}$$

The fitting value of pressure can be computed as:

$$p_M = \left(\frac{p_D}{\Delta p}\right)_M \tag{7.38}$$

The pressure drawdown in the pressure drawdown test is:

$$\Delta p = p_i - p_w(t) \tag{7.39}$$

The pressure drop in the pressure buildup test is:

$$\Delta p = p_w(\Delta t) - p_w(\Delta t = 0) \tag{7.40}$$

And:

$$(C_D e^{2S})_M = 10^{\frac{p'_{\text{max}} - 0.4409}{0.4273}} \tag{7.41}$$

The formation properties can be obtained after fitting.

The permeability is:

$$K = 1.842 \times 10^{-3} \frac{Q\mu B}{h} p_M \tag{7.42}$$

And:

$$\frac{Kh}{\mu} = 1.842 \times 10^{-3} QBp_M \tag{7.43}$$

The wellbore storage coefficient is:

$$C = \frac{7.2\pi Kh}{\mu} \frac{1}{t_M} \tag{7.44}$$

The dimensionless wellbore storage coefficient is:

$$C_D = \frac{C}{2\pi\phi C_t h R_w^2} \tag{7.45}$$

The skin factor is:

$$S = \frac{1}{2} \ln \frac{(C_D e^{2S})_M}{C_D} \tag{7.46}$$

The units of all the parameters in this section are listed as follows: p — pressure, MPa; μ — viscosity, mPa·s; K — permeability, mD; Q — flow rate, m^3/d; t — time, h; h — thickness, m; R_w — well radius, m; C — wellbore storage coefficient, m^3/MPa; C_t — total compressibility, 1/MPa; ϕ — porosity, decimal; S — skin factor.

7.3 Application to production engineering

7.3.1 *Introduction*

If the flow is consistent with Darcy's law when the single-phase liquid is flowing steadily, then the production rate and pressure difference will have a linear relationship, and then the bottom hole pressure and production rate will also have a linear relationship. When multiphase flow or multiple oil layers are involved, the relationship between bottom hole pressure and production rate is no longer linear.

The oil and gas flowing from the reservoir to the bottom hole of the well is the object of study of flow mechanics in porous media, and the oil and gas flowing from the bottom hole of the well to the wellhead on the ground surface is the object of study of production engineering.

When oil and gas flow from the reservoir to the bottom of the well, the pressure distribution and production variation can be obtained from the flow law of the reservoir. The variation of the bottom hole pressure with flow rate can be used to analyze the artificial lifting design. IPR curve (Inflow Performance Relationship) in oil production engineering is a practical method to analyze well productivity.

7.3.2 *Concepts*

Production engineering: *it is the general term to describe engineering techniques taken to production and injection wells in the reservoir according to oil and gas development objectives. It includes artificial lift, injection, fracturing, acidification, paraffin cleaning, sand control, etc.*

IPR curve: *it is short for Inflow Performance Relationship curve, and refers to the relationship between oil well production and bottom hole pressure. It can reflect oil supply capacity of the well as shown in Figure 7.4.*

Productivity equation: *it is used to express the relationship between pressure difference or bottom hole pressure and flow rate,*

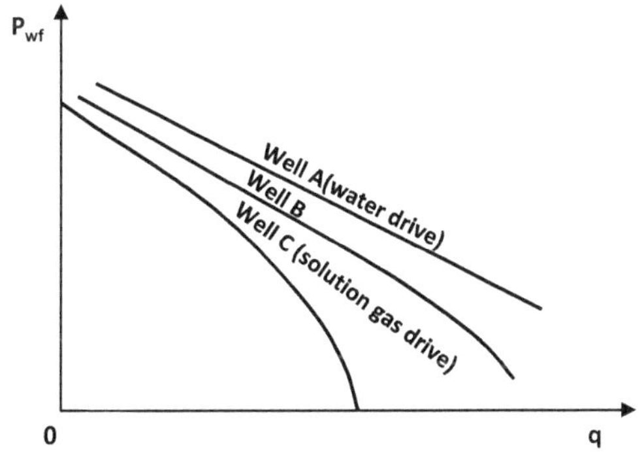

Figure 7.4. Typical well inflow performance curve

and can be written as:

$$Q_o = J(p_R - p_{wf}) \tag{7.47}$$

where Q_o *is the oil well production;* J *is the oil productivity index;* p_R *is the current formation pressure; and* p_{wf} *is the bottom hole pressure.*

Vogel equation: *it is a production equation used to represent a well under the dissolved gas drive process. The equation is:*

$$\frac{Q_o}{Q_{o\max}} = 1 - 0.2\frac{p_w}{p_R} - \left(\frac{p_w}{p_R}\right)^2 \tag{7.48}$$

where Q_{omax} *is the maximum oil well production rate.*

7.3.3 *Analysis*

7.3.3.1 *Inflow performance of a single-phase liquid*

The production rate of a well in the center of a circular reservoir with supply boundary is:

$$Q_o = \frac{2\pi K_o h(p_R - p_w)}{\mu_o B_o\left(\ln\frac{R_e}{R_w} - \frac{1}{2} + S\right)} \tag{7.49}$$

The production rate of a well in the center of a circular reservoir with closed boundary is:

$$Q_o = \frac{2\pi K_o h(p_R - p_w)}{\mu_o B_o\left(\ln\frac{R_e}{R_w} - \frac{3}{4} + S\right)} \tag{7.50}$$

For oil well production with non-circular closed boundary, it can be corrected according to the drainage area and the location of the oil well, as shown in Table 7.1. And the way to do that is to put $\frac{R_e}{R_w} = X$ in the equation, and thus, the flow rate of oil well can be obtained as:

$$Q_o = J(p_R - p_w) \tag{7.51}$$

The IPR curve can be drawn as shown in Figure 7.5. The IPR curve is a straight line in this case, the intercept on the y-axis is

Table 7.1. Correction coefficients for different shapes of drainage area and well locations

Drainage area shape and well location	X	Drainage area shape and well location	X
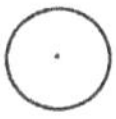	$\frac{R_e}{R_w}$	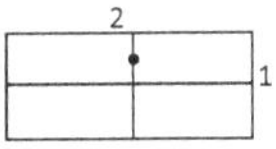	$\frac{0.966A^{1/2}}{R_w}$
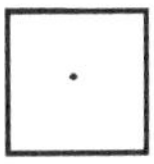	$\frac{0.571A^{1/2}}{R_w}$	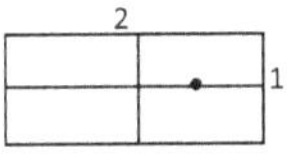	$\frac{1.44A^{1/2}}{R_w}$
	$\frac{0.565A^{1/2}}{R_w}$	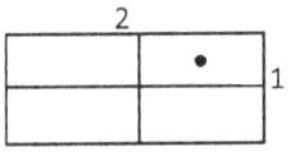	$\frac{2.206A^{1/2}}{R_w}$
	$\frac{0.604A^{1/2}}{R_w}$	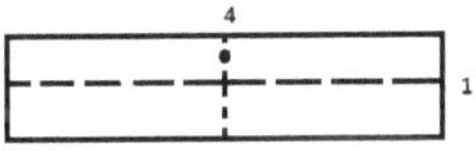	$\frac{1.925A^{1/2}}{R_w}$
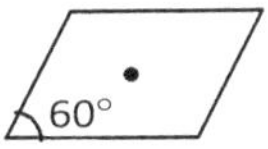	$\frac{0.61A^{1/2}}{R_w}$	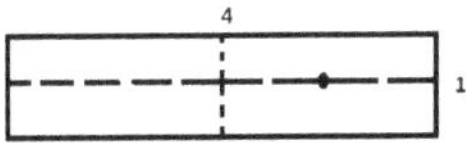	$\frac{6.59A^{1/2}}{R_w}$
	$\frac{0.678A^{1/2}}{R_w}$	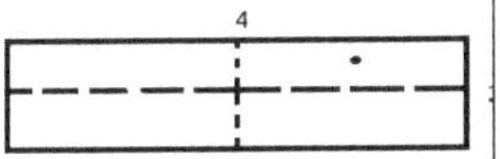	$\frac{9.36A^{1/2}}{R_w}$
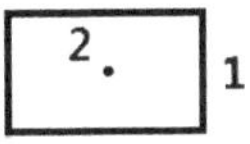	$\frac{0.668A^{1/2}}{R_w}$	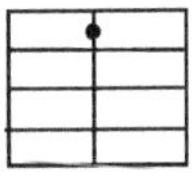	$\frac{1.724A^{1/2}}{R_w}$
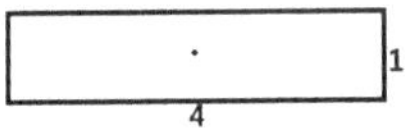	$\frac{1.368A^{1/2}}{R_w}$	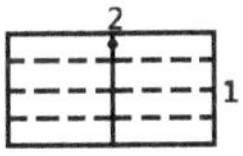	$\frac{1.794A^{1/2}}{R_w}$
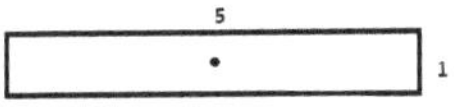	$\frac{2.066A^{1/2}}{R_w}$	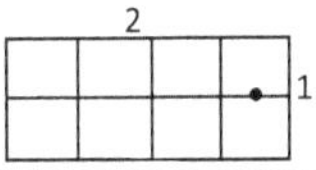	$\frac{4.072A^{1/2}}{R_w}$
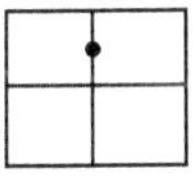	$\frac{0.884A^{1/2}}{R_w}$	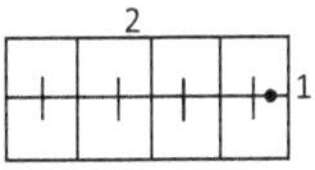	$\frac{9.523A^{1/2}}{R_w}$
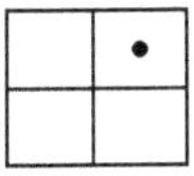	$\frac{1.485A^{1/2}}{R_w}$	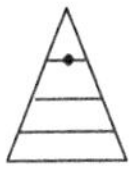	$\frac{10.135A^{1/2}}{R_w}$

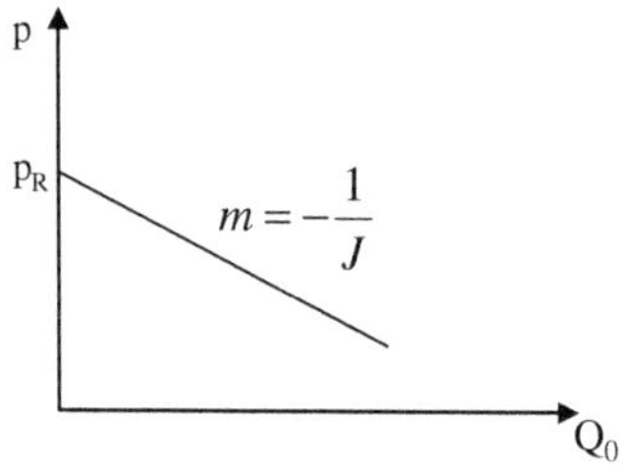

Figure 7.5. IPR curve of the single-phase liquid

the reservoir pressure, and the negative inverse of the slope is the productivity index.

7.3.3.2 *Inflow performance of oil and gas two-phase fluids*

The oil production rate of the oil and gas two-phase flow has already been discussed in the previous section as:

$$Q_o = \frac{2\pi Kh}{\ln \frac{R_e}{R_w}} \int_{p_w}^{p_e} \frac{K_{ro}}{\mu_o B_o} dp \tag{7.52}$$

It is difficult to use this equation with the integral form of the pseudo-pressure function in the IPR curves, so IPR curves are often drawn based on the field production data in well performance analysis.

(1) Vogel curve

In 1968, Vogel calculated the solution gas flooding in various reservoirs and obtained the Vogel equation:

$$\frac{Q_o}{Q_{o\max}} = 1 - 0.2\frac{p_w}{p_R} - 0.8\left(\frac{p_w}{p_R}\right)^2 \tag{7.53}$$

If the reservoir's current formation pressure p_R is known, IPR curves can be drawn with a pair of measured production Q_o and flow pressure p_w as shown in Figure 7.6.

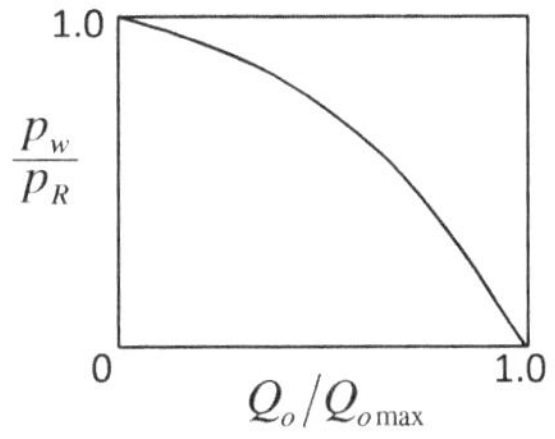

Figure 7.6. Dimensionless IPR curve of oil and gas two-phase fluids (Vogel curve)

(2) Modified Vogel curve

When an imperfect well is considered, it needs to be corrected. The flow efficiency of the oil well is defined as:

$$FE = \frac{p_R - p_w - \Delta p_{sk}}{p_R - p_w} \tag{7.54}$$

where Δp_{sk} is the additional pressure drop caused by the skin factor of the imperfect well, that is, the difference between the ideal and actual bottom hole pressure, and the equation is:

$$\Delta p_{sk} = p'_w - p_w \tag{7.55}$$

From the production rate of the perfect and imperfect wells, the expression of skin factor S can be obtained as:

$$S = \left(\frac{K_o}{K_s} - 1\right) \ln \frac{R_s}{R_w} \tag{7.56}$$

where K_s is the permeability in damaged zone; and R_s is the radius of damage zone.

Then:

$$\Delta p_{sk} = \frac{Q_o \mu_o B_o}{2\pi K_o h} S \tag{7.57}$$

The skin factor S can be obtained by well testing.

The flow efficiency FE and ideal bottom hole pressure p'_w can be obtained. The Vogel curve can be modified by using flow efficiency,

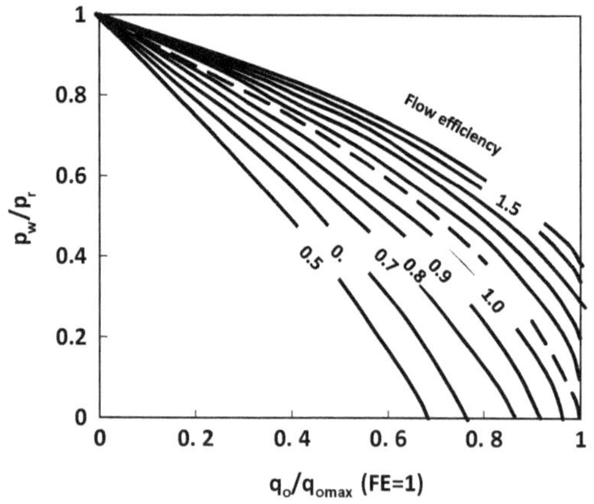

Figure 7.7. IPR curve considering imperfect oil wells (Standing curve)

and the modified equation is:

$$\frac{Q_o}{Q_{o\max(FE=1)}} = 1 - 0.2\frac{p'_w}{p_R} - 0.8\left(\frac{p'_w}{p_R}\right)^2 \tag{7.58}$$

The IPR curve is shown in Figure 7.7.

7.3.3.3 *Inflow performance of oil, gas and water three-phase fluids*

The IPR curve of the oil, gas and water three-phase fluids is shown in Figure 7.8.

Curve A is IPR curve when the fractional flow of water is $f_w = 0$, which is called the oil IPR curve.

Curve B is IPR curve when the fractional flow of water is $f_w = 100$, which is called the water IPR curve.

Curve C is IPR curve at a certain water cut, which is called the comprehensive IPR curve of oil, gas and water three-phase fluids.

The IPR curve of oil, gas and water three-phase fluids is essentially a weighted average according to water cut. In practice, when calculating the productivity index, the total liquid production rate can be the weighted average based on the production rate. When the production rate and flow pressure are predicted, they can be

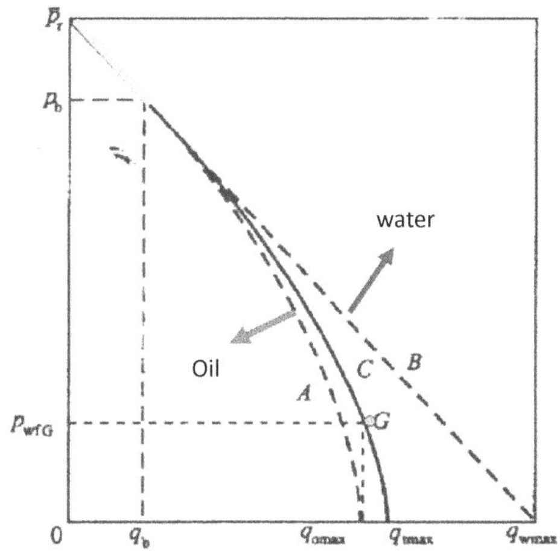

Figure 7.8. IPR curve of oil, gas and water three-phase fluids

weighted by flow pressure. That is:

$$q_t = (1 - f_w)q_{oil} + f_w q_{water} \tag{7.59}$$

$$P_{wf} = (1 - f_w)P_{wf(oil)} + f_w P_{wf(water)} \tag{7.60}$$

7.4 Application to reservoir engineering

7.4.1 *Introduction*

Flow mechanics in porous media is the basic theory to study the flow law of oil and gas in reservoirs. Reservoir engineering is the application of flow theory to develop oil and gas reservoirs efficiently. In addition to flow theory, reservoir engineering combines the geological, oil recovery, economic and other scientific knowledge together to form a complex management system. This section briefly introduces the application of flow mechanics in porous media to reservoir engineering.

7.4.2 *Concepts*

Reservoir engineering: *it is a comprehensive discipline that takes oil and gas reservoirs as the research object, uses the flow mechanics in porous media, petrophysics, petroleum geology and other displines* *as the theoretical basis, and mathematics, computer science, economics and other disciplines as the research tools.*

Relative mobility: *it is the ratio of relative permeability to viscosity, i.e.*

$$\lambda_r = \frac{K_r}{\mu} \tag{7.61}$$

where λ_r *is the relative mobility;* K_r *is the relative permeability; and* μ *is the viscosity of reservoir fluid.*

Apparent viscosity: *it is also known as effective viscosity. When oil and water flow in two phases, the equation of apparent viscosity is:*

$$\lambda_r^{-1} = \frac{1}{\lambda_{or} + \lambda_{wr}} \tag{7.62}$$

where λ_r^{-1} *is the apparent viscosity;* λ_{or} *is the relative mobility of oil; and* λ_{wr} *is the relative mobility of water;*

Areal well pattern: *it refers to well patterns with water injection wells and oil production wells that are evenly distributed with a certain geometry and density. The reservoir is divided into many small development units for water injection. There are mainly: direct line drive well pattern, staggered line drive well pattern, four-spot well pattern, five-spot well pattern, seven-spot well pattern, nine-spot well pattern, inverted nine-spot well pattern and so on.*

7.4.3 *Analysis*

7.4.3.1 *One-dimensional water flooding*

The equation of fractional flow has already been derived as:

$$f_w = \frac{Q_w}{Q_w + Q_o} \quad \text{or} \quad f_w = \frac{v_w}{v_w + v_o} \tag{7.63}$$

In reservoir engineering, the expression of water cut with respect to mobility is

$$f_w = \frac{\lambda_w}{\lambda_w + \lambda_o} = \frac{1}{1 + \frac{\lambda_{or}}{\lambda_{wr}}} \tag{7.64}$$

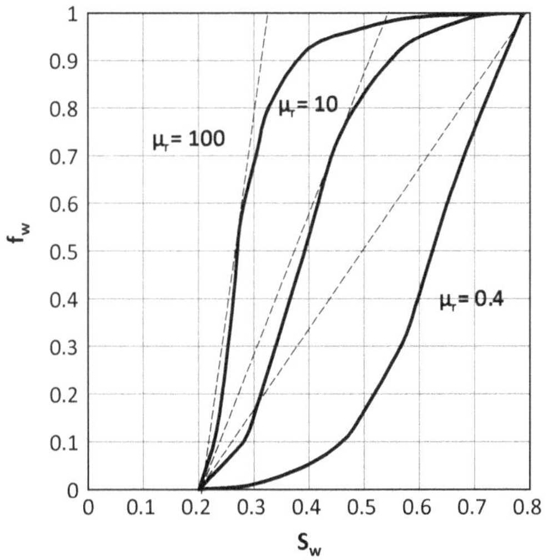

Figure 7.9. Relationship between saturation and fractional flow under different oil-water viscosity ratios

where $\lambda_o = \frac{kk_{ro}}{\mu_o}$ is the mobility of oil; $\lambda_w = \frac{kk_{rw}}{\mu_w}$ is the mobility of water; $\lambda_{or} = \frac{k_{ro}}{\mu_o}$ is the relative mobility of oil; and $\lambda_{wr} = \frac{k_{rw}}{\mu_w}$ is the relative mobility of water.

As shown in Figure 7.9, the water cut curves are made for different oil-water viscosity ratios. From the water saturation value at the saturation front, it can be found that the larger the oil-water viscosity ratio, the smaller the front saturation, and the worse the displacement effect. Three scenarios are described here.

① When the oil-water viscosity ratio is 100, there is a big difference between oil and water, and the flow capacity of water is greater than that of oil. The water breakthrough occurs quickly, and the performance of the oil displacement by water is poor.

② When the oil-water viscosity ratio is 10, the oil-water difference is relatively small, and the displacement effect becomes better.

③ When the oil-water viscosity ratio is 0.4, the viscosity of water is greater than that of oil, the front saturation is generally close to 1, and displacement effect tends to be close to piston-like water flooding. The oil displacement effect is good.

The application of B-L equation in flow mechanics in porous media is generally to keep the water injection rate unchanged, that is, the cumulative injection remains unchanged.

Sometimes, in order to keep the pressure difference between the wells unchanged, the flow law of constant pressure difference water flooding needs to be considered.

The production rate at the initial moment can be obtained by Darcy's law:

$$Q_o = \frac{KK_{ro}(S_{wc})A}{\mu_o}\frac{\Delta p}{L} \tag{7.65}$$

For the oil phase, the production rate before the water breakthrough moment can be:

$$q_o = -\frac{kk_{ro}A}{\mu_o}\left(\frac{dP}{dx}\right) \tag{7.66}$$

For the water phase, the production rate before the water breakthrough moment can be:

$$q_w = -\frac{kk_{rw}A}{\mu_w}\left(\frac{dP}{dx}\right) \tag{7.67}$$

The total liquid production rate before the water breakthrough is:

$$q_t = -\left(\frac{k_{ro}}{\mu_o} + \frac{k_{rw}}{\mu_w}\right)k\mathrm{A}\left(\frac{dP}{dx}\right) \tag{7.68}$$

Let $\lambda_r = \frac{k_{ro}}{\mu_o} + \frac{k_{rw}}{\mu_w}$ be the total relative mobility, and the above equation is transformed into:

$$q_t = -\lambda_r k\mathrm{A}\left(\frac{dP}{dx}\right) \tag{7.69}$$

It can be rewritten as:

$$q_t = \frac{kA(P_i - P_p)}{\int_0^L \lambda_r^{-1}dx} \tag{7.70}$$

The key to obtain the solution after water breakthrough is to get the average apparent viscosity, which requires the apparent viscosity before water breakthrough.

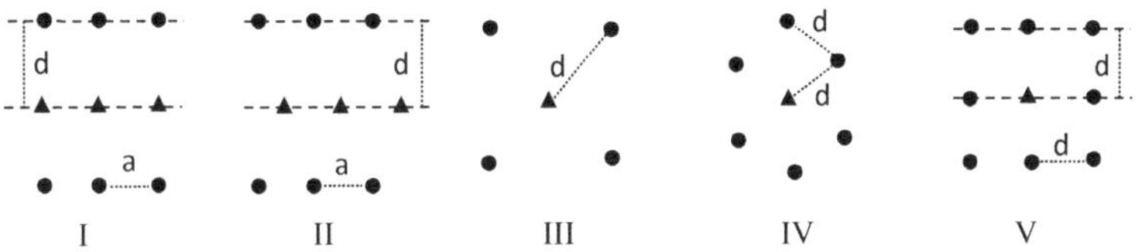

Figure 7.10. Different areal well patterns (I: direct line drive; II: staggered line drive; III: inverted five-spot; IV: inverted seven-spot; V: inverted nine-spot)

7.4.3.2 *Productivity of areal well pattern*

As shown in Figure 7.10, the injection rate with different areal well patterns can be summarized as follows.

(1) Direct line drive ($d \geq a$)

$$Q_{in} = \frac{0.1178KK_{rw}(S_{or})h\Delta p}{\mu_w\left(\lg\frac{a}{R_w} + 0.682\frac{d}{a} - 0.798\right)} \tag{7.71}$$

(2) Staggered line drive ($d \geq a$)

$$Q_{in} = \frac{0.1178KK_{ro}(S_{wc})h\Delta p}{\mu_o\left(\lg\frac{a}{R_w} + 0.682\frac{d}{a} - 0.798\right)} \tag{7.72}$$

(3) Inverted five-spot

$$Q_{in} = \frac{0.1178KK_{ro}(S_{wc})h\Delta p}{\mu_o\left(\lg\frac{d}{R_w} - 0.2688\right)} \tag{7.73}$$

(4) Inverted seven-spot

$$Q_{in} = \frac{0.1571KK_{ro}(S_{wc})h\Delta p}{\mu_o\left(\lg\frac{d}{R_w} - 0.2472\right)} \tag{7.74}$$

(5) Inverted nine-spot

$$Q_{in} = \frac{0.1178KK_{ro}(S_{wc})h\Delta p_{ic}}{\mu_o\left(\frac{1+R}{2+R}\right)\left(\lg\frac{d}{R_w} - 0.1183\right)} \tag{7.75}$$

or

$$Q_{in} = \frac{0.1178KK_{ro}(S_{wc})h\Delta p_{is}}{\mu_o\left[\left(\frac{3+R}{2+R}\right)\left(\lg\frac{d}{R_w} - 0.1183\right) - \frac{0.301}{2+R}\right]} \tag{7.76}$$

where R is the flow rate ratio between the corner well and the side well; while Δp_{ic} and Δp_{is} are the difference of flow pressure between injection wells and that between the corner well and side well respectively.

The above production equation is derived with the assumption that the mobility ratio is 1 and the fluid is incompressible. When the mobility ratio is not 1 or the well is watered out, the calculation method will change accordingly.

7.4.3.3 *Material balance equation*

The mass conservation law is applied to the derivation of the continuity equation, which states that the mass difference between the fluid flowing into and out of the REV is equal to the mass difference caused by the change of porosity and fluid density in the REV. The continuity equation is also called the mass conservation equation.

The mass balance equation in this section is derived based on the volumetric balance, but the "volume" needs to be converted into the volume of the original reservoir condition, that is, the cumulative volumetric production rate expressed in the original reservoir condition is equal to the summation of the volume of fluid expansion and invasion caused by the pressure drop in the reservoir.

The material balance equation does not consider the flow process, but it can establish the relationship between flow rate and pressure difference, which can also be used to study the driving mechanism of reservoir and predict the dynamics.

In the process of reservoir development, the decrease in reservoir pressure will lead to the invasion of edge water, gas cap expansion,

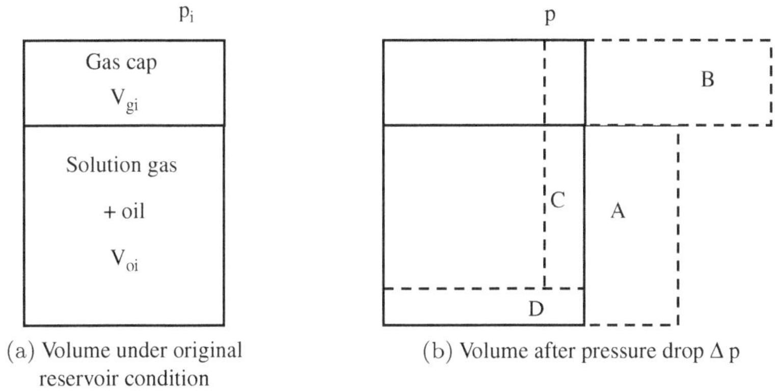

Figure 7.11. Volume changes after reservoir pressure changes

dissolved gas separation and expansion, and the release of these energies will drive the oil to flow out of the ground, which needs to be considered comprehensively when the material balance equation is established.

If the actual reservoir is completely filled with oil and gas under the original formation pressure, with the pressure drop Δp, the solid lines in Figure 7.11(a) represent the original pore volume. A is the increased volume due to the expansion of the oil and dissolved gas, B is the increased volume due to the expansion of the gas cap, C is the decreased pore volume due to the bound water expansion and pore volume decrease, and D is the decreased pore volume caused by the edgc water invasion, as shown in Figure 7.11(b).

If the total production rate of oil, gas and water is expressed as the production rate under the reservoir condition, then:

$$Underground\ \ Production = A + B + C + D \tag{7.77}$$

Thus, the general equation of material balance is established as:

$$N = \frac{N_p(B_t - R_{si}B_g) - (W_e + W_i - W_p) + G_pB_g}{(B_t - B_{ti}) + mB_{ti}\frac{B_g - B_{gi}}{B_{gi}} + B_{ti}\frac{(1+m)}{1-S_{wc}}(C_f + C_wS_{wc})\Delta p} \tag{7.78}$$

It can be rewritten as:

$$N_p(B_t - R_{si}B_g) + G_pB_g + W_p$$
$$= N\left[(B_t - B_{ti}) + mB_{ti}\frac{B_g - B_{gi}}{B_{gi}} + B_{ti}\frac{(1+m)}{1-S_{wc}}(C_f + C_wS_{wc})\Delta p\right] + (W_e + W_i) \tag{7.79}$$

where N_p is the cumulative oil production (at the reservoir condition); G_p is the cumulative gas production (at the reservoir condition), $G_n = N_nR_n$; R_p is the cumulative gas-oil ratio; W_p is the cumulative production rate (at the reservoir condition); N is the original oil reserves (at the reservoir condition), $N = V\varphi(1 - S_{wc})/B_{oi}$; V is the reservoir volume; W_e is the natural water inflow (either surface or reservoir condition, because $B_w = 1$); and W_i is the water injection (either surface or reservoir condition, because $B_w = 1$);

$$m = \frac{V_{gi}}{V_{oi}}; \quad B_{ti} = B_{oi}B_t = B_o + (R_{si} - R_s)B_g; \quad \Delta p = p_i - p;$$

where the subscript i is the original pressure state.

For different driving types of reservoirs, the material balance equation can be simplified.

(1) Unsaturated closed reservoir with elastic drive

The assumptions are $W_e = W_i = W_p = 0$ and $R_p = R_s = R_{si}$:

$$N_pB_o = NB_{oi}C_T\Delta p = N \cdot X \tag{7.80}$$

$$C_T = C_o + \frac{C_wS_{wc} + C_f}{1 - S_{wc}} \tag{7.81}$$

Let $X = B_{oi}C_T\Delta p$, then:

$$N_pB_o = N \cdot X \tag{7.82}$$

where $X \sim N_pB_o$ is the line that goes through the origin, and the slope is N.

(2) Unsaturated reservoir with the elastic water drive

The assumptions are $m = 0$; $R_p = R_s = R_{si}$; and $W_i = 0$:

$$N_pB_o = NB_{oi}C_T\Delta p + (W_e - W_p) \tag{7.83}$$

It can be rewritten as:

$$\frac{N_pB_o + W_p}{B_{oi}C_T\Delta P} = N + \frac{W_e}{B_{oi}C_T\Delta P} \tag{7.84}$$

Let $X = \frac{W_e}{B_{oi}C_T\Delta P}$, $Y = \frac{N_pB_o+W_p}{B_{oi}C_T\Delta P}$, then:

$$Y = N + X \tag{7.85}$$

where $X \sim Y$ is the straight line with intercept N.

(3) Unsaturated reservoir with natural water and injected water drive

The assumptions are $m = 0$ and $R_p = R_s = R_{si}$:

$$N_pB_o = NB_{oi}C_T\Delta p + (W_e + W_i - W_p) \tag{7.86}$$

It can be rewritten as:

$$\frac{N_pB_o + W_p - W_i}{B_{oi}C_T\Delta P} = N + \frac{W_e}{B_{oi}C_T\Delta P} \tag{7.87}$$

Let $X = \frac{W_e}{B_{oi}C_T\Delta P}$, $Y = \frac{N_pB_o+W_p-W_i}{B_{oi}C_T\Delta P}$, then:

$$Y = N + X \tag{7.88}$$

$X \sim Y$ is the straight line with intercept N.

(4) Saturated reservoir, no gas cap, no bottom water, solution gas drive

The assumptions are $W_e = W_p = W_i = 0$; $m = 0$; and $R_p \neq R_s \neq R_{si} \neq 0$:

$$N_p[B_t - (R_{si} - R_p)B_g] = N\frac{C_f + C_w S_{wc}}{1 - S_{wc}} B_{ti}\Delta p + N(B_t - B_{ti}) \tag{7.89}$$

Ignoring the elastic effects of liquid and rock, it changes to:

$$N_p[B_t - (R_{si} - R_p)B_g] = N(B_t - B_{ti}) \tag{7.90}$$

Let $X = B_t - B_{ti}$, $Y = N_p[B_t - (R_{si} - R_p)B_g]$, then:

$$Y = N \cdot X \tag{7.91}$$

$X \sim Y$ is the straight line with intercept N.

(5) Saturated reservoir, gas cap, no bottom water, solution gas drive

The assumptions are $W_e = W_i = W_p = 0$; $m \neq 0$; and $R_p \neq R_s \neq R_{si} \neq 0$:

$$\begin{aligned} N_p[B_t - (R_{si} - R_p)B_g]N\frac{1+m}{1 - S_{wc}} B_{ti}(C_f + C_w S_{wc})\Delta p \\ +N[(B_t - B_{ti}) + mB_{ti}\frac{B_g - B_{gi}}{B_{gi}}] \end{aligned} \tag{7.92}$$

Ignoring the elastic effects of liquid and rock, it changes to:

$$N_p[B_t - (R_{si} - R_p)B_g] = N\left[(B_t - B_{ti}) + mB_{ti}\frac{B_g - B_{gi}}{B_{gi}}\right] \tag{7.93}$$

Let $X = (B_t - B_{ti}) + mB_{ti}\frac{B_g - B_{gi}}{B_{gi}}$, $Y = N_p[B_t - (R_{si} - R_p)B_g]$, then:

$$Y = N + X \tag{7.94}$$

$X \sim Y$ is the straight line with intercept N.

(6) Saturated reservoir, gas cap, edge and bottom water, solution gas drive

The assumptions are $W_i = 0$; $m \neq 0$; and $R_p \neq R_s \neq R_{si} \neq 0$:

$$N_p[B_t - (R_{si} - R_p)B_g] = N\frac{1+m}{1-S_{wc}}B_{ti}(C_f + C_w S_{wc})\Delta p$$
$$+N\left[(B_t - B_{ti}) + mB_{ti}\frac{B_g - B_{gi}}{B_{gi}}\right] + (W_e - W_p) \qquad (7.95)$$

Ignoring the elastic effects of liquid and rock, it changes to:

$$\frac{N_p[B_t - (R_{si} - R_p)B_g] + W_p}{(B_t - B_{ti}) + mB_{ti}\frac{B_g - B_{gi}}{B_{gi}}} = N + \frac{W_e}{(B_t - B_{ti}) + mB_{ti}\frac{B_g - B_{gi}}{B_{gi}}} \qquad (7.96)$$

Let

$$X = \frac{W_e}{(B_t - B_{ti}) + mB_{ti}\frac{B_g - B_{gi}}{B_{gi}}} \qquad (7.97)$$

$$Y = \frac{N_p[B_t - (R_{si} - R_p)B_g] + W_p}{(B_t - B_{ti}) + mB_{ti}\frac{B_g - B_{gi}}{B_{gi}}} \qquad (7.98)$$

Then:

$$Y = N + X \qquad (7.99)$$

$X \sim Y$ is the straight line with intercept N.

Bibliography

Ahmed, T. and McKinney, P. (2004). *Advanced Reservoir Engineering* (Gulf Professional Publishing, USA).

Bear, J. (1972). *Dynamics of Fluids in Porous Media* (Elsevier, USA).

Buckley, S. and Leverett, M. (1942). Mechanism of fluid displacement in sands, *Transactions of the AIME*, 146, pp. 107–116.

Craft, B., Hawkins, M., and Terry, R. (1959). *Applied Petroleum Reservoir Engineering*, 2nd Ed. (Prentice Hall, USA).

Crichlow H. (1977). *Modern Reservoir Engineering: A Simulation Approach* (Prentice Hall, USA).

De Swaan, O. (1976), Analytic solutions for determining naturally fractured reservoir properties by well testing, *Society of Petroleum Engineers Journal*, 16, pp. 117–122

Helfferich, F.G. (1981). Theory of multicomponent multiphase displacement in porous media, *Society of Petroleum Engineers Journal*, 21, pp. 51–62.

Horner, D. (1951). Pressure Build-up in Wells, *3rd World Petroleum Congress*, World Petroleum Congress, pp. 503–521.

Javadpour, F. (2009). Nanopores and apparent permeability of gas flow in mudrocks (shales and siltstone), *Journal of Canadian Petroleum Technology*, 48, pp. 16–21.

Kazemi, H. (1969). Pressure transient analysis of naturally fractured reservoirs with uniform fracture distribution, *Society of Petroleum Engineers Journal*, 9, pp. 451–462.

Klinkenberg, L.J. (1941). The permeability of porous media to liquids and gases, *Drilling and Production Practice*, American Petroleum Institute, pp. 200–214.

Langmuir, I. (1918). The adsorption of gases on plane surfaces of glass, mica and platinum, *Journal of the American Chemical Society*, 40, pp. 1361–1403.

Matthews, C., Brons, F., and Hazebroek, P. (1954). A method for the determination of average pressure in a bounded reservoir, *Transactions of the AIME*, 201, pp. 182–191.

Moridis, G., Collett, T.S., Pooladi-Darvish, M., Hancock, S.H., Santamarina, C., Boswell, R., Kneafsey, T., Rutqvist, J., Kowalsky, M., Reagan, M., Sloan, E., and Sum, A. (2011). Challenges, uncertainties, and issues facing gas production from gas-hydrate deposits, *SPE Reservoir Evaluation & Engineering*, 14, pp. 76–112.

Prats, M. (1985). *Thermal Recovery.* (Society of Petroleum Engineers, USA).

Savins, J.G. (1969), Non-Newtonian flow through porous media, *Industrial & Engineering Chemistry*, 61, pp. 18–47.

Scheidegger, A.E. (1961). General theory of dispersion in porous media, *Journal of Geophysical Research*, 66, pp. 3273–3278.

Silin, D. and Kneafsey, T. (2012). Shale gas: nanometer-scale observations and well modelling, *Journal of Canadian Petroleum Technology*, 51, pp. 464–475.

Tester, J.W., Anderson, B.J., Batchelor, A.S., Blackwell, D.D., DiPippo, R., Drake, E.M., Garnish, J., Livesay, B., Moore, M.C., Nichols, K., Petty, S., Toksoz, M.N., Veatch, R.W., Baria, R., Augustine, C., Murphy, E., Negraru, P., and Richards, M. (2007). Impact of enhanced geothermal systems on US energy supply in the twenty-first century, *Philosophical Transactions of the Royal Society A*, 365, pp.1057–1094.

Warren, J. and Root, P. (1963). The behavior of naturally fractured reservoirs, *Society of Petroleum Engineers Journal*, 3, pp. 245–255.

Yu, W. and Sepehrnoori, K. (2014). Simulation of gas desorption and geomechanical effects for unconventional gas reservoirs, *Fuel*, 116, pp. 455–464.

Index

Milton Keynes UK
Ingram Content Group UK Ltd.
UKHW010629150324
439410UK00004B/23

9 789811 219528